STP 1044

Advances in Thermoplastic Matrix Composite Materials

Golam M. Newaz, editor

ASTM
1916 Race Street
Philadelphia, PA 19103

Library of Congress Cataloging-in-Publication Data

Advances in thermoplastic matrix composite materials/Golam M. Newaz, editor.
    "ASTM publication code number 04-010440-33"—T.p. verso.
    "Papers presented at the symposium of the same name held in Bal Harbour, Florida 19–20 Oct. 1987 and sponsored by ASTM Committee D-30 on High Modulus Fibers and Their Composites"—Foreword.
    Includes bibliographical references.
    1. Thermoplastic composites—Congresses. I. Newaz, Golam M., 1954- . II. ASTM Committee D-30 on High Modulus Fibers and Their Composites. III. Series: ASTM special technical publication: 1044.
    TA418.9.C6A323   1989   620.1′92396—dc20   89–17634 CIP

NOTE

The Society is not responsible, as a body,
for the statements and opinions
advanced in this publication.

Peer Review Policy

Each paper published in this volume was evaluated by three peer reviewers. The authors addressed all of the reviewers' comments to the satisfaction of both the technical editor(s) and the ASTM Committee on Publications.

The quality of the papers in this publication reflects not only the obvious efforts of the authors and the technical editor(s), but also the work of these peer reviewers. The ASTM Committee on Publications acknowledges with appreciation their dedication and contribution of time and effort on behalf of ASTM.

Printed in Chelsea, MI
September 1989

Foreword

This publication, *Advances in Thermoplastic Matrix Composite Materials,* contains papers presented at the symposium of the same name held in Bal Harbour, Florida 19–20 Oct. 1987. The symposium was sponsored by ASTM Committee D-30 on High Modulus Fibers and Their Composites. Golam M. Newaz, Battelle Memorial Institute, presided as symposium chairman and was editor of this publication.

Contents

Overview   1

PROCESSING

Process Modeling of Thermoplastic Matrix Composites—C-L. LEUNG AND T. T. LIAO   5

Resin Impregnation During the Manufacture of Thermoplastic Matrix Composite
Materials—J. P. COULTER AND S. I. GÜÇERI   14

The Effect of Processing on Autohesive Strength Development in Thermoplastic Resins
and Composites—J. C. HOWES, A. C. LOOS, AND J. A. HINKLEY   33

Thermoplastic Coating of Carbon Fibers—B. W. GANTT, D. D. EDIE, G. C. LICKFIELD,
M. J. DREWS, AND M. S. ELLISON   50

Pultrusion Process Modeling—R. M. HACKETT AND S. N. PRASAD   62

THERMAL AND MECHANICAL BEHAVIOR

Thermoplastic Matrix Composites: Finite-Element Analysis of Mode I and Mode II
Failure Specimens—R. J. BANKERT, N. D. LAMBROPOULOS, M. S. SHEPHARD,
AND S. S. STERNSTEIN   73

A Study of Graphite/PEEK Under High Temperatures—J. RAMEY AND A. PALAZOTTO   91

Load-Frequency Effect on Fatigue Life of IMP6/APC-2 Thermoplastic Composite
Laminates—E. DAN-JUMBO, S. G. ZHOU, AND C. T. SUN   113

Response of Notched AS4/PEEK Laminates to Tension/Compression Loading—
R. A. SIMONDS AND W. W. STINCHCOMB   133

Investigation into Thermal Conductivity of Composite Materials for Electronics
Packaging—R. R. HOFFMAN, R. P. TYE, AND J. G. CHERVENAK   146

Stress-State Effects on the Viscoelastic Response of Polyphenylene Sulfide (PPS) Based
Thermoplastic Composites—F. A. MYERS   154

Fiber Orientation and Fracture Morphology in Short Fiber-Reinforced
Thermoplastics—H. T. HAHN, K. L. JERINA, AND P. BURRETT   183

Mechanical Properties of the Carbon Fiber/PEEK Composite APC-2/AS-4 for
Structural Applications—D. R. CARLILE, D. C. LEACH, D. R. MOORE, AND
N. ZAHLAN   199

Effects of Moisture Content on the Mechanical Properties of Polyphenylene Sulfide
Composite Materials—S. V. HOA, S. LIN, AND J. R. CHEN   213

Moisture Effects on Mode I Interlaminar Fracture Toughness of a Graphite Fiber
Thermoplastic Matrix Composite—J. P. LUCAS AND B. C. ODEGARD   231

DELAMINATION

Interlaminar Fracture Toughness of Thermoplastic Composites—J. A. HINKLEY,
N. J. JOHNSTON, AND T. K. O'BRIEN   251

Delamination Growth Under Cyclic Loading at Elevated Temperature in
Thermoplastic Composites—G. M. NEWAZ, A. LUSTIGER, AND J-Y. YUNG   264

Delamination Resistance of Stitched Thermoplastic Matrix Composite Laminates—
K. B. SU   279

Indexes   301

Overview

In recognition of the considerable interest in advanced thermoplastic composites in the aerospace area, Committee ASTM D30 on High Modulus Fibers and Their Composites organized the symposium on Advances in Thermoplastic Matrix Composite Materials held in Bal Harbour, Florida 19–20 Oct. 1987.

Thermoplastic composites are known to provide high impact toughness, delamination fracture energy, and environmental resistance. These qualities are complemented by the flexibility in processing and unlimited shelf life of the composites. However, the materials are relatively more expensive and the knowledge base is just beginning to develop. The symposium was organized to provide a forum to discuss the advantages and disadvantages of these materials in the context of processing and material response. Design and analysis were not emphasized, primarily because the application of these materials for advanced aircraft is just evolving. Currently, major efforts are underway in materials development, processing, and characterization of advanced thermoplastic composites. The symposium captured many of these issues and had a large audience from industry, government, and universities. The program was broad enough to incorporate a few papers on short-fiber reinforced thermoplastic composites for nonaerospace applications, although the emphasis of the symposium was on continuous fiber structural composites.

The papers presented in the symposium can be divided into three major technical areas: processing, thermal and mechanical behavior, and delamination. They are similarly grouped in this special technical publication. The papers related to processing are concerned with process modeling, influence of processing on interfacial strength development, and on coating of carbon fibers for application in thermoplastic composites. Papers on thermal and mechanical behavior address experimental aspects, mechanical properties, thermal characteristics, failure mechanisms, effect of adverse environment, and fatigue. Papers on delamination discuss delamination toughness characterization and delamination growth modeling. A few new concepts are forwarded in this area. Also discussed are morphological issues as they relate to delamination.

In all, a broad overview of state-of-the art research in advanced thermoplastic composites is well represented in this book. The purpose of this publication was not only to inform the technical community of some recent accomplishments in the field, but also to raise questions which will form the basis for future investigations. This discussion and consideration of unresolved issues is expected to advance the thermoplastic matrix composite technology, however small the effort of the ASTM symposium may seem.

I would like to thank the ASTM staff for their dedication to make the symposium a success and to bring out the technical publication in a timely manner. Also, I would like to acknowledge Dr. Clarence Wolf of the McDonnell-Douglas Research Laboratories, Professor Thomas Hahn of the Pennsylvania State University, Dr. Jeffrey Hinkley of NASA-Langley, and Dr. George Sendeckyj of the Air Force Wright Aeronautical Laboratories for their assistance in conducting the sessions. Much thanks to all the reviewers who contributed their valuable time to review the papers. Finally, the authors are congratulated for their contributions to this volume.

Golam M. Newaz

Battelle Memorial Institute, Columbus, OH;
symposium chairman and editor

Processing

Chuk-Ling Leung[1] and Tony T. Liao[1]

Process Modeling of Thermoplastic Matrix Composites

REFERENCE: Leung, C-L. and Liao, T. T., **"Process Modeling of Thermoplastic Matrix Composites,"** *Advances in Thermoplastic Matrix Composite Materials, ASTM STP 1044,* G. M. Newaz, Ed., American Society for Testing and Materials, Philadelphia, 1989, pp. 5–13.

ABSTRACT: Heat transfer is perhaps the most critical parameter in the processing of thermoplastic matrix composites. Temperature distribution affects the quality and microstructure of the final product in that flow, compaction, melting, and crystallization are thermally controlled processes. Using readily available thermal properties of the composites, a model that predicts the temperature profile within a composite laminate during processing was developed. Graphite-reinforced thermoplastic composites of various thicknesses, up to 80 plies, were fabricated and used to validate the analytical thermal profiles. When compared with empirical measurements, the heat transfer model was shown to generate highly accurate temperature distributions within the composites under various processing conditions. When correlated with the dynamic viscosities of the polymer at various temperatures, t-T-V profiles of the composites during processing can then be obtained.

KEY WORDS: heat transfer, modeling, thermoplastic composites, dynamic viscosity, thermal profiles

The principle of processing management is essentially the utilization of a monitor in a closed loop control of the production process through the program time (t) versus temperature (T), external pressure (P), and internal degassing vacuum (V) imposed on the part being processed. Information for "real time" management of this t-T-P-V program for production process control can be derived from (*a*) prior processability testing and (*b*) current processing monitor data from the part. Item (*b*) can be conceptually divided into two parts: first, the gathering of the in-process data; second, the comparison of these dynamic data and information with the predicted behavior of the materials under similar situations. This task addresses the development of such an analytical model for the heat transfer from the processing equipment, for example, a heated press, to the thermoplastic composite.

Heat transfer is perhaps the most critical parameter in the processing of high-temperature thermoplastics. Temperature distribution affects the quality and microstructure of the final product in that flow, compaction, melting, and crystallization (or chemical reaction) are thermally controlled processes.

For the conventional processing of thermoplastic composites whereby the composite is heated between two platens of a hydraulic press, temperature is changing with time, and heat sources may be present within the composite. The energy balance may thus be made

Energy conducted from top face + face generated within composite = change in internal energy + energy conducted out bottom face

[1]Manager and technical specialist, respectively, Polymer Synthesis and Processing Department, Rockwell International Science Center, Thousand Oaks, CA 91360.

For the vast majority of commercial thermoplastic materials, no chemical reaction is involved during processing. Also assume that the total heat of crystallization of the polymer is negligible when compared to the heat energy released by the platens. Furthermore, assume that heat loss or gain through the thickness of the composite by convection and radiation is small since the thickness is much less than the length and width of the composite. This reduces the problem to that of a one-dimensional heat conduction across the thickness of the composite, the equation for which can be approximated as

$$k \Delta^2 T = \rho C_p (\partial T / \partial t) \tag{1}$$

or

$$\Delta^2 T = 1 / \alpha (\partial T / \partial t) \tag{2}$$

where

k = heat conductivity,
ρ = density,
C_p = heat capacity,
α = heat diffusivity $\left(= \dfrac{k}{\rho C_p} \right)$,
T = temperature, and
t = time.

A number of analytical solutions for the one-dimensional heat transfer through a thermoplastic laminate has been published over the last few years. For example, Seferis [1] developed a model that is derived from a complex series solution whose evaluation becomes quite difficult for different geometries and processing procedures in which the boundary conditions vary with time. For such situations, a most fruitful approach is based on finite-difference techniques. Finite differences are used to approximate differential increments in the temperature and space coordinates; the smaller the increments, the more closely the true temperature distribution will be approximately.

If the thickness is further divided into equal increments, each of which is called a nodal point, the temperature at each of the nodal points can be established. To approximate differential increments in temperature and position, a method called backward finite difference is used [2]. Thus Eq 2 can be approximated by

$$\frac{T_{m+1}^{P+1} + T_{m-1}^{P+1} - 2T_m^{P+1}}{(\Delta x)^2} = \frac{1}{\alpha} \frac{T_m^{P+1} - T_m^P}{\Delta t} \tag{3}$$

or

$$T_m^P = -\frac{\alpha \Delta t}{\Delta x^2} \left(T_{m+1}^{P+1} + T_{m-1}^{P+1} \right) + \left(1 + \frac{2\alpha \Delta t}{\Delta x^2} \right) T_m^{P+1} \tag{4}$$

where

T_m^P = temperature at nodal m at time p,
T_m^{P+1} = temperature at nodal m at time $p + 1$,
Δx = increment in x direction, and
Δt = time increment.

To calculate the temperature T_m^{P+1}, Eq 3 or 4 is solved iteratively for the entire nodal system, that is, thickness of the composite, using the temperatures of the heated press platens as boundary conditions.

The quantities that constitute the thermal diffusivity of the composite such as thermal conductivity, density, and heat capacity are all functions of temperature. However, for simplicity, these quantities are assumed to be constants. Thus the thermal diffusivity of the composite can be assumed to be a constant, the value of which can be derived from the physical properties of the neat resin matrix and fiber, as described below.

The thermal properties of the neat resin [3,4] and graphite fibers are tabulated in Table 1. Because heat was assumed to be conducted perpendicularly across the fibers, the value for transverse heat conductivity was used.

Thermal conductivity of the composite was obtained by averaging the values for the neat resin and graphite fibers [1], that is,

$$k_c = 1/(X_{vm}/k_m + X_{vf}/k_f) \tag{5}$$

where

k_c = composite thermal conductivity,
k_m = matrix resin thermal conductivity,
k_f = fiber thermal conductivity,
X_{vm} = matrix volume fraction, and
X_{vf} = fiber volume fraction.

Similarly, the heat capacity of a composite can be obtained by a mass-averaged heat capacities of the matrix and reinforcements, that is,

$$C_{pc} = X_{mm}C_{pm} + (1 - X_{mm})C_{pf} \tag{6}$$

where

C_{pc} = composite heat capacity,
C_{pm} = matrix resin heat capacity,
C_{pf} = fiber heat capacity, and
X_{mm} = mass fraction of matrix resin.

Assuming a void-free composite, the density of a composite can be derived from a volume average of the respective densities of the matrix and the fibers [5], that is,

$$\rho_c = X_{vm}d_m + (1 - X_{vm})\rho_f \tag{7}$$

TABLE 1—Physical properties of PEEK APC2.

Material	Density, g/C³	Resin, weight %	Resin, vol. %	Heat Capacity, J/g/K	Thermal Conductivity, J/s/m/K
PEEK	1.263	100	100	1.339	0.251
C Fiber	1.79			1.255	0.427
APC2	1.53	40	48	1.297	0.318

where

ρ_c = composite density,
ρ_m = density of matrix resin,
ρ_f = density of fibers, and
X_{vm} = volume fraction of matrix resin.

Finally, the thermal diffusivity of the composite can be calculated from the heat conductivity, heat capacity, and density as

$$\alpha_c = k_c/(d_c C_{pc}) \tag{8}$$

Based on the above equations, an analytical model capable of predicting the thermal profile of a composite as a function of time and position was developed and written as computer code on a personal computer.

Experimental Procedures

Carbon fiber reinforced poly (etheretherketone) prepregs (APC2) were obtained from Imperial Chemical Industries (ICI) and cut into 4 by 4-in. squares. Unidirectional composites consisting of 16, 32, 48, 64, and 80 plies, respectively, were stacked.

Thermocouples were inserted at the top and bottom as well as in the center of the layup. The layup was then compacted under 1500 psi (10 MPa) from room temperature to 370°C and then cooled by either air or water circulation within the press platens. Two cool-down rates were examined: a slow rate by turning off the power supply to the press, and a faster rate by circulating compressed air through the press platens. Temperatures at the surfaces of the composite during the operation were digitized and input into the computer code at 1-min intervals as boundary conditions, while temperatures at the middle of the composite were recorded and compared with those generated by the analytical model.

Dynamic viscosity of neat film-grade poly (etheretherketone) was measured on a Rheometric mechanical spectrometer using the 25-mm parallel plates at 10 rad/s, 10% strain, at 5°C/min heat-up and cool-down rates.

Results and Discussion

Heat-up

Figures 1A–C show the increase in temperature at the center of each of the composites at various times during the heat-up (compaction) phase of the cycle. The figures show that the calculated time-temperature profiles track the experimental profiles remarkably well. Indeed, as illustrated in Fig. 1B, even when electrical power was inadvertently interrupted and then restored, the computed sample temperatures during that period were exactly as those measured by the thermocouples. Only in comparatively thick laminates are deviations between experimental and calculated temperatures observed.

Cool-down

Cool-down was achieved with or without air circulation within the press platens while the composite was still under pressure. Correlations between empirical and calculated time-temperature profiles of the composites are shown in Figs. 2A–C (with air assist) and Figs. 3A–C (without air assist). Again, highly accurate predictive results are observed. In cooling without air circulation, the hydraulic press automatically injects cooling water after a certain temperature

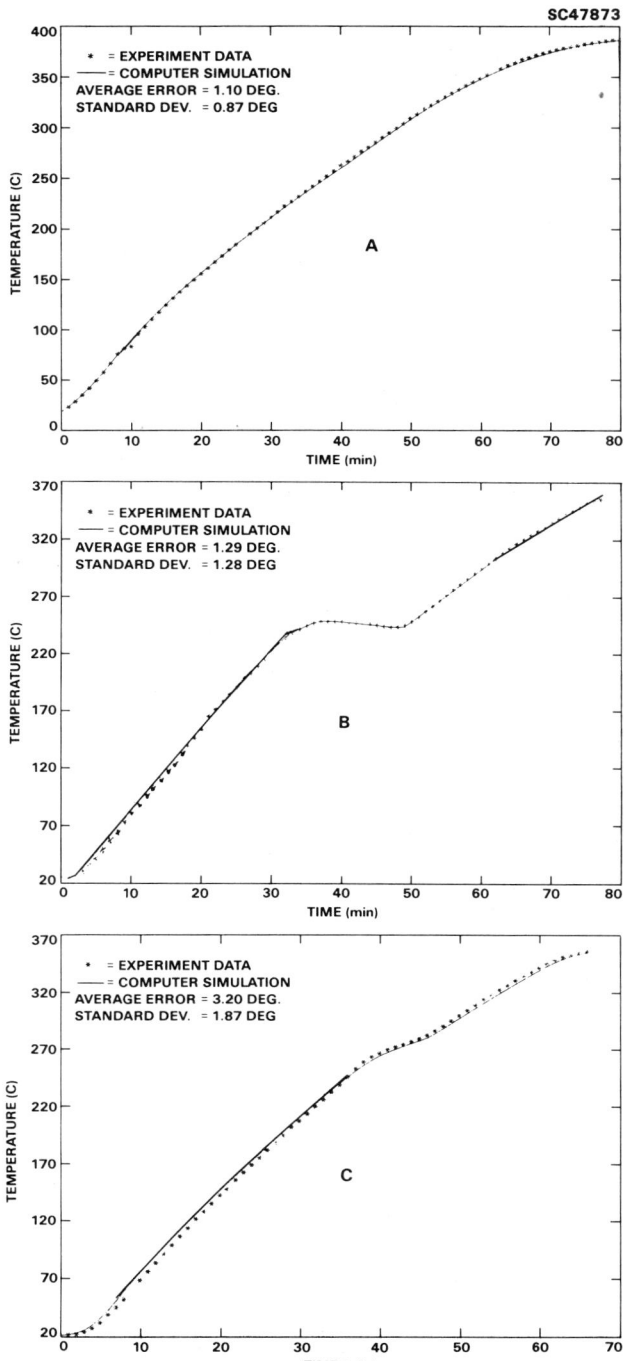

FIG. 1—*Comparison of experimental with predicted temperature profiles for:* (A) *32,* (B) *64 and* (C) *80 plies PEEK-APC2 laminates during heat-up.*

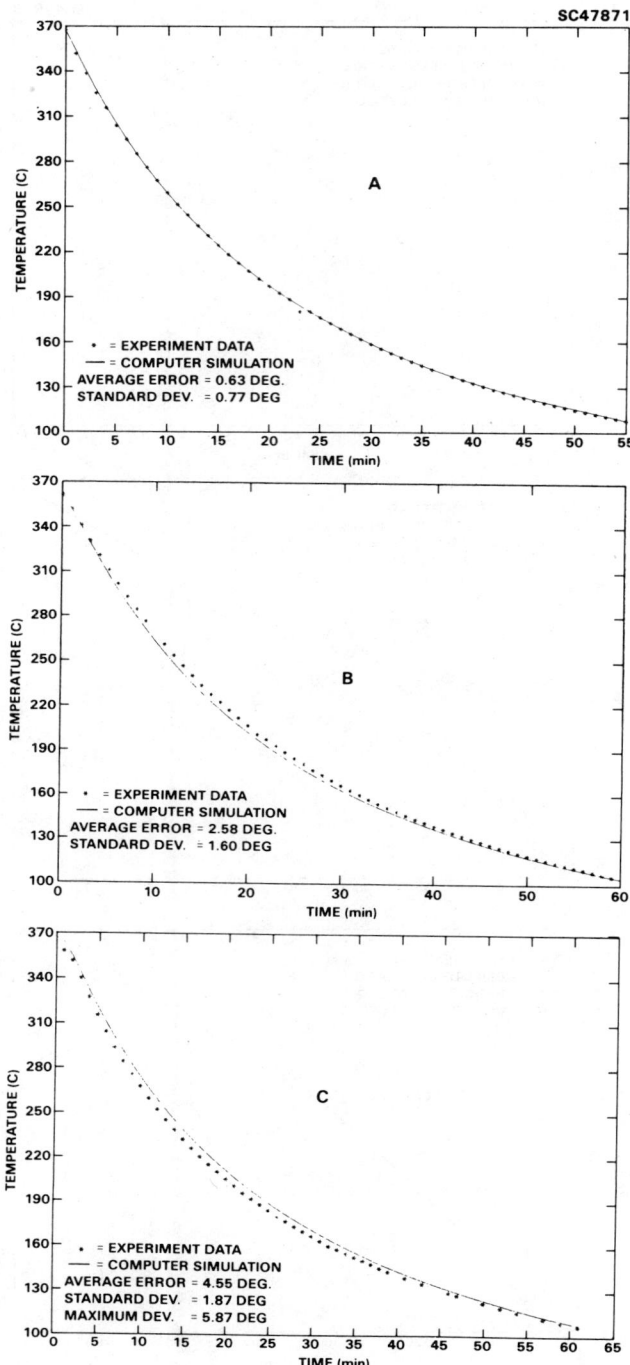

FIG. 2—*Comparison of experimental with predicted temperature profiles for:* (A) *32,* (B) *64 and* (C) *80 plies PEEK-APC2 laminates during cool-down with air assist.*

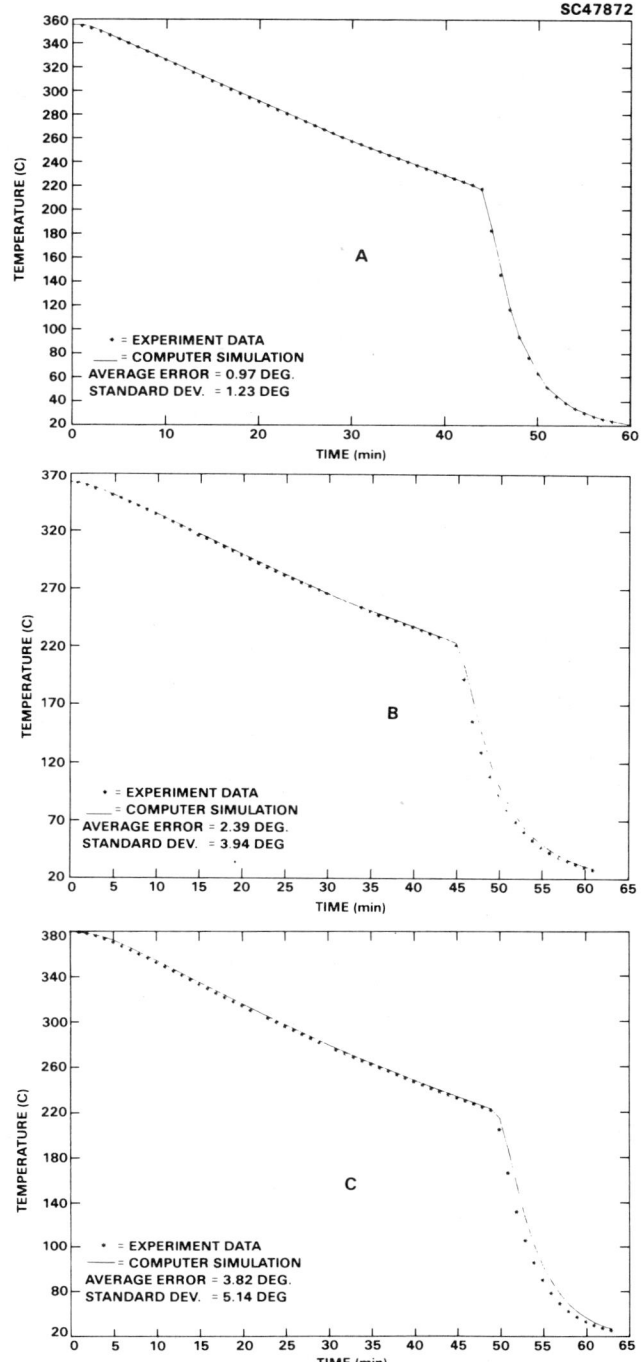

FIG. 3—*Comparison of experimental with predicted temperature profiles for:* (A) *32,* (B) *64 and* (C) *80 plies PEEK-APC2 laminates during cool-down without air assist.*

TABLE 2—*Deviation between calculated and measured temperatures (in °C).*

No. of Plies	Heat-Up Phase	Cool-Down Phase	
		No Air	Forced Air
16	1.10	. . .	. . .
32	1.66	0.97	0.63
48	1.29	1.80	1.69
64	2.48	2.39	2.58
80	3.20	3.82	4.55

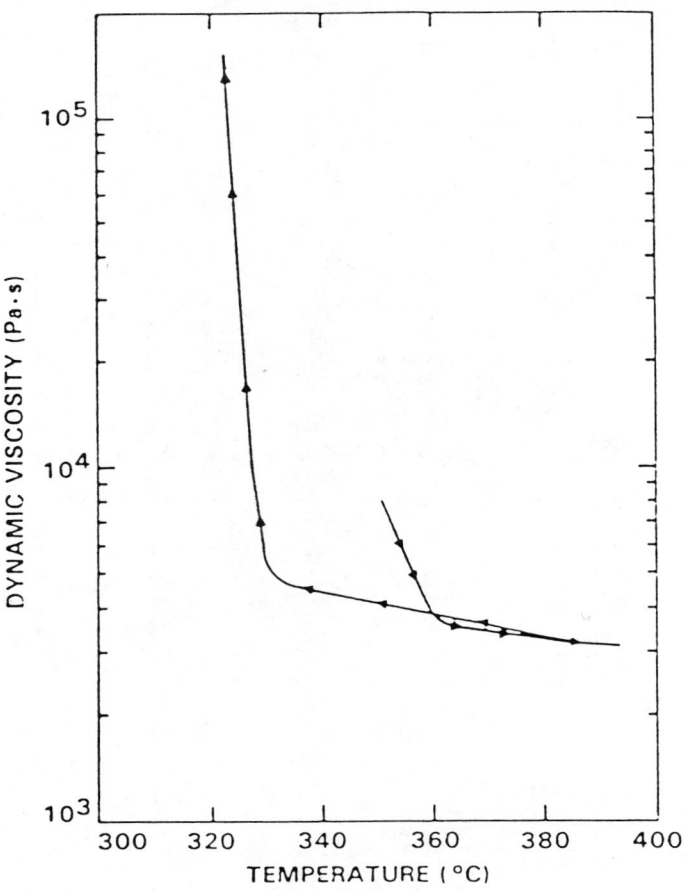

FIG. 4—*Dynamic viscosity of film-grade PEEK resin, 5°C/min heating and cooling, 10% strain, 10 rad/s, 25-mm parallel plates.*

is reached, as shown by the sharp downturn in Figs. 3A–C. As is the case in the heat-up phase, these rapid changes in processing temperatures were handled extremely well by the analytical model.

To gauge the accuracy of the computed results and to determine the effect of sample thickness on the accuracy, deviations of the computed results from the measured temperatures were calculated and tabulated. The results are shown in Table 2.

The data show that there is a rapid rise in the deviation as the laminate thickness increases beyond 80 plies. This is due to the fact that the one-dimensional heat transfer model does not adequately consider the effect of heat dissipation through the sides. However, even at 80 plies, the temperature deviations are within the acceptability of composite manufacturing. The discrepancy will likely decrease as the surface area of the composite increases (relative to the thickness). Therefore, the model provides for highly accurate and important time-temperature profiles in the processing of thermoplastic composites.

Dynamic Viscosity

Dynamic viscosity of film-grade PEEK is shown in Fig. 4. The minimum viscosity reached in PEEK at the upper processing temperature of 390°C is shown to be 3×10^3 Pa · s (3×10^4 poise). The high viscosity of molten PEEK may impede the consolidation of composite laminates in certain processing methods. By comparison, the minimum viscosity for epoxies is on the order of 1 Pa · s (1×10^1 poise). A detailed thermal profile of the composite as predicted by the model can be used with Fig. 4 to establish the viscosities of the laminate at any distance from the heated platens, thereby enabling a satisfactory compaction of the laminate. This will be of immense assistance in the production of satisfactory thermoplastic matrix composites.

Summary and Conclusion

This paper describes a simple yet highly accurate heat transfer model for the processing of thermoplastic matrix composites. Using easily obtainable thermal data and a one-dimensional heat flow model, the predicted temperature distribution within a thick composite coincides very closely with that of experimental values.

Acknowledgments

This research was performed under Rockwell International's independent research and development program.

References

[1] Seferis, J. C. and Velisaris, C. N., *Proceedings*, 31st International SAMPE Symposium, April 1986, p. 1236.
[2] Holman, J. P., in *Heat Transfer*, McGraw-Hill, New York, 1976.
[3] Blundell, D. J. and Osborn, B. N., *Polymer*, Vol. 24, No. 953, 1983.
[4] Data Sheet from Imperial Chemical Industries, 1980.
[5] Nielsen, L. E., *Journal of Composite Materials*, Vol. 1, No. 100, 1967.

John P. Coulter[1] and Selçuk I. Güçeri[2]

Resin Impregnation During the Manufacture of Thermoplastic Matrix Composite Materials

REFERENCE: Coulter, J. P. and Güçeri, S. I., **"Resin Impregnation During the Manufacture of Thermoplastic Matrix Composite Materials,"** *Advances in Thermoplastic Matrix Composite Materials, ASTM STP 1044,* G. M. Newaz, Ed., American Society for Testing and Materials, Philadelphia, 1989, pp. 14–32.

ABSTRACT: A numerical investigation was performed into the resin flow that occurs during the manufacture of thermoplastic matrix composite materials. Specific attention was directed towards modeling the fiber/matrix consolidation that occurs during resin film stacking/compression molding processes. The understanding of these processes is important to the field of composite materials because it is desirable to know the proper processing conditions to apply during manufacturing to obtain completely consolidated, void-free thermoplastic resin composite parts.

The numerical study was based on the assumption of two-dimensional Darcy flow through stationary porous media. Fiber motion was neglected and quasi-steady state flow conditions were assumed. Inhomogeneity and anisotropy of fibrous preforms were allowed for. Due to the degree of geometrical complexity inherent to impregnation processes, the computational technique of boundary-fitted coordinate systems encompassing numerical grid generation was used and was found to be suitable for solving flow problems involving moving boundaries. The boundary conditions chosen for the investigation include constant applied pressure at the inlet, no-slip flow conditions along all walls, and vanishing shear stress along the impregnation front. Resin front/mold wall contact point movement was accomplished using either a no-slip based relocation algorithm or an imposed orthogonality method.

Results are presented in the forms of temporal impregnation front positions and neat resin depth distributions for two sample process configurations. From the analysis of this information, processing parameters are suggested with which the successful manufacturing of thermoplastic composite materials can be accomplished. The need for further experimental and analytical work in this area is also discussed.

KEY WORDS: composites manufacture, resin impregnation, resin film stacking, compression molding, computational fluid dynamics, porous media, numerical grid generation

Nomenclature

$A_{p,k}$, $A_{q,k}$	Mesh concentration amplitudes
B_w, B_{fs}	Boundary point concentration constants
$C_{p,k}$, $C_{q,k}$	Mesh concentration decay factors
$C_{x\xi}$, $C_{x\eta}$	Geometrical coefficients
$C_{y\xi}$, $C_{y\eta}$	Geometrical coefficients
$C_{xx\xi} \cdots C_{xx\xi\eta}$	Geometrical coefficients
$C_{yy\xi} \cdots C_{yy\xi\eta}$	Geometrical coefficients
$C_{xy\xi} \cdots C_{xy\xi\eta}$	Geometrical coefficients

[1]Office of Naval Research graduate fellow and Ph.D. candidate, Department of Mechanical Engineering, University of Delaware, Newark, DE 19716.

[2]Professor, Department of Mechanical Engineering, University of Delaware, Newark, DE 19716.

D_{11}, D_{12}, D_{13} Governing equation coefficients
i, j Computational domain node indices
J Jacobian of the transformation $= (x_\xi y_\eta - x_\eta y_\xi)$
k, k_p, k_q Mesh concentration locating variables
K_x, K_y Fibrous preform principle direction permeabilities, m^2
M, N Mesh size variables
p Thermodynamic pressure, Pa
P, Q Mesh concentration functions
s, s_t Curvilinear boundary distance and length, m
u, v x and y direction velocities, m/s
x, y Physical domain coordinates, m

Greek Symbols

α, β, γ Geometric functions ($\alpha = x_\eta^2 + y_\eta^2$, $\beta = x_\xi x_\eta + y_\xi y_\eta$, $\gamma = x_\xi^2 + y_\xi^2$)
ξ, η Computational domain coordinates
θ Impregnation front orientation angle, radians
μ Molecular viscosity, poise
σ_{xx} Applied pressure, Pa
τ Shear stress, Pa
ψ Stream function, m^2/s

With the acceptance of advanced composite materials as viable components for a wide range of applications, the challenge to effectively produce composite parts has gained significant importance in the manufacturing industry. The traditional composites manufacturing processes are typically labor intensive and, therefore, are not well suited to mass production. Among the newly emerging fabrication techniques that are in the early stages of development, resin transfer molding and resin film stacking/compression molding appear to have good potential for a significant amount of use in producing parts at high volume rates. A process of major importance during both of these manufacturing methods is the impregnation of neat resin into a preformed unwetted fibrous network. In resin transfer molding, the resins which are commonly used are thermosetting systems of relatively low viscosity. Reasonably large impregnation depths can be achieved. Resin film stacking/compression molding, on the other hand, is typically characterized by the use of higher viscosity thermoplastic resin systems with relatively small penetration depths. In both cases, the full understanding of the resin impregnation during the processes is necessary to establish a science base and optimize the use of these methods. The present study addresses this need for an understanding of these impregnation processes, with an emphasis on the manufacturing of thermoplastic composites using the resin film stacking/compression molding process.

In a realistic manufacturing process, there exist both small-scale and large-scale factors that influence the impregnation of resin into a fibrous network, making investigations based on both microscopic and macroscopic flow theories equally important. Investigations which have focused on the small-scale resin/fiber interactions that occur during composites manufacturing have been carried out by several investigators [1–3]. The present study, on the other hand, is based on a macroscopic approach. Assuming that no macroscopic fiber motion occurs, the impregnation of a resin of either thermoplastic or thermosetting nature into a fibrous preform can be considered as flow into a porous medium. Excellent overviews of this subject are available in the literature, which contain pertinent information [4–8], and a concise review of single phase flows through porous media was completed by Dullien [9]. In addition to these works, there exists an abundance of literature concerning flows through porous materials with particular applications to the petroleum, agricultural, filtration, and heating and refrigeration industries.

A significant amount of qualitative information pertaining to flow through porous materials has been recently obtained analytically using boundary layer theories, perturbation analyses, and the method of matched asymptotic expansion [10–15]. Quantitative information, however, has mostly been obtained experimentally, much of this being individual application specific. As far as numerical investigations are concerned, reasonable success has been observed in using several of the currently popular computational approaches to simulate single and multicomponent flows through isotropic porous materials [16–19].

In the area of composites manufacturing, several workers have used the porous medium flow approach to model the flow of resin during traditional autoclave curing processes. Particularly noteable are the works of Springer [20,21], Lee and Springer [22], Loos and Springer [23], Gutowski [24], Gutowski et al. [25–28], and Williams et al. [29,30]. In these studies, liquid-free surface effects were not considered. Analytical and experimental studies in which free surface effects were considered have been performed by Williams et al. [31], Miller and Clark [32,33], and Adams et al. [34]. To date, the only numerical studies in this area were reported by Martin and Son [35] and Coulter and Güçeri [36]. Martin and Son simulated two-dimensional mold filling processes by considering Darcy law type flows and modifying the finite-element code POLYFLOW [37]. These authors also included some related experimental results. Coulter and Güçeri simulated two-dimensional impregnation processes using the numerical method based on boundary-fitted coordinate systems encompassing numerical grid generation. The finite difference based code TGIMPG that was based on this approach is available through the Center for Composite Materials at the University of Delaware. In its initial form, this code could simulate impregnation processes with prescribed injection rates only. Following the initial development of their model, Coulter et al. performed an experimental investigation to test and verify the computational simulation [38]. From this study it was concluded that the package gave good qualitative results and that further numerical analysis related to resin-front/mold-wall contact point movement was necessary. This further expansion of the method of contact point movement within the TGIMPG model was undertaken during the present study. In addition to this, the code was implemented with the capabilities to simulate constant applied pressure processes as well as processes involving impregnation into preforms which are both anisotropic and nonhomogeneous in nature. A treatment of the neat resin region which is common during resin film stacking/compression molding processes was also included.

Problem Formulation

A schematic diagram of a sample resin film stacking/compression molding configuration is shown in Fig. 1. The formulation that is used to model such processes during the present study is based on two initial assumptions. First, it is assumed that the resistance to flow at any given time during the process is caused by the existence of the fibrous preform only. The simplification that follows from this assumption is that the pressure drop across the neat resin region above the preform can be assumed to be negligible. Thus at each point in time during the impregnation process, the applied platen pressure is exerted directly on the resin entering the preform at the neat resin/fibrous preform interface. The subsequent vanishing of the neat resin region then is predicted using a conservation of mass principle, taking into account the fiber volume fraction of the preform. The second assumption chosen for the present study is that the process could be modeled using a quasisteady state approach. This is based on the fact that due to the high-resin viscosities and low-preform permeabilities common to resin film stacking/compression molding processes, the resultant impregnation flow velocities are very small. Assuming this type of behavior, the two-dimensional macroscopic flow of an isothermal Newtonian viscous fluid through a porous medium such as a fibrous preform can be modeled using Darcy's laws, which for a cartesian x-y coordinate system neglecting body forces are

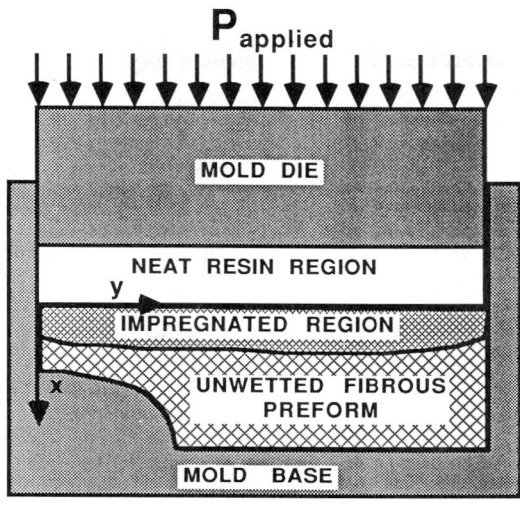

FIG. 1—*Sample resin film stacking/compression molding configuration.*

$$\frac{\partial p}{\partial x} + \frac{\mu}{K_x} u = 0 \tag{1}$$

$$\frac{\partial p}{\partial y} + \frac{\mu}{K_y} v = 0 \tag{2}$$

In the above equations, K_x and K_y represent the preform permeabilities in the x and y directions, respectively. These quantities are usually obtained by experimentation. Using the stream function-velocity relations

$$u = \frac{\partial \psi}{\partial y} \tag{3}$$

$$v = -\frac{\partial \psi}{\partial x} \tag{4}$$

Equations 1 and 2 can be combined in a manner resulting in the elimination of the pressure terms. Performing this operation yields a single equation with the stream function as the dependent variable, which is

$$\frac{\mu}{K_y} \frac{\partial^2 \psi}{\partial x^2} + \frac{\mu}{K_x} \frac{\partial^2 \psi}{\partial y^2} + \left[\frac{1}{K_x} \left(\frac{\partial \mu}{\partial y} \right) - \frac{\mu}{K_x^2} \left(\frac{\partial K_x}{\partial y} \right) \right] \frac{\partial \psi}{\partial y}$$

$$+ \left[\frac{1}{K_y} \left(\frac{\partial \mu}{\partial x} \right) - \frac{\mu}{K_y^2} \left(\frac{\partial K_y}{\partial x} \right) \right] \frac{\partial \psi}{\partial x} = 0 \tag{5}$$

The above expression takes into consideration the fact that both the principle direction permeabilities and the viscosity may vary throughout the flow domain.

Boundary Conditions

Due to the elliptic nature of the governing differential Eq 5, it is necessary to specify conditions along the boundaries of the entire flow domain. The boundary conditions which were applied to the impregnated flowfield region during each quasisteady state timestep are shown in Fig. 2 and can be considered in the following three groups:

Neat Resin Region/Fibrous Preform Interface—As stated previously, the neat resin/fibrous preform interface was modeled as the inlet to the impregnated region flowfield. Thus, it was along this boundary that a constant pressure loading condition was applied. Assuming that the resin is Newtonian and that Stoke's hypothesis is justified, the relationship between resultant or applied normal stress, σ_{xx}, and velocity is

$$\sigma_{xx} = -p + 2\mu \frac{\partial u}{\partial x} \tag{6}$$

Introducing the stream function via Eq 3, this equation can be written as

$$\sigma_{xx} = -p + 2\mu \frac{\partial^2 \psi}{\partial x \partial y} \tag{7}$$

Equation 7, then, serves as the boundary condition for stream function which was satisfied along the entire neat resin/fibrous preform interface. Due to the fact that the mode of applied loading in a compression molding process is compression, the value of σ_{xx} used was always less than zero and corresponded identically to the desired applied platen pressure.

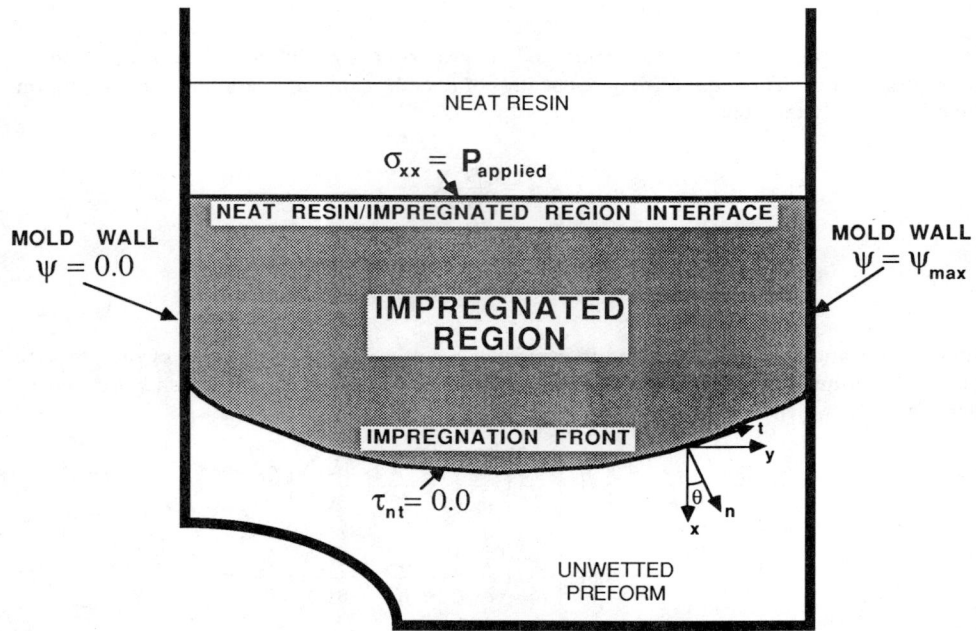

FIG. 2—*Boundary conditions applied during each quasi-steady state timestep.*

In using the constant loading boundary condition, Eq 7, it becomes necessary either to know the resultant thermodynamic pressure distribution along the neat resin/fibrous preform inter-face where the condition is applied or to relate this pressure distribution to a surrounding stream function distribution, thereby eliminating pressure from Eq 7. Due to the belief that the coupled determination of pressure along with stream function would lead to computational complexities of lesser magnitude than the elimination of pressure from Eq 7, the boundary condition was applied as is and pressure was included in the formulation as a dependent vari-able. By integrating and combining Darcy's laws, Eqs 1 and 2, an expression relating thermody-namic pressure drop to stream function distribution can be derived, which is

$$p_A = \left\{ \frac{(\mu_B + \mu_A)(y_A - y_B)}{2(K_{y,B} + K_{y,A})} \right\} \left[\left(\frac{\partial \psi}{\partial x} \right)_B + \left(\frac{\partial \psi}{\partial x} \right)_A \right]$$

$$- \left\{ \frac{(\mu_B + \mu_A)(x_A - x_B)}{2(K_{x,B} + K_{x,A})} \right\} \left[\left(\frac{\partial \psi}{\partial y} \right)_B + \left(\frac{\partial \psi}{\partial y} \right)_A \right] + p_B \qquad (8)$$

The application of this equation to determine the pressure at any point A, which could for example be a point along the neat resin/fibrous preform interface, requires that the pressure at a nearby point B is known. Since the only boundary along which pressure is known during impregnation processes is the impregnation front, where pressure is assumed to be identically zero, a well-posed system of pressure equations can be obtained only through the continuous application of Eq 8 throughout the flowfield, starting at points which are next to the impregna-tion front and ending at the nodal points along the neat resin/fibrous preform interface. This produces a complete set of equations suitable for obtaining the thermodynamic pressure distri-bution along the boundary where the constant applied load, Eq 7, is to be applied.

Solid Boundaries—Along the mold walls the stream function was set to constant values. Along the lower wall, the wall corresponding to the computational domain boundary $\eta = 1$, this constant stream function value was set identically to zero. The upper mold wall, corresponding to the computational domain boundary $\eta = N$, was assigned a constant stream function value which was determined by the flow rate of the resin entering the preform at that instant in time. The liquid/solid contact points of the problem were treated as wall points and therefore as-signed these same upper and lower wall constant stream function values.

Resin Impregnation Front—One of the primary concerns in this class of problems is the treatment of the moving boundary. The boundary condition that was chosen for application to the resin impregnation front during the present study was that of vanishing shear stress at all points on this surface. In terms of the physical coordinate direction velocities, this condition can be stated as

$$\tau_{nt} = \mu \left[\left(\frac{\partial u}{\partial x} - \frac{\partial v}{\partial y} \right) \sin 2\theta - \left(\frac{\partial u}{\partial y} + \frac{\partial v}{\partial x} \right) \cos 2\theta \right] = 0 \qquad (9)$$

where

θ = angle between the outward normal at a point along the free surface and the x-axis.

This representation is shown in Fig. 2. Following the introduction of stream function, this equa-tion becomes

$$\cos 2\theta \left(\frac{\partial^2 \psi}{\partial x^2} \right) + 2 \sin 2\theta \left(\frac{\partial^2 \psi}{\partial x \partial y} \right) - \cos 2\theta \left(\frac{\partial^2 \psi}{\partial y^2} \right) = 0 \qquad (10)$$

It is this equation which was applied at all free surface nodal points excluding the liquid solid contact points during the calculation of each quasi-steady state timestep solution.

Solution Technique

One of the computational challenges of the present class of problems is the common existence of geometrical complexity. This is caused by the irregularity of mold walls and the movement of the impregnation front, which is not known a priori. In order to account for this geometrical complexity, the computational technique based on boundary-fitted coordinate systems (BFCS) encompassing numerical grid generation was selected. This method has been developed primarily during the last decade, and a significant amount of information pertaining to the technique has been reported by Thompson et al. [39–43], Hauser and Taylor [44], and Güçeri [45]. This technique combines the geometrical flexibility of finite-element methods with the computational simplicity of finite difference methods. The BFCS technique previously had been applied to a wide range of thermofluid problems by many investigators including Yost [46], Coulter and Güçeri [47,48], Coulter et al. [49], Wei and Güçeri [50], Muralidhar and Güçeri [51] and has been found to be particularly suitable for moving boundary type problems as reported by Coulter and Güçeri [36], Beyeler et al. [52,53], Subbiah et al. [54], Projahn et al. [55], and Reiger et al. [56]. The underlying concept of the BFCS technique is the mapping of the irregular problem geometry in the physical x-y domain into a regular geometry, such as a rectangular region, in a computational ξ, η domain as shown in Fig. 3. Unlike in the case of other types of transformations where the physical and computational coordinates are related by algebraic expressions, the relations between the physical and computational coordinates are given in the form of dif-

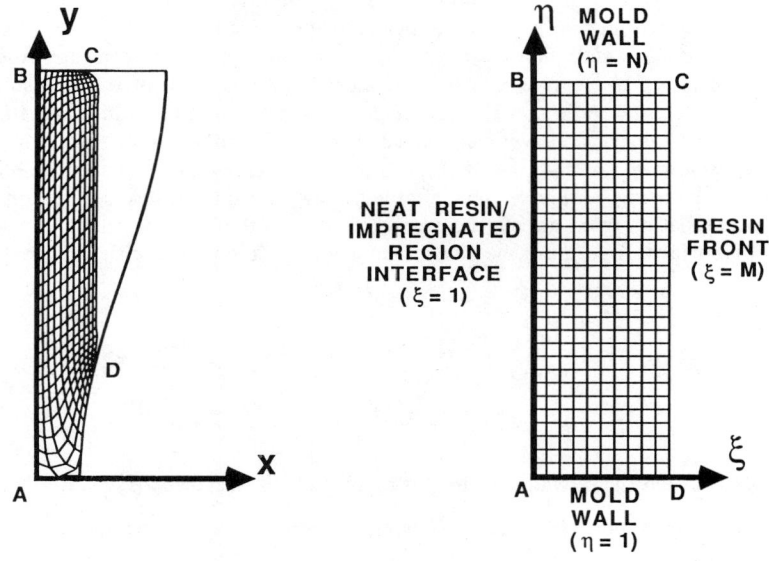

PHYSICAL   DOMAIN   COMPUTATIONAL   DOMAIN

FIG. 3—*Physical and computational domains of the boundary-fitted coordinate system transformation.*

ferential equations. In general, these generating equations can be of elliptic, hyperbolic, or parabolic types. In the present study, elliptic type transformation equations of the Poisson form were chosen as follows

$$\frac{\partial^2 \xi}{\partial x^2} + \frac{\partial^2 \xi}{\partial y^2} = P(\xi, \eta) \tag{11}$$

$$\frac{\partial^2 \eta}{\partial x^2} + \frac{\partial^2 \eta}{\partial y^2} = Q(\xi, \eta) \tag{12}$$

These elliptic equations generate inherently uniform meshes and have the characteristic of smoothening the boundary slope discontinuities in the physical domain. In addition, Poisson equations with the nonhomogeneous terms $P(\xi, \eta)$ and $Q(\xi, \eta)$ allow for grid control to regulate nodal point concentration in desired regions. When Eqs 11 and 12 are rearranged so as to yield corresponding expressions with the physical (x, y) coordinates as the dependent variables, the result is

$$\alpha \frac{\partial^2 x}{\partial \xi^2} - 2\beta \frac{\partial^2 x}{\partial \xi \partial \eta} + \gamma \frac{\partial^2 x}{\partial \eta^2} + J^2 \left(P \frac{\partial x}{\partial \xi} + Q \frac{\partial x}{\partial \eta} \right) = 0 \tag{13}$$

$$\alpha \frac{\partial^2 y}{\partial \xi^2} - 2\beta \frac{\partial^2 y}{\partial \xi \partial \eta} + \gamma \frac{\partial^2 y}{\partial \eta^2} + J^2 \left(P \frac{\partial y}{\partial \xi} + Q \frac{\partial y}{\partial \eta} \right) = 0 \tag{14}$$

Equations 13 and 14, with Dirichlet-type boundary conditions, can be solved in an explicit iterative manner to produce the desired transformation from any simply connected domain of general irregular shape into a rectangular domain containing square unit cells in the computational domain where traditional finite difference techniques can be used to solve the problem. The Dirichlet boundary conditions that must be provided are the x and y positions of the nodal points chosen to define the shape in the physical domain. As can be seen in Fig. 3, boundary point concentration along and interior nodal point concentration towards the mold walls and free surface was applied during the present study. Boundary point concentration was accomplished using expressions of the form

$$s = s_t \left[\frac{\tan^{-1} \left\{ B_w \left(\frac{i-1}{M-1} \right) \right\}}{\tan^{-1} \{ B_w \}} \right] \tag{15}$$

along the mold walls BC and AD and

$$s = s_t \left\{ 0.5 + \left[\frac{\tan^{-1} \left\{ B_{fs} \left[\left(\frac{2j-2}{N-1} \right) - 1 \right] \right\}}{2 \tan^{-1} \{ B_{fs} \}} \right] \right\} \tag{16}$$

along the impregnation front CD, where

$s = $ curvilinear distance along the boundary segment,
$s_t = $ maximum value of s,
B_w and $B_{fs} = $ constant concentration coefficients,

i and j = nodal point indices,

M = number of nodal points in the ξ direction ($i = 1, 2, \ldots M$), and

N = nodal point numbers in the η direction ($j = 1, 2, \ldots N$).

B_w and B_{fs} values of 1.5 and 2.0, respectively, were used to produce the concentration shown in Fig. 3. Interior nodal point concentration was achieved by using functions of the following form as initially suggested by Thompson et al. [40].

$$P(\xi, \eta) = -\sum_{k=1}^{k_p} A_{p,k} \, \text{sign}[\xi - \xi_{0,k}] \exp[-C_{p,k}|\xi - \xi_{0,k}|] \tag{17}$$

$$Q(\xi, \eta) = -\sum_{k=1}^{k_q} A_{q,k} \, \text{sign}[\eta - \eta_{0,k}] \exp[-C_{q,k}|\eta - \eta_{0,k}|] \tag{18}$$

Equations 17 and 18 accomplish the concentration of nodal points towards the curvilinear coordinate lines corresponding to the computational coordinates ξ_0 and η_0, respectively. In these equations, the coefficients $A_{p,k}$, $A_{q,k}$ and $C_{p,k}$, $C_{q,k}$ are the chosen amplitudes and decay factors of interior point concentration, respectively. The grid in Fig. 3 was generated using concentration towards the coordinate lines corresponding to the computational coordinates $\xi = 1$, $\xi = M$, $\eta = 1$, and $\eta = N$. Amplitude and decay factor values of $A_{p,\xi=1} = 20.0$, $A_{p,\xi=M} = 1.0 \times 10^6$, $C_{p,\xi=1} = C_{p,\xi=M} = 0.1$ and $A_{q,\eta=1} = A_{q,\eta=N} = 1.0 \times 10^5$. $C_{q,\eta=1} = C_{q,\eta=N} = 0.08$ were used.

In addition to the transformation of geometrical domain as discussed above, a transformation of the governing equations and boundary conditions must also be performed. Once this is done, the new set of transformed governing equations and boundary conditions can be solved over the computational domain using traditional finite difference methods. The direct one-to-one correspondence between nodal points in the physical and computational domains then gives the problem solution in the physical domain, which represents the flow solution for any given quasi-steady state timestep during the impregnation process.

Quasi-Steady State Timestep Solution

As stated in the previous section, the solution of the present class of problems using BFCS required that the governing stream function Eq 5, boundary condition Eqs 7 and 10, and supporting pressure Eq 8 be transformed into computational domain equations. Upon performing a chain rule differentiation, these expressions take forms as follows. The governing stream function Eq 5 becomes

$$[C_{xx\xi} + D_{11}C_{yy\xi} + D_{12}C_{y\xi} + D_{13}C_{x\xi}] \frac{\partial \psi}{\partial \xi}$$

$$+ [C_{xx\eta} + D_{11}C_{yy\eta} + D_{12}C_{y\eta} + D_{13}C_{x\eta}] \frac{\partial \psi}{\partial \eta}$$

$$+ [C_{xx\xi\xi} + D_{11}C_{yy\xi\xi}] \frac{\partial^2 \psi}{\partial \xi^2} + [C_{xx\eta\eta} + D_{11}C_{yy\eta\eta}] \frac{\partial^2 \psi}{\partial \eta^2}$$

$$+ [C_{xx\xi\eta} + D_{11}C_{yy\xi\eta}] \frac{\partial^2 \psi}{\partial \xi \partial \eta} = 0 \tag{19}$$

where

$$D_{11} = \frac{K_y}{K_x} \qquad (20)$$

$$D_{12} = \frac{K_y}{J\mu K_x}\left[\frac{\partial\mu}{\partial\eta}\frac{\partial x}{\partial\xi} - \frac{\partial\mu}{\partial\xi}\frac{\partial x}{\partial\eta}\right] - \frac{K_y}{JK_x^2}\left[\frac{\partial K_x}{\partial\eta}\frac{\partial x}{\partial\xi} - \frac{\partial K_x}{\partial\xi}\frac{\partial x}{\partial\eta}\right] \qquad (21)$$

$$D_{13} = \frac{1}{J\mu}\left[\frac{\partial\mu}{\partial\xi}\frac{\partial y}{\partial\eta} - \frac{\partial\mu}{\partial\eta}\frac{\partial y}{\partial\xi}\right] - \frac{1}{JK_y}\left[\frac{\partial K_y}{\partial\xi}\frac{\partial y}{\partial\eta} - \frac{\partial K_y}{\partial\eta}\frac{\partial y}{\partial\xi}\right] \qquad (22)$$

The constant applied pressure boundary condition Eq 7 transforms into

$$\sigma_{xx} = -p + 2\mu\left\{C_{xy\xi}\frac{\partial\psi}{\partial\xi} + C_{xy\eta}\frac{\partial\psi}{\partial\eta} + C_{xy\xi\xi}\frac{\partial^2\psi}{\partial\xi^2} + C_{xy\eta\eta}\frac{\partial^2\psi}{\partial\eta^2} + C_{xy\xi\eta}\frac{\partial^2\psi}{\partial\xi\partial\eta}\right\} \qquad (23)$$

with the pressure equation supporting this boundary condition taking the form:

$$p_A = \left\{\frac{(\mu_B + \mu_A)(y_A - y_B)}{2(K_{y,B} + K_{y,A})}\right\}\left[\left(C_{x\xi}\frac{\partial\psi}{\partial\xi} + C_{x\eta}\frac{\partial\psi}{\partial\eta}\right)_{,B} + \left(C_{x\xi}\frac{\partial\psi}{\partial\xi} + C_{x\eta}\frac{\partial\psi}{\partial\eta}\right)_{,A}\right]$$

$$- \left\{\frac{(\mu_B + \mu_A)(x_A - x_B)}{2(K_{x,B} + K_{x,A})}\right\}\left[\left(C_{y\xi}\frac{\partial\psi}{\partial\xi} + C_{y\eta}\frac{\partial\psi}{\partial\eta}\right)_{,B} + \left(C_{y\xi}\frac{\partial\psi}{\partial\xi} + C_{y\eta}\frac{\partial\psi}{\partial\eta}\right)_{,A}\right]$$

$$+ p_B \qquad (24)$$

Lastly, the vanishing shear stress boundary condition of Eq 10 to be applied along the impregnation front becomes:

$$\cos 2\theta\left[(C_{xx\xi} - C_{yy\xi})\frac{\partial\psi}{\partial\xi} + (C_{xx\eta} - C_{yy\eta})\frac{\partial\psi}{\partial\eta} + (C_{xx\xi\xi} - C_{yy\xi\xi})\frac{\partial^2\psi}{\partial\xi^2}\right.$$

$$+ (C_{xx\eta\eta} - C_{yy\eta\eta})\frac{\partial^2\psi}{\partial\eta^2} + (C_{xx\xi\eta} - C_{yy\xi\eta})\frac{\partial^2\psi}{\partial\xi\partial\eta}\right]$$

$$+ 2\sin 2\theta\left[C_{xy\xi}\frac{\partial\psi}{\partial\xi} + C_{xy\eta}\frac{\partial\psi}{\partial\eta} + C_{xy\xi\xi}\frac{\partial^2\psi}{\partial\xi^2} + C_{xy\eta\eta}\frac{\partial^2\psi}{\partial\eta^2} + C_{xy\xi\eta}\frac{\partial^2\psi}{\partial\xi\partial\eta}\right] = 0 \qquad (25)$$

The C's in the above equations are geometrical transformation coefficients and are listed in Fig. 4. Following a second-order accurate descretization process, Eqs 19, 23, 24, and 25, along with the conditions of constant stream function along the mold walls and zero pressure along the impregnation front, form a linear system of $2 \times M \times N$ equations in $2 \times M \times N$ unknowns, the unknowns being the stream function and thermodynamic pressure values at each of the nodal points in the impregnated region flowfield. Due to the numerical stability characteristics of the system, the use of a fully implicit finite difference solution technique was necessary to generate results.

$C_{xx\xi} = \frac{1}{J^2}\left[y_\eta y_{\xi\eta} - y_\xi y_{\eta\eta} + \frac{J_\eta}{J} y_\xi y_\eta - \frac{J_\xi}{J} y_\eta^2\right]$	$C_{xx\eta} = \frac{1}{J^2}\left[y_\xi y_{\xi\eta} - y_\eta y_{\xi\xi} + \frac{J_\xi}{J} y_\eta y_\eta - \frac{J_\eta}{J} y_\xi^2\right]$
$C_{xx\xi\xi} = \frac{1}{J^2} y_\eta^2$	$C_{xx\eta\eta} = \frac{1}{J^2} y_\xi^2$
$C_{xx\xi\eta} = \frac{-2}{J^2} y_\xi y_\eta$	$C_{yy\xi} = \frac{1}{J^2}\left[x_\eta x_{\xi\eta} - x_\xi x_{\eta\eta} + \frac{J_\eta}{J} x_\xi x_\eta - \frac{J_\xi}{J} x_\eta^2\right]$
$C_{yy\eta} = \frac{1}{J^2}\left[x_\xi x_{\xi\eta} - x_\eta x_{\xi\xi} + \frac{J_\xi}{J} x_\xi x_\eta - \frac{J_\eta}{J} x_\xi^2\right]$	$C_{yy\xi\xi} = \frac{1}{J^2} x_\eta^2$
$C_{yy\eta\eta} = \frac{1}{J^2} x_\xi^2$	$C_{yy\xi\eta} = \frac{-2}{J^2} x_\xi x_\eta$
$C_{xy\xi} = \frac{1}{J^2}\left[x_\xi y_{\eta\eta} - x_\eta y_{\xi\eta} + \frac{J_\xi}{J} x_\eta y_\eta - \frac{J_\eta}{J} x_\xi y_\eta\right]$	$C_{xy\eta} = \frac{1}{J^2}\left[x_\eta y_{\xi\xi} - x_\xi y_{\xi\eta} + \frac{J_\eta}{J} x_\xi y_\xi - \frac{J_\xi}{J} x_\eta y_\xi\right]$
$C_{xy\xi\xi} = \frac{-1}{J^2} x_\eta y_\eta$	$C_{xy\eta\eta} = \frac{-1}{J^2} x_\xi y_\xi$
$C_{xy\xi\eta} = \frac{1}{J^2}[x_\xi y_\eta + x_\eta y_\xi]$	$C_{x\xi} = \frac{1}{J} y_\eta$
$C_{x\eta} = \frac{-1}{J} y_\xi$	$C_{y\xi} = \frac{-1}{J} x_\eta$
$C_{y\eta} = \frac{1}{J} x_\xi$	

FIG. 4—*Geometrical coefficients in the transformed governing equations.*

Resin Movement

Impregnation Front Movement

In the interim between successive quasi-steady state timestep solutions, movement of the resin impregnation front within the mold was carried in a manner shown in Figs. 5 and 6. First, the resultant free surface node velocities (the velocities at nodes 2, 3, . . . in Figs. 5 and 6) were obtained from the previous timestep solution. These velocity values were then used, along with a predetermined time step value, to relocate the free surface nodes which were not contact points. This is indicated by the movement of the previous timestep impregnation front nodes 2, 3, . . . in Figs. 5 and 6 to the positions 2′, 3′, . . . along the new resin front. New impregnation front/ mold wall contact points were determined using either a nonslip or imposed orthogonality relocation method. During nonslip based contact point movement, shown in Fig. 5, new impregnation front/mold wall contact points, such as point 1′ in the figure, were determined by locating intersections between cubic spline curve fits of the mold walls (the points W1, W2, . . .) and a cubic spline curve fit of the newly determined resin front using the contact point from the previous timestep (a spline of the points 1, 2′, 3′, . . .). In some cases, this contact point relocation was found to bring about the elimination from the flow domain of newly determined resin front nodal points such as the point 2′ in the figure which are adjacent to the previous contact point. When this occurred, it was necessary to add replacement points along the newly determined free surface so as to maintain the invariance of the number of nodes in the computational domain η direction. The resultant new resin impregnation front after this redistribution of nodal points is shown as the points 1′, 2″, 3″, . . . in Fig. 5.

The imposed orthogonality type of contact point relocation is depicted in Fig. 5. This method imposes the condition that the resultant new impregnation front and the mold walls meet orthogonally. Mathematically, this condition consists of the determination of the point along the mold wall which satisfies the relation

$$\beta = \frac{\partial x}{\partial \xi}\frac{\partial x}{\partial \eta} + \frac{\partial y}{\partial \xi}\frac{\partial y}{\partial \eta} = 0.0 \qquad (26)$$

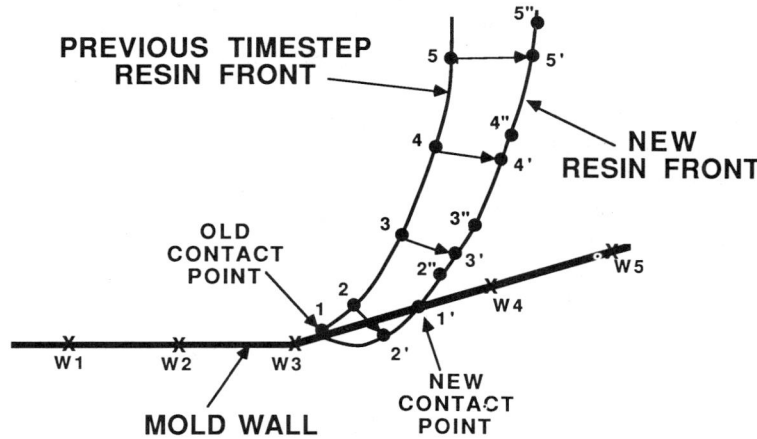

FIG. 5—*Methods of impregnation front movement, no-slip based method.*

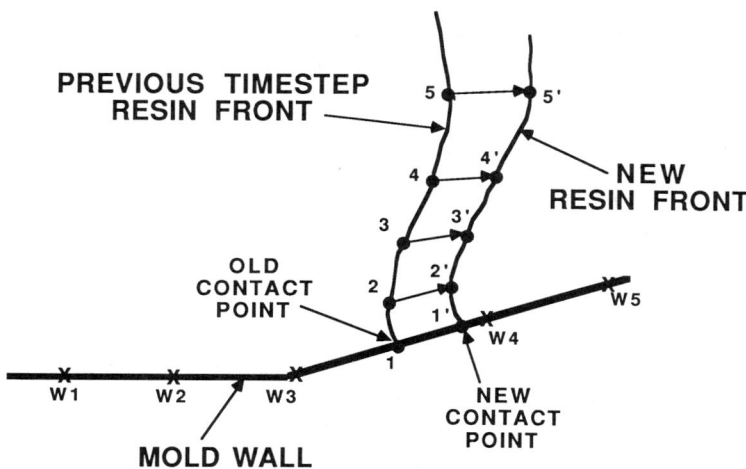

FIG. 6—*Methods of impregnation front movement, imposed orthogonality method.*

The type of contact point relocation used during a simulation was based on the relative magnitudes of the tangential velocities in near wall regions as suggested by Coulter et al. [38]. This suggestion was motivated by experimental findings which showed that, due to the existence of fiber terminations near walls, local regions of higher preform permeability result and allow a slip-like phenomenon to occur to a small degree. This small amount of slip, however, was found only to be significant near walls where the local velocities were small in comparison with the velocities in the remainder of the flowfield. The numerical simulation of this behavior was accomplished as follows. If the near wall tangential resin velocity was found to be above a chosen percentage of the maximum velocity in the flowfield, the contact point relocation method based on the no-slip criterion was used. Alternatively, the imposed orthogonality method was used when the resultant near wall tangential velocities fell below the specified percentage of the maximum flowfield velocity.

The successive growth in the physical size of the fluid flowfield as time progressed during impregnation simulations brought about a need for increasing the number of computational nodal points in the ξ direction as the number of timesteps performed during each simulation grew. This mesh size enlargement did not produce any significant computational difficulties because of the quasi-steady state nature of the processes studied which effectively uncoupled solutions of successive timesteps. As for the selection of proper timesteps for mesh size enlargement, a new column of mesh nodes was added to the flowfield every time the average absolute physical distance between free surface and inlet source nodal points increased by a predetermined percentage of the average initial fluid configuration width.

Mold Die Movement

In order to simulate realistic resin film stacking/compression molding processes, a treatment of the mold die movement corresponding to the vanishing of the neat resin region during impregnation was necessary. During the present study, the movement of the die was modeled using a mass conservation approach, taking into consideration the fiber volume fraction of the preform, V_f, which was assumed to be known. An initial neat resin region depth was assumed at the start of each simulation. This resin depth was then adjusted following each timestep solution calculation by determining the average velocity of the resin entering the preform at the neat resin/fibrous preform interface, $u_{\text{avg interface}}$, and then calculating the instantaneous mold die velocity, u_{die}, using the relationship

$$u_{\text{die}} = u_{\text{avg interface}}(1.0 - V_f) \tag{27}$$

The multiplication of this instantaneous mold die velocity by the chosen timestep then indicated the incremental distance of travel of the mold die during that timestep.

Results and Discussion

As mentioned earlier, the present formulation and solution methodology was partially validated for impregnation processes characterized by prescribed resin injection rate in a previous study involving low viscosity resins [38]. Unfortunately, no data has yet been obtained for cases involving both prescribed applied pressure loading conditions and higher viscosity thermoplastic resins. In order to qualitatively validate the present model, however, a test case of resin impregnation into a fibrous preform encased by a rectangular mold was considered. The mold was chosen to be 0.2 m wide and the preform to be 0.02 m deep. The preform was assumed to be isotropic with permeability values of $K_x = K_y = 1.0 \times 10^{-10}$ m^2 (≈ 100 Darcy). A constant resin viscosity of $\mu = 2000$ poise was chosen along with a constant applied pressure of $p_{\text{applied}} = 1.5 \times 10^6$ Pa. In order to start the simulation, an initial impregnation depth of 0.0025 m was necessary. This depth, along with an assumption of an incompressible preform with a fiber volume fraction of $V_f = 50\%$ led to an initial neat resin region depth of 0.00875 m. A sequence of impregnation configurations which resulted for this case is presented in Fig. 7. The forced orthogonality type of contact point movement was used throughout the simulation. As expected, the impregnation of resin into the preform occurred rapidly in the early stages of the process and slowed down as the depth of penetration into the preform increased. The time of complete preform wetting was $t_{\text{fill}} = 283$ s. The temporal distribution of neat resin region depth corresponding to this case is shown in Fig. 8. This distribution also shows the decay in the rate of resin impregnation as the process proceeded. In performing this simulation, an initial mesh size of (6 $\times$ 41) was chosen which was enlarged to a final size of (13 $\times$ 41) as the impregnated

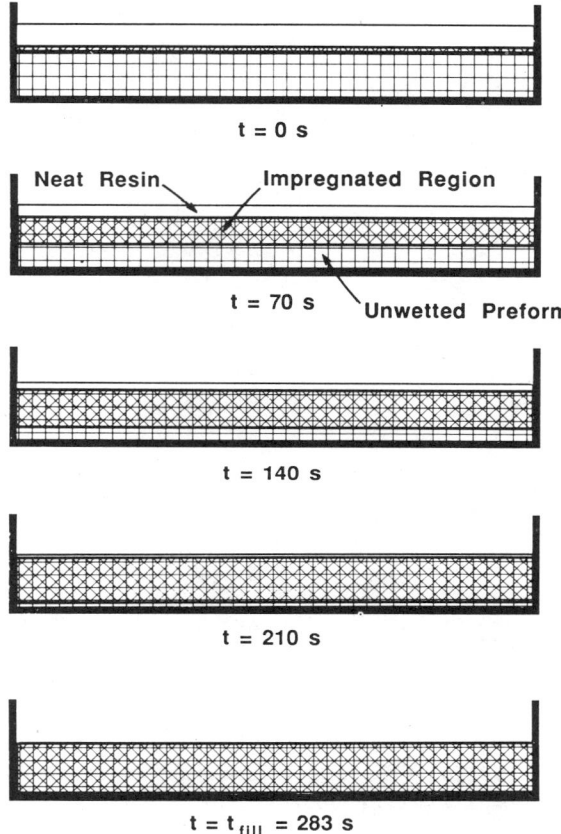

FIG. 7—*Sequence of configurations during resin impregnation into a fibrous preform encased by a rectangular mold.*

region grew. Timestep increments between 2 and 10 s were used resulting in a total of 48 quasi-steady state time step calculations. Computationally, between 3 min and 40 s and 10 min and 29 s of CPU time were required to perform calculations for each timestep on a Vax 11/785 mainframe computer.

The geometrical capability of the TGIMPG package is exhibited by the simulation of resin impregnation into a preform in a mold with a sinusoidal base shown in Figs. 9 and 10. The mold was chosen to be 0.05 m wide and the preform was given a depth ranging from 0.005 to 0.015 m. An initial impregnated region depth of 0.0025 m was chosen, which along with a fiber volume fraction assumption of $V_f = 50\%$ resulted in an initial neat resin region depth of 0.00375 m. The resin viscosity and preform permeability values used were the same as in the rectangular mold case presented above. The applied pressure at the neat resin/impregnated region interface was again chosen as $p_{applied} = 1.5 \times 10^6$ Pa. A contact point movement type parameter of 10% of the maximum flowfield velocity was selected, which resulted in the use of no-slip based contact point movement along both walls throughout the simulation. A sequence of impregnation configurations which resulted is shown in Fig. 9 and the corresponding temporal neat resin region depth distribution is presented in Fig. 10. Again, the rate of impregnation was seen to

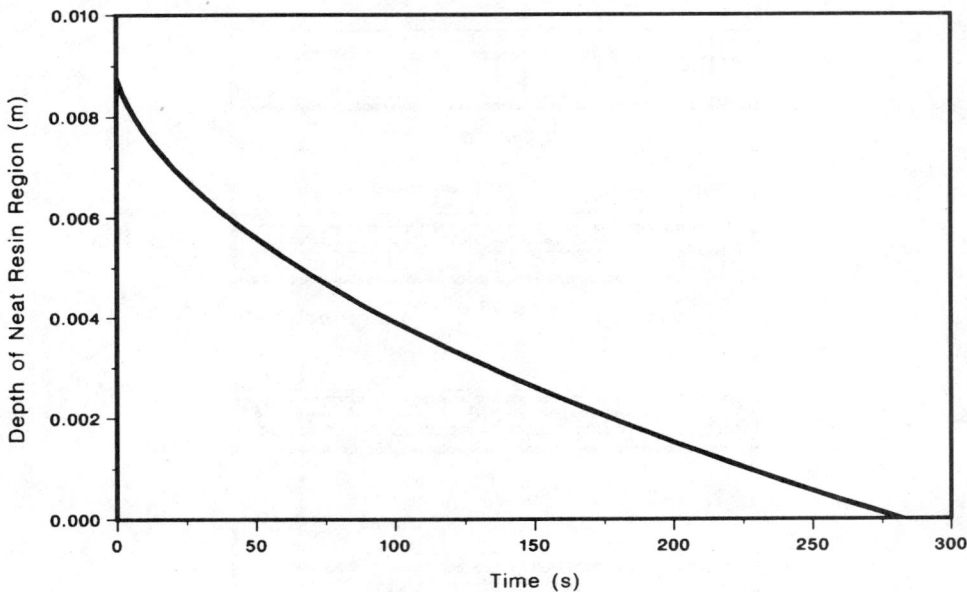

FIG. 8—*Temporal distribution of neat resin region depth during resin impregnation into a fibrous preform encased by a rectangular mold.*

decay with time as the process proceeded. The total preform wetting time for this case was 150 s. Mesh sizes ranging from (6 × 31) to (17 × 31) were used along with a timestep increment range of 0.5 to 5.0 s. The quasi-steady state timestep calculations required CPU times between 1 min and 16 s and 20 min and 1 s. A total of 49 timestep calculations were performed before the complete impregnation of the preform was achieved.

Conclusions

The present investigation consisted of the initial development of a numerical model to simulate resin impregnation processes which commonly occur during the manufacturing of thermoplastic matrix composite materials. The resultant computational code, TGIMPG, was shown to produce reasonable qualitative results for two sample resin film stacking/compression molding processes. The additional features of the code, including the capability to accommodate nonhomogeneous anisotropic fibrous preforms, variable viscosity resins, and prescribed resin injection rate processes, were not exhibited. These features either have been shown in previous publications or will be discussed in detail in future reports. Overall, the success of the TGIMPG code has shown that the numerical technique of boundary-fitted coordinate systems using numerical grid generation is applicable to flow problems involving irregular and/or changing geometry flowfields.

Turning to the area of thermoplastic matrix composites manufacturing, it has been suggested that the impregnation of unwetted fibrous preforms with high viscosity thermoplastic resin systems using a resin film stacking/compression molding process is possible under reasonable applied pressure loadings. Future experimental work is necessary before the numerical model developed can be rigorously tested. This experimentation could consist of the performing of fully monitored resin film stacking/compression molding. Investigations directed towards the deter-

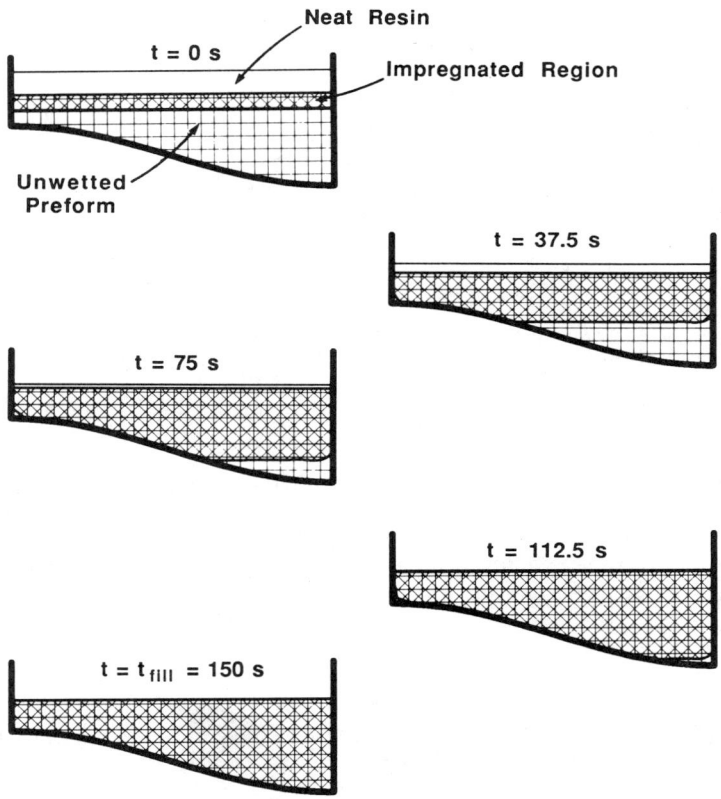

FIG. 9—*Sequence of configurations during resin impregnation into a fibrous preform encased by a mold with a sinusoidal base.*

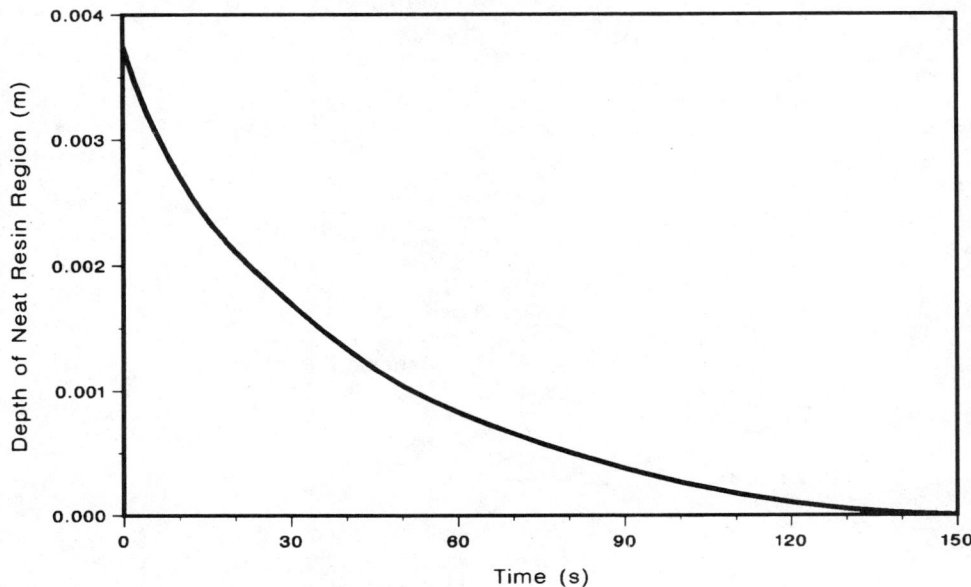

FIG. 10—*Temporal distribution of neat resin region depth during resin impregnation into a fibrous preform encased by a mold with a sinuosoidal base.*

mining of permeability values of fibrous preforms under compressive loading is also needed. With these studies, the validity of the present model can be further tested. The result, in either the current or a modified form, will be a software package that will provide the composites industry with a predictive capability for understanding thermoplastic composite manufacturing processes, as well as increasing the design efficiency, manufacturing reliability, and cost effectiveness.

References

[1] Larson, R. E. and Higdon, J. J. L., "Microscopic Flow Near the Surface of Two-Dimensional Porous Media. Part 1: Axial Flow," *Journal of Fluid Mechanics,* Vol. 166, 1986, pp. 449–472.

[2] Peterson, B. K., Walton, J. P. R. B., and Gubbins, K. E., "Microscopic Studies of Fluids in Pores: Computer Simulation and Mean-Field Theory," *International Journal of Thermophysics,* Vol. 6, No. 6, 1985, pp. 585–593.

[3] Behrens, R. A., *Transient Domain Free Surface Flows and Their Applications to Mold Filling,* Technical Report No. CCM-83-14, Center for Composite Materials, University of Delaware, Newark, DE, 1983.

[4] Bear, J., *Dynamics of Fluids in Porous Media,* Elsevier, New York, 1972.

[5] Scheidegger, A. E., *The Physics of Flow Through Porous Media,* 3rd ed., University of Toronto Press, Toronto, 1974.

[6] Dullien, F. A. L., *Porous Media: Fluid Transport and Pore Structure,* Academic Press, New York, 1979.

[7] Slattery, J. C., *Momentum, Energy, and Mass Transfer in Continua,* McGraw-Hill, New York, 1961.

[8] Bejan, A., "Principles of Convection Through Porous Media," *Convection Heat Transfer,* John Wiley and Sons, New York, 1984, pp. 343–387.

[9] Dullien, F. A. L., "Single Phase Flow Through Porous Media and Pore Structure," *The Chemical Engineering Journal,* Vol. 10, 1975, pp. 1–34.

[10] Pan'ko, S. V., "Exact Solutions of Boundary-Value Problems for Nonlinear Flow in Porous Media," *Fluid Dynamics,* Vol. 20, No. 3, 1985, pp. 427–432.

[11] Vafai, K., "Convective Flow and Heat Transfer in Variable-Porosity Media," *Journal of Fluid Mechanics,* Vol. 147, 1984, pp. 233–259.

[12] Panfilov, M. B., "A Symptotic Form of the Solution to the Problem of Multicomponent Mixture Flow Through a Porous Medium with a Boundary Layer," *Fluid Dynamics,* Vol. 20, No. 4, 1985, pp. 574–580.

[13] Tyvand, P. A., "Decay of a Disturbed Free Surface in an Anisotropic Porous Medium," *Journal of Hydrology,* Vol. 83, Nos. 3-4, 1986, pp. 367–371.

[14] De La Cruz, V., Spanos, T. J. T., and Sharma, R. C., "Stability of a Steam-Water Front in a Porous Medium," *Canadian Journal of Chemical Engineering,* Vol. 63, No. 5, 1985, pp. 735–745.

[15] Advani, S. H., Torok, J. S., and Lee, J. K., "General Solutions for Pistonlike Displacement of Compressible Fluids in Porous Media," *Journal of Energy Resources Technology,* Vol. 107, No. 4, 1985, pp. 523–526.

[16] Levi, B. I. and Shakirov, Kh. G., "Numerical Solution of Flow of Multicomponent Mixtures in Porous Media," *Fluid Dynamics,* Vol. 20, No. 4, 1985, pp. 581–590.

[17] Prasad, V., "Natural Convection in Porous Media—An Experimental and Numerical Study for Vertical Annular and Rectangular Enclosures," Ph.D. thesis, University of Delaware, Newark, DE, 1983.

[18] Levin, D. and Tal, A., "Boundary Collocation Method for the Solution of a Flow Problem in a Complex Three-Dimensional Porous Medium," *International Journal of Numerical Methods in Fluids,* Vol. 6, No. 9, 1986, pp. 611–622.

[19] Ewing, R. E., "Finite Element Methods for Nonlinear Flows in Porous Media," *Computational Methods Applied to Mechanical Engineering,* Vol. 51, Nos. 1-3, 1985, pp. 421–439.

[20] Springer, G. S., "Resin Flow During the Cure of Fiber Reinforced Composites," *Journal of Composite Materials,* Vol. 16, 1982, pp. 400–410.

[21] Springer, G. S., "Modeling the Cure Process of Composites," *Proceedings of the 31st International SAMPE Symposium,* Las Vegas, NV, April 1986, pp. 776–787.

[22] Lee, W. I. and Springer, G. S., "Microwave Curing of Composites," *Journal of Composite Materials,* Vol. 18, 1984, pp. 387–409.

[23] Loos, A. C. and Springer, G. S., "Curing of Epoxy Matrix Composites," *Journal of Composite Materials,* Vol. 17, No. 2, 1983, pp. 135–169.

[24] Gutowski, T. G., "A Resin Flow/Fiber Deformation Model for Composites," *SAMPE Quarterly,* Vol. 16, No. 4, 1985, pp. 58–64.

[25] Gutowski, T. G., Cai, J., Kingery, J., and Wineman, S. J., "Resin Flow/Fiber Deformation Experiments," *SAMPE Quarterly,* Vol. 17, No. 4, 1986, pp. 54–58.

[26] Gutowski, T. G., Wineman, S., and Cai, Z., "Applications of the Resin Flow/Fiber Deformation Model," *Proceedings of the 31st International SAMPE Symposium,* Las Vegas, NV, April 1986, pp. 245–254.

[27] Gutowski, T. G., Tadahiko, M., and Zhong, C., "The Consolidation of Laminate Composites," *Journal of Composite Materials,* Vol. 21, No. 2, 1987, pp. 172–187.

[28] Gutowski, T. G., Cai, Z., Bauer, S., Boucher, D., Kingery, J., and Wineman, S., "Consolidation Experiments for Laminate Composites," *Journal of Composite Materials,* 1987, submitted.

[29] Williams, J., Donnellan, T., and Trabocco, R., "A Predictive Model for Resin Flow During Composite Processing," Technical Report NADC-85164-60, Naval Air Development Center, Warminster, PA, 1985.

[30] Williams, J., Donnellan, T., and Trabocco, R., "Experimental Verification of the NADC Composite Resin Flow Model," Technical Report NADC-86048-60, Naval Air Development Center, Warminster, PA, 1986.

[31] Williams, J. G., Morris, C. E. M., and Ennis, B. C., "Liquid Flow Through Aligned Fiber Beds," *Polymer Engineering and Science,* Vol. 14, No. 6, 1974, pp. 413–419.

[32] Miller, B. and Clark, D. B., "Liquid Transport Through Fabrics: Wetting and Steady-State Flow, Part i: A New Experimental Approach," *Textile Research Journal,* Vol. 48, 1978, pp. 150–155.

[33] Clark, D. B. and Miller, B., "Liquid Transport Through Fabrics: Wetting and Steady-State Flow, Part ii: Fabric Wetting," *Textile Research Journal,* Vol. 48, 1978, pp. 256–260.

[34] Adams, K. L., Miller, B., and Rebenfeld, L., "Forced In-Plane Flow of an Epoxy Resin in Fibrous Networks," *Polymer Engineering and Science,* Vol. 26, No. 20, 1986, pp. 1434–1441.

[35] Martin, G. Q. and Son, J. S., "Fluid Mechanics of Mold Filling for Fiber Reinforced Plastics," *Advanced Composites: The Latest Developments, Proceedings of the ASM/ESD Second Conference on Advanced Composites,* Dearborn, MI, November 1986, pp. 149–157.

[36] Coulter, J. P. and Güçeri, S. I., "Resin Impregnation During the Manufacturing of Composite Materials Subject to Prescribed Injection Rate," *International Journal of Reinforced Plastics and Composites,* Vol. 7, No. 3, 1987, pp. 200–219.

[37] Crochet, M. J., Davies, A. R., and Walters, K., *Numerical Simulation of Non-Newtonian Flow,* Elsevier Science Publishing Co., New York, 1984.

[38] Coulter, J. P. and Güçeri, S. I., "Resin Impregnation During the Manufacturing of Composite Materials: Theory and Experimentation," *Composites Science and Technology,* 1987.

[39] Thompson, J. F., Thames, F. C., and Mastin, C. W., "Automatic Numerical Generation of Boundary-Fitted Curvilinear Coordinate System for Field Containing Any Number of Arbitrary Two-Dimensional Bodies," *Journal of Computational Physics,* Vol. 15, 1974, pp. 219–319.

[40] Thompson, J. F., Thames, F. C., and Mastin, C. W., "Tomcat—A Code for Numerical Generation of Boundary-Fitted Curvilinear Coordinate Systems on Fields Containing Any Number of Arbitrary Two-Dimensional Bodies," *Journal of Computational Physics,* Vol. 24, 1977, pp. 274–302.

[41] Thompson, J. F., Warsi, Z. U. A., and Mastin, C. W., "Boundary-Fitted Coordinate Systems for Numerical Solution of Partial Differential Equations—A Review," *Journal of Computational Physics,* Vol. 47, 1982, pp. 1–108.

[42] Thompson, J. F., *Numerical Grid Generation,* Elsevier Science Publishing Co., Inc., New York, 1982.

[43] Thompson, J. F., *A Survey of Grid Generation Techniques in Computational Fluid Dynamics,* Technical Report AIAA Paper 83-0447, AIAA 21st Aerospace Sciences Meeting, 1983.

[44] Hauser, J. and Taylor, C. D., *Numerical Grid Generation in Computational Fluid Dynamics,* Pineridge Press, Swansea, UK, 1986.

[45] Güçeri, S. I., "Finite Difference Methods in Polymer Processing," *Fundamentals of Computer Modeling for Polymer Processing,* C. L. Tucker, Ed., Hanser Publishing Company, 1988.

[46] Yost, B. A., "The Analysis of Fluid Flow/Solidification Problems in Arbitrarily Shaped Domains," Ph.D. thesis, University of Delaware, Newark, DE, 1984.

[47] Coulter, J. P. and Güçeri, S. I., "Laminar and Turbulent Natural Convection in Irregularly Shaped Enclosures," Technical Report CCM-86-08, Center for Composite Materials, University of Delaware, Newark, DE, 1986.

[48] Coulter, J. P. and Güçeri, S. I., "Laminar and Turbulent Natural Convection Within Irregularly Shaped Enclosures," *Numerical Heat Transfer,* Vol. 12, 1987, pp. 211–227.

[49] Coulter, J. P., Gilmore, S. D., and Güçeri, S. I., "Tgflow—A Software Package for the Analysis of Laminar Fluid Flow," *Numerical Grid Generation in Computational Fluid Dynamics,* J. Hauser and C. Taylor, Eds., Pineridge Press, Swansea, UK, 1986, pp. 515–526.

[50] Wei, S. S. and Güçeri, S. I., "Laminar and Turbulent Recirculating Flows Through Irregular Domains," *International Journal of Heat and Fluid Flow,* 1987, submitted.

[51] Muralidhar, K. and Güçeri, S. I., "Comparative Study of Two Numerical Procedures for Free-Convection Problems," *Numerical Heat Transfer,* Vol. 9, 1986, pp. 631–638.

[52] Beyeler, E. P. and Güçeri, S. I., "Tgsold—A Software to Simulate Solidification in Irregularly Shaped Domains," *Numerical Grid Generation in Computational Fluid Dynamics,* J. Hauser and C. Taylor, Eds., Pineridge Press, Swansea, UK, 1986, pp. 655–666.

[53] Beyeler, E. P. and Güçeri, S. I., "Two-Dimensional Solidification in Irregularly Shaped Domains," *Journal of Heat Transfer,* 1986, submitted.

[54] Subbiah, S., Trafford, D. L., and Güçeri, S. I., "Non-isothermal Flow of Polymers into Two-Dimensional, Thin Cavity Molds: A Numerical Grid Generation Approach," *International Journal of Heat and Mass Transfer,* 1987, submitted.

[55] Projahn, U., Rieger, H., and Beer, H., "Numerical Analysis of Laminar Natural Convection Between Concentric and Eccentric Cylinders," *Numerical Heat Transfer,* Vol. 4, 1981, pp. 131–146.

[56] Rieger, H., Projahn, U., and Beer, H., "Analysis of the Heat Transport Mechanisms During Melting Around a Horizontal Circular Cylinder," *International Journal of Heat and Mass Transfer,* Vol. 25, 1982, pp. 137–147.

Jeremy C. Howes,[1] Alfred C. Loos,[1] and Jeffrey A. Hinkley[2]

The Effect of Processing on Autohesive Strength Development in Thermoplastic Resins and Composites

REFERENCE: Howes, J. C., Loos, A. C., and Hinkley, J. A., **"The Effect of Processing on Autohesive Strength Development in Thermoplastic Resins and Composites,"** *Advances in Thermoplastic Matrix Composite Materials, ASTM STP 1044,* G. M. Newaz, Ed., American Society for Testing and Materials, Philadelphia, 1989, pp. 33–49.

ABSTRACT: The effects of processing on autohesive bond strength development in polysulfone resin and graphite-polysulfone composites were investigated. Autohesive bond strength development in polysulfone resin was observed by measuring the refracture toughness of precracked compact tension specimens that were healed at a given temperature and contact time. The fractured specimens, when completely healed, achieved the original toughness of the virgin resin. Results of the compact tension (CT) tests, when corrected for nonisothermal effects, compared favorably with diffusion models explaining crack healing and welding of amorphous polymers. Interply strength development in graphite-polysulfone unidirectional composites was measured as a function of healing temperature and contact time using a double cantilever beam (DCB) interlaminar fracture toughness test. The critical strain energy release rate of the refractured composites did not show a strong time or temperature dependence as observed in the neat resin tests. Furthermore, only 80 to 90% of the undamaged fracture energy can be recovered. The fracture mechanisms were determined to be different in the healed DCB specimen from those in the virgin specimen due to resin flow at the crack plane upon healing.

KEY WORDS: autohesion, self-diffusion, polysulfone, thermoplastic, composites, process modeling, composite processing, polymers

The use of advanced fiber-reinforced composites has increased significantly in recent years. High specific strength and stiffness make composites candidate materials for many aerospace and space applications. However, fiber-reinforced composites using thermosetting matrix resins, such as epoxies, have low damage tolerance and low service temperatures when compared to the more traditional aerospace materials. To overcome these shortcomings, there is great interest in the use of thermoplastic resins as matrix materials for fiber-reinforced composites.

Thermoplastic resins are generally high toughness materials and subsequently can improve the damage tolerance of composites. However, the mechanisms describing consolidation and interply bonding during processing of thermoplastic matrix composites are quite different than the mechanisms observed during cure of thermosetting matrix composites. Extreme tow height nonuniformity and lack of flow make thermoplastic prepregs more difficult to process than thermosetting prepregs. Unlike thermosetting resins, which rely on low viscosity and high flow of the resin to coalesce the ply interfaces, thermoplastic matrix prepregs must be physically deformed to cause intimate contact and coalescence of the ply interfaces.

The mechanisms explaining intimate contact and consolidation in thermoplastic composites

[1]Graduate research assistant and associate professor, respectively, Department of Engineering Science and Mechanics, Virginia Polytechnic Institute and State University, Blacksburg, VA 24061.

[2]Chemical engineer, Polymeric Materials Branch, NASA Langley Research Center, Hampton, VA 23665.

have been established as viscoelastic deformation and autohesive bonding, respectively [1]. These mechanisms are not well quantified and present processing models developed for thermosetting resin composites cannot be applied directly to thermoplastic matrix composites. Processing cycles for thermoplastic composites currently are derived empirically by trial and error. These methods do not necessarily lead to processing cycles that result in fully consolidated composite structures with strong interply bonds. In order to improve the processing theory for continuous fiber-reinforced thermoplastic matrix composites, the processing parameters, temperature, pressure, and time must be related to the overall state of consolidation in the composite.

In development of a processing model, the physical processes that occur during production of thermoplastic composites must be understood fully. The mechanism controlling interply bond formation (consolidation) has been recognized as autohesion or self-diffusion. Numerous theories have been developed describing autohesion in neat resins, and tests to determine autohesive strength are available. Autohesion in fiber-reinforced thermoplastic prepregs is not well understood due to the complications of the fiber/matrix interface. Thus, the objective of the present study was to develop test methods that can be used to characterize autohesive strength development in amorphous thermoplastic matrix resins and fiber-reinforced amorphous thermoplastic prepregs, and to model the results for incorporation into future processing theories.

In previous studies of thermoplastic composite processing, it has been established that individual prepreg plies consolidate by interply bonding [1]. The resulting bond strength is a function of the processing parameters, temperature, pressure, and time to which the interface is subjected. The mechanism governing the formation of interply bonds has been established as autohesion of self-diffusion [2]. The following is a brief description of the autohesive phenomenon as it is presented in the literature [2–16].

Autohesive strength is controlled by two mechanisms: (1) intimate contact between the interfacial surfaces, and (2) diffusion of the macromolecules across the interface [4,7,9,12].

Figure 1 shows the phenomenon of autohesion for an amorphous thermoplastic polymer. At time zero, the two surfaces are pressed together. Providing the temperature is high enough (normally above the glass transition temperature, T_g), the surfaces will deform viscoelastically, come into contact, and wet (Fig. 1a). The polymer chains will begin to diffuse across the interface due to random thermal motions. After time has passed, the chains will have partially diffused across the interface and entangled with molecular chains on the other side of the interface, thus giving the interface strength (Fig. 1b). Following a long period of time, the polymer chains will have penetrated and entangled into the adjacent interface so that the interface is no

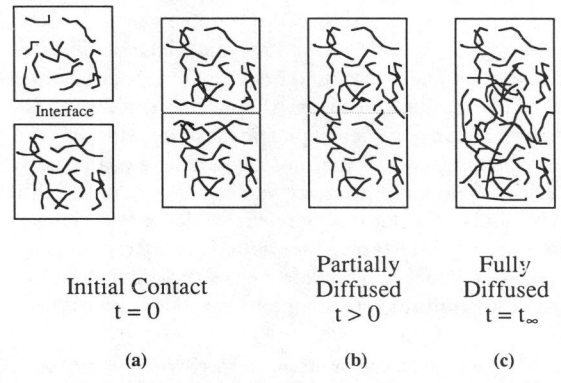

Initial Contact
t = 0

Partially
Diffused
t > 0

Fully
Diffused
$t = t_\infty$

(a)

(b)

(c)

FIG. 1—*Schematic representation of the autohesive phenomenon.*

longer distinguishable from the bulk polymer. At this point, the interface is considered completely healed (Fig. 1c).

Either wetting or diffusion can account for significant portions of the interfacial strength. Diffusion is conditional upon the surfaces being in intimate contact, as the molecules cannot move across open space [8]. Theories describing polymer diffusion are based on de Gennes' theory of molecular motion [3]. Wool [4,8], Wool and O'Connor [5–7], Prager and Tirrell [9], and Jud et al. [10] have developed theories explaining strength development of a polymer-polymer interface and crack healing in thermoplastic polymers.

Destructive mechanical tests commonly are used to characterize autohesion of high polymers. In the mechanical tests, two polymer surfaces normally are pressed together at a given temperature for a specified length of time. The fracture stress or fracture energy of the interface then is measured using the appropriate test.

Tack measurements were used by Skewis [11], Bauer [13], and Rhee and Andries [14] to measure interfacial tensile strength of two surfaces that have been pressed together under pressure for a specified time. The tack tests were performed at room temperature, well above T_g of the elastomers studied.

Fracture toughness tests were performed by Wool and O'Connor [6,7] and Jud et al. [10]. Wool and O'Connor studied healing of elastomers using double cantilever beam toughness tests, and healing of PMMA and polystyrene using Izod impact tests. The specimens were healed above the T_g of the polymer in question. Jud et al. used a compact tension fracture toughness test to measure autohesion in healed PMMA specimens. The specimens were healed in a hot press at temperatures above the T_g of PMMA (approximately 100°C).

Hamed [15] investigated tack formation in elastomers using a T-peel test. A T-peel test measures the energy required per unit area to separate the two surfaces. The polymer was dissolved in a solvent and spread on a flexible base. Two strips of the polymer/base material were pressed together for a measured time and then the strips were separated in a tension testing machine.

Interfacial tests were performed by Wool and O'Connor [7], Boenig et al. [16], Dara and Loos [1], and Bothe and Rehage [12]. Wool and O'Connor used notched tensile bars of hydroxy-terminated polybutadiene (HTPB) that were fractured and healed to evaluate fracture stress, fracture strain, and fracture energy as a function of healing time. Boenig et al. used the ASTM Test Methods for Rubber Properties in Tension (D 412-51T), type C tension test to measure tack in urethran elastomers. Dara and Loos used a parallel plate plastometer fitted with a tensile/compressive load cell to measure autohesion in polysulfone resin. The specimens were bonded at elevated temperatures and tested mechanically at the bond temperature in a nitrogen-purged atmosphere. Bothe and Rehage used a through-the-thickness tension test at room temperature to test autohesion in polybutadiene (BR), crosslinked acrylonitrile-butadiene copolymer (NBR), ethylenepropylene copolymer (EPM), and polychlorobutadiene (CR). They studied the effects of contact pressure, contact time, polymer structure, and strain rate.

A close examination of the literature reveals that only a few studies have addressed autohesion in thermoplastic resins suitable for use as matrix materials in advanced composites. Furthermore, autohesive bonding of the ply interfaces during formation of a thermoplastic composite laminate has received little attention. Thus, the goal of this investigation was to develop mechanical tests that can be used to characterize autohesive bond strength development in thermoplastic resins and thermoplastic composites.

Experimental Information

Mechanical tests were performed to measure autohesive bond strength development in neat resin and fiber-reinforced composite specimens. A compact tension (CT) fracture toughness test was used to measure interfacial strength development in neat resin samples. All neat resin tests were done using Union Carbide Corp. Udel P1700 polysulfone resin with a glass transition temperature of 194°C. Interply strength development in thermoplastic composites was measured using a double cantilever beam (DCB) fracture toughness test. Composite specimens were fabricated from prepreg consisting of Hercules AS4 graphite fiber impregnated with P1700 polysulfone resin.

Compact Tension Toughness Test

Compact tension specimens were cut from a 6.36-mm (0.25-in.) thick sheet of annealed polysulfone and machined to a width of 31.8 mm (1.25 in.) and a height of 30.5 mm (1.2 in.). The specimens were sized to satisfy the requirements for both thickness and crack length as specified in the ASTM Test Method for Plane-Strain Fracture Toughness of Metallic Materials (E 399-81). Sharp, naturally arrested cracks were introduced by driving a new, chilled razor blade into a saw cut.

The CT specimens were fractured at a crosshead speed of 0.5 mm/min (0.02 in./min). Critical stress intensity factors were calculated using the following formulas

$$K_c = \frac{(P_c Y)}{(b W^{1/2})} \tag{1}$$

where

P_c = peak load recorded by the chart recorder,
b = specimen thickness,
W = specimen width, and
Y = geometrical factor calculated as follows

$$Y = \frac{(2 + X)(0.886 + 4.64X - 13.32X^2 + 14.72X^3 - 5.6X^4)}{(1 - X)^{3/2}} \tag{2}$$

The parameter X is defined as

$$X = a/W \tag{3}$$

where

a = crack length.

Values of X should fall within the range of $0.2 \leq a/W \leq 1.0$.

The critical strain energy release rate, G_{IC}, was calculated from the critical stress intensity factor by assuming that the material is linearly elastic and in a state of plane strain as follows [17]

$$G_{IC} = \frac{K_C^2(1 - \nu^2)}{E} \tag{4}$$

where

E = tensile modulus, and
ν = Poisson's ratio.

Fractured CT specimens were healed using the following procedure. A 0.0127-mm (0.0005-in.) thick sliver of Kapton was placed in the crack end to ensure that the same crack plane was fractured after healing. The CT specimen was wrapped in Kapton and placed in a fixture which was preheated in a forced air oven at the desired healing temperature. External pressure was applied to the specimen by placing a dead weight on top of the specimen. After healing for the specified period of time, the specimen was removed from the oven and cooled under pressure to room temperature. The specimen was refractured and the healed critical stress intensity factor and critical strain energy release rate were calculated.

Double Cantilever Beam Composite Test

Unidirectional composite specimens were fabricated from AS4/P1700 polysulfone prepreg tape. The 12-ply thick specimens were compression molded in a 76-mm (3-in.) square steel mold. During lay-up, a 25-mm (1-in.) wide piece of Kapton film was placed along one edge of the specimen at midplane for crack initiation. The specimens were processed at 371°C (700°F) and 6.9 MPa (1000 psi) for 15 min and cooled to ambient temperature under pressure. After processing, the laminae were C-scanned for defects and cut into 12.7-mm (0.5-in.) wide coupons. Aluminum tabs were bonded to the top and bottom surfaces at the precracked end of each DCB specimen.

The DCB specimens were fractured at room temperature in tension using a crosshead speed of 0.5 mm/min (0.02 in./min). Peak load, crack length, and the chart recording of load versus time were recorded for calculation of the critical strain energy release rates, G_{IC}, using the compliance calibration method [18]. For most specimens, maximum load was followed by stable, slow crack growth. In these cases, the crosshead was stopped and 1 min allowed for the crack to stop growing. The crack length was marked on both sides of the specimen and the specimen unloaded. This procedure was repeated until the crack had grown to within 13 mm (0.5 in.) of the end of the sample. Typically, eight to ten measurements were obtained from each specimen. A total of twelve specimens were tested.

The fractured DCB specimens were placed in a special alignment fixture, and the fixture was placed between the platens of a hot press that was preheated to the desired temperature. The platens were closed, and a load of 445 N (100 lb.) was applied to the specimen. After healing at the desired temperature and pressure for the specified length of time, the fixture was removed from the press and cooled to ambient temperature. The critical strain energy release rate of the healed specimen was measured using the aforementioned test procedures.

Results

Compact Tension Tests

Autohesive bond strength development in polysulfone resin was measured using a compact tension (CT) fracture toughness test. In the CT test, a crack was allowed to propagate only far enough to give a suitable amount of data without complete fracture of the specimen. Thus, the same surface that was broken was healed, eliminating the wetting and alignment problems that often occur in interfacial tension tests [19].

To ensure that data recorded at different temperatures are comparable, the healed specimens must have the same interfacial wetting functions regardless of temperature. One method of obtaining this is to apply enough pressure so that the surfaces come into intimate contact and wet immediately. This pressure is called saturation pressure and varies in accordance with temperature [12].

To determine the saturation pressure, different pressures were applied to precracked CT specimens at the beginning of healing. The pressure was estimated by dividing the dead weight load applied to the specimen by the area of the fractured surface being healed. The effect of pressure on the measured refracture toughness is shown in Fig. 2. The symbols represent the mean refractured toughness, and the error bars represent ±1 standard deviation from the mean.

Results show that the refracture toughness reaches a plateau value with initial application of pressure and remains fairly constant for pressures less than 80 kPa. The plateau indicates that the saturation pressure has been reached, resulting in extremely rapid interfacial contact and wetting of the fractured surfaces. At pressures greater than 80 kPa, the refracture toughness decreases due to excessive deformation of the specimens. A pressure of 46 kPa was used for all remaining CT tests.

The results of the CT healing tests are shown in Fig. 3 for polysulfone specimens healed at 196, 200, 205, 213, 225, and 245°C. The symbols represent the mean of at least five measurements at each time, and temperature condition and the error bars represent ±1 standard deviation from the mean. The dashed line represents the fracture toughness of virgin polysulfone. A mean value of 2.26 MPa · m$^{1/2}$ with a standard deviation from the mean of ±0.242 MPa · m$^{1/2}$ was measured. At all temperatures, except for 200 and 205°C, the CT specimens regained the original fracture toughness of undamaged polysulfone, indicating complete healing of the interface. Additional time at temperature would most likely result in complete healing of the CT specimens at 200 and 205°C.

A plot of refracture toughness versus the fourth root of contact time will be linear if healing is isothermal and wetting of the crack interface is instantaneous [6–10]. However, the healed fracture toughness data do not pass through the origin as reported in previous investigations [10], and there is a considerable time lag between the beginning of the healing process and the point where the fracture toughness increases. The reason for the time lag lies in the experimental procedure used to heal the specimens in the present investigation. Healing was performed in a forced-air convection oven preheated to the desired temperature. Due to the finite surface heat

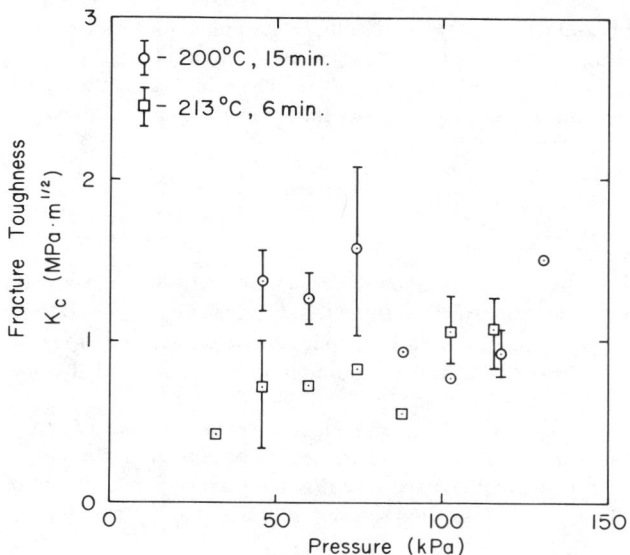

FIG. 2—*Fracture toughness versus healing pressure.*

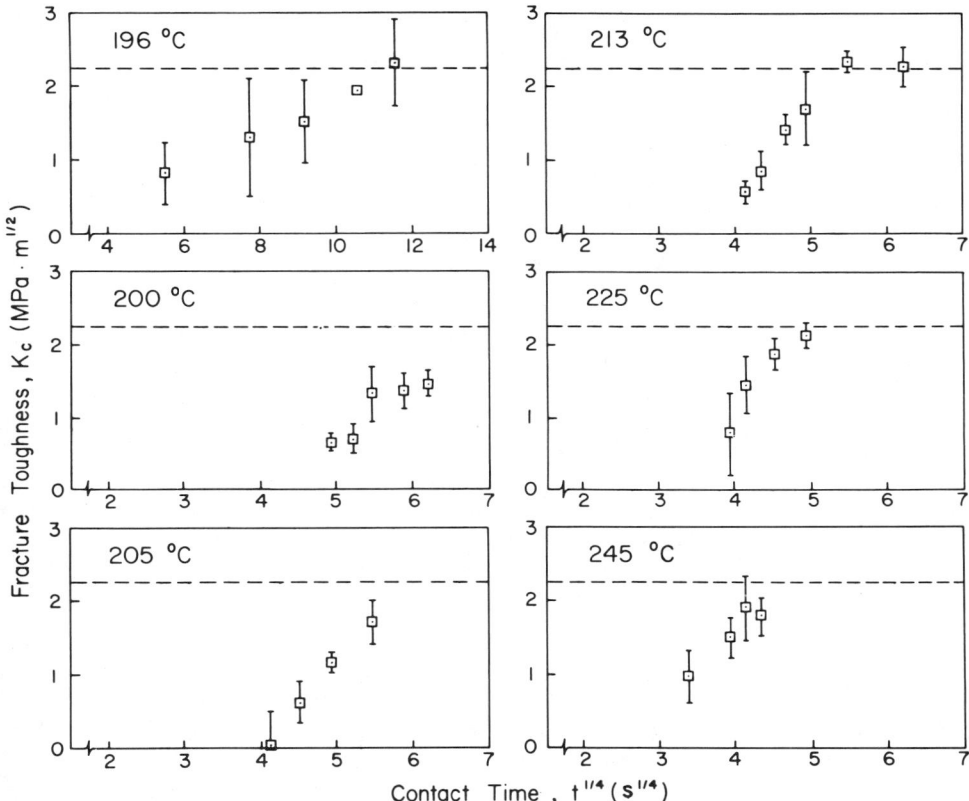

FIG. 3—*Fracture toughness versus fourth root of contact time. Note different time scale for 196°C.*

transfer coefficient between the oven fluid and the specimen, the time required for the specimen to reach the healing temperature was a significant portion of the total healing time. Since healing begins at T_g, the specimens were healed nonisothermally.

In order to allow comparisons to be made between neat resin CT tests and composite healing data, critical strain energy release rates were calculated using Eq 4. The data were nondimensionalized by dividing the calculated strain energy release rate, G_{IC}, by the critical strain energy release rate for the undamaged polysulfone specimens, G_{IC_∞}, as follows

$$R = \frac{G_{IC}}{G_{IC_\infty}} \tag{5}$$

where

$R =$ the healing function.

Shown in Fig. 4 is a summary plot of the healing function, R, versus the square root of time. According to the healing model of Wool and O'Connor [6], if healing is isothermal and interfacial wetting is instantaneous, a plot of the healing function versus square root of time should be

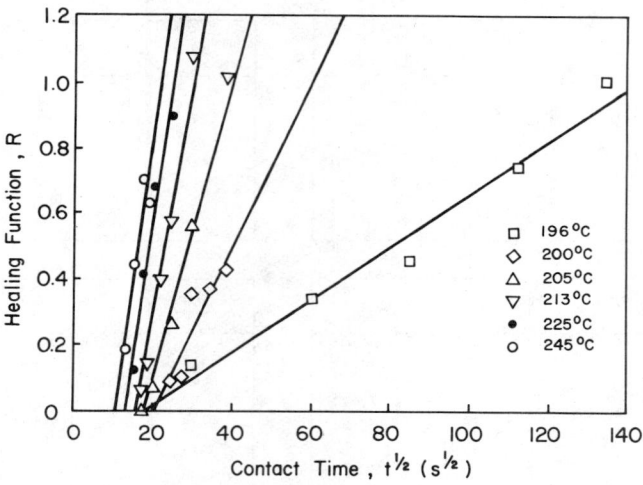

FIG. 4—*Healing function,* R, *versus square root of contact time. Symbols represent the mean of the data. Solid lines are linear regression curve fits to the data.*

a straight line that passes through the origin. The healing data obtained in the present investigation appear to follow the straight line relationship but do not pass through the origin. Furthermore, above 205°C, the slope of each curve is nearly constant. These differences are due to the nonisothermal healing effects.

In order to measure temperature as a function of time during the healing tests, six thermocouples were embedded at various locations inside a CT specimen. One thermocouple was placed in each of the following locations: the top surface, bottom surface, front edge, and rear edge of the specimen. The other two thermocouples were embedded in the crack plane and the crack tip. The CT specimen was placed in the oven and exposed to the same processing cycles used in the CT healing tests. Each thermocouple was sampled at 30-s intervals by a Fluke scanning digital thermometer.

Results of the measurements showed that the temperature inside the CT specimen was uniform during healing. The maximum temperature difference across the crack plane was less than 10°C and occurred only at the beginning of the healing process. The temperature difference between the surface of the specimen and the oven fluid was significantly greater than the temperature difference between any two points inside the specimen.

A negligible internal resistance (NIR) heat transfer solution was compared with measured data. In the NIR solution it is assumed that the internal thermal resistance of the system is small compared to the external thermal resistance between the surface of the system and the surrounding medium. Thus, the temperature of the body may be taken as uniform for any instant of time.

The temperature response of the specimen can be written as [20]

$$\frac{T - T_f}{T_i - T_f} = \exp\left(\frac{-hA_s}{\rho C_p V} t\right) \tag{6}$$

where

h = surface heat transfer coefficient,
A_s = exposed surface area,
T = specimen temperature,
T_f = fluid temperature,
T_i = initial temperature,
ρ = density,
V = volume of the body,
C_p = specific heat of the body, and
t = time.

For polysulfone CT specimens, using a heat transfer coefficient h of 32 W/(m$^2 \cdot$ °C), ρ of 1.24 Mg-m^{-3} [21], C_p of 1.13 kJ/kg-°C [22], V of 6.145 $\cdot$ 10^{-6} m^3, and A_s of 2.725 $\cdot$ 10^{-3} m^2, we obtain the following

$$\theta = \frac{T - T_f}{T_i{}' - T_f} = e^{(-t/98.743)} \tag{7}$$

The heat transfer coefficient was obtained by using Eq 6 to calculate temperature versus time for different values of the heat transfer coefficient until the calculated temperature matched thermocouple data. Figure 5 shows the correlation between Eq 7 and the thermocouple data for a heat transfer coefficient of 32 W/(m$^2 \cdot$ °C). The NIR solution gave very good correlation with the measured results. However, since the CT specimens are quite thick (6.35 mm), a more detailed heat transfer analysis which includes the thermal resistance of polysulfone may be warranted.

To fully describe and model the autohesive phenomenon for thermoplastic resins, it was recognized that the temperature dependence must be determined. As discussed previously, the low convective heat transfer coefficient between the oven fluid and the CT specimen required a finite amount of time to heat the specimen to the oven set point temperature. Thus, the CT specimens were healed nonisothermally. Furthermore, only data measured after the specimen reached the specified isothermal oven temperature can be used to determine the temperature dependence. To determine the time required for the CT specimens to reach the oven temperature, the NIR heat transfer expression, Eq 7, was solved for time, t, as follows

$$t = -98.743 \, \ell n \left(\frac{T - T_f}{T_i - T_f} \right) \tag{8}$$

The time to reach 0.5°C below oven setpoint temperature, T_f, was calculated for the CT specimens and verified by the thermocouple data. A temperature 0.5°C below the setpoint temperature was selected rather than the actual setpoint temperature because the NIR expression predicts infinite time to reach the actual oven setpoint temperature. The half degree approximation agreed well with the thermocouple data.

Healing data obtained after the specimen reached the oven setpoint temperature, T_f, were isothermal and can be used in the diffusion model developed by Wool and O'Connor [6, 7] to determine the temperature dependence. If wetting is instantaneous and if the instantaneous wetting load at initial time is negligible, then the healing function defined in Eq 5 can be written as follows:

$$R = C(T)t^{1/2} \tag{9}$$

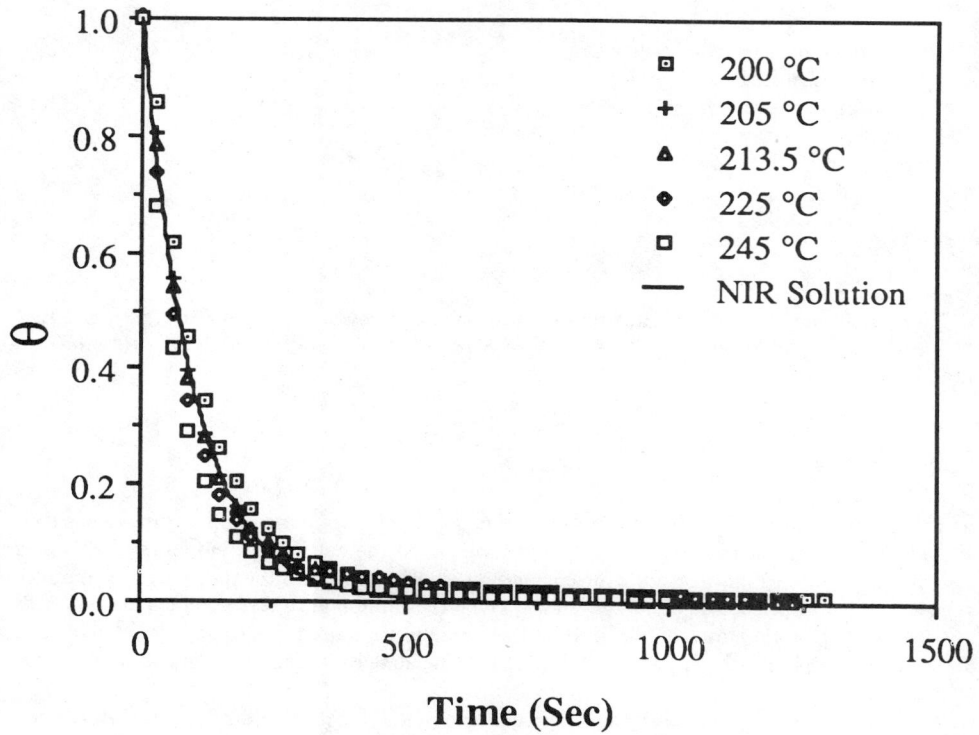

FIG. 5—*Nondimensionalized temperature,* Θ *versus time. Symbols represent measured temperature of the crack plane at different oven setpoint temperatures. Solid line represents the negligible internal resistance heat transfer solution.*

where $C(T)$ = temperature dependent parameter proportional to the polymer self-diffusion coefficient.

In order to determine $C(T)$, nonisothermal data points were removed from the plot of the healing function versus square root of contact time and a linear least squares curve was fit to the isothermal data at each temperature condition (Fig. 6). Measurement of the slope of each curve gave $C(T)$. Table 1 shows the time to obtain the oven setpoint temperature, measured values of $C(T)$, and the correlation coefficient from the least squares fit to the isothermal data. Determination of $C(T)$ at temperatures above 213°C was precluded by the fact that the data were not isothermal.

Wool and O'Connor [6] stated that the self-diffusion coefficient should follow a WLF temperature dependence, providing that the mode of failure remains the same between samples healed at different temperatures. Ferry [23] states that the WLF relationship is accurate at temperatures between the glass transition temperature, T_g, and fifty degrees above the glass transition temperature, $T_g + 50°C$, due to the fact that free volume changes control the mechanical properties of polymers in this range. The temperature of the tests in this study falls within this range.

The WLF relationship can be written as follows [23]

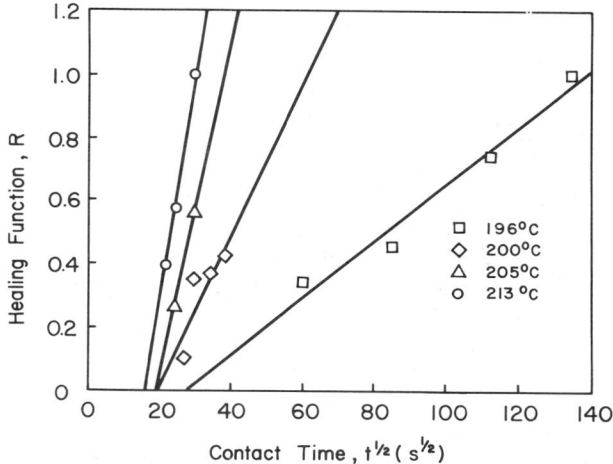

FIG. 6—*Healing function, R, versus square root of contact time. Symbols represent the mean of the data. Solid lines represent linear regression curve fits to the data.*

TABLE 1—*Time to reach oven temperature, C(T), and correlation coefficients from the least squares fit to the isothermal rehealing data.*

Temperature, °C	Time to T_f, s	$C(T)$	Correlation
196.0	579.7	0.0091	0.98
200.0	581.9	0.0239	0.83
205.0	584.6	0.0543	1.00
213.0	589.0	0.0752	1.00
225.0	594.6	...	...
245.0	603.8	...	...

$$\log a_T = \frac{-C_1(T - T_r)}{C_2 + T - T_r} \qquad (10)$$

where

$$a_T = \text{shift factor,}$$
$$C_1 \text{ and } C_2 = \text{constants,}$$
$$T = \text{temperature, and}$$
$$T_r = \text{reference temperature.}$$

The value a_T is defined as

$$a_T = \frac{C_{S,0}T}{C_S T_r} \qquad (11)$$

where

$C_{S,0}$ = property being measured at the reference temperature, and
C_S = property at temperature T.

In the present investigation, the reference temperature, T_r, was taken as 196°C and C_S represents $C(T)$ at the healing temperature T taken from Table 1.

The constants C_1 and C_2 were determined by plotting $1/\log(a_T)$ versus $1/(T - T_r)$. A least squares linear curve fit to the data gives values of 1.55 and 10.95 for C_1 and C_2, respectively (Fig. 7).

A nonisothermal healing model was developed by combining the isothermal healing model Eq 9 with the NIR heat transfer model Eq 7. The model was developed using a WLF temperature dependence for determination of $C(T)$.

A numerical scheme (stepwise) was developed in which the healing equation was solved incrementally for small time steps. A flowchart of the solution process is shown in Fig. 8. At the beginning of healing, the specimen is at ambient temperature and the initial value of the healing function is zero. Time is incremented by a small time step, Δt, and the new temperature of the sample is calculated using the NIR heat transfer solution. The former and new temperatures are averaged over the time step, and $C(T)$ is calculated using this temperature. The incremental healing, ΔR, corresponding to Δt at the averaged temperature is calculated and added to the previous value of R. Then time is incremented and the calculations continue until the time reaches a set flag.

Figure 9 shows a comparison between the results of the nonisothermal healing model using a WLF temperature dependence and the measured CT data. The model accurately predicts the onset of healing at all temperatures. Furthermore, the model predicts the degree of healing at long healing times for all temperatures tested. The amount of nonisothermal healing can be ascertained from the plots. At 196°C the healing curve is linear, with a slope corresponding to $C(T)$ (see Fig. 6 and Table 1). These results indicate that most of the healing occurred isothermally at the oven temperature. On the other hand, at 245°C the healing curve does not approach a straight line, indicating that most of the healing was nonisothermal at temperatures below the oven temperature.

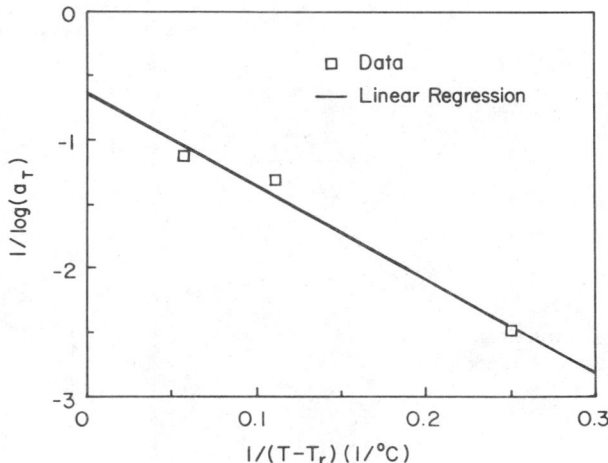

FIG. 7—*Determination of WLF constants for the temperature dependency of autohesion. Symbols represent data. Solid line represents a linear regression curve fit to the data.*

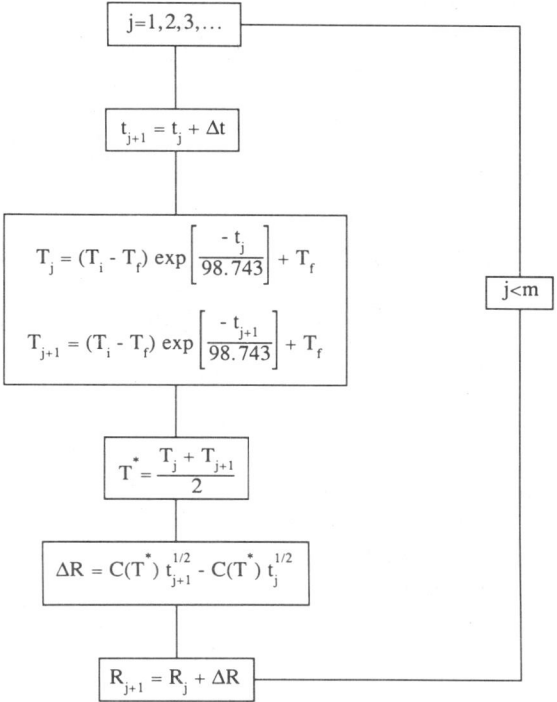

FIG. 8—*Flowchart of the stepwise solution process.*

Double Cantilever Beam Composite Test

A double cantilever beam (DCB) interlaminar toughness test was used to examine interply strength development in unidirectional graphite-polysulfone composites. The test measured the critical strain energy release rate of a healed interlaminar crack.

Extensive fiber bridging caused the critical strain energy release rate to increase with crack length. A two-fold increase in G_{IC} with crack extension was commonly observed for both virgin and healed specimens [24]. Hence, only the initial value of the critical strain energy release rate, measured during first load application, is reported for each specimen.

Following the same nondimensional scheme used for the CT specimens, the healing function is defined as

$$R = \frac{G_{IC}}{G_{IC_\infty}} \tag{5}$$

where G_{IC_∞} = initial critical strain energy release rate for the undamaged specimens. A mean value of 394 N/m with a standard deviation from the mean of ±61.6 N/m was measured.

The healing function, R, versus the square root of time is plotted in Fig. 10 for healing temperatures of 213, 225, and 245°C. The square symbols represent data obtained from specimens that were initially fractured and healed only once, while the circle symbols represent data obtained from specimens that were fractured and rehealed more than one time.

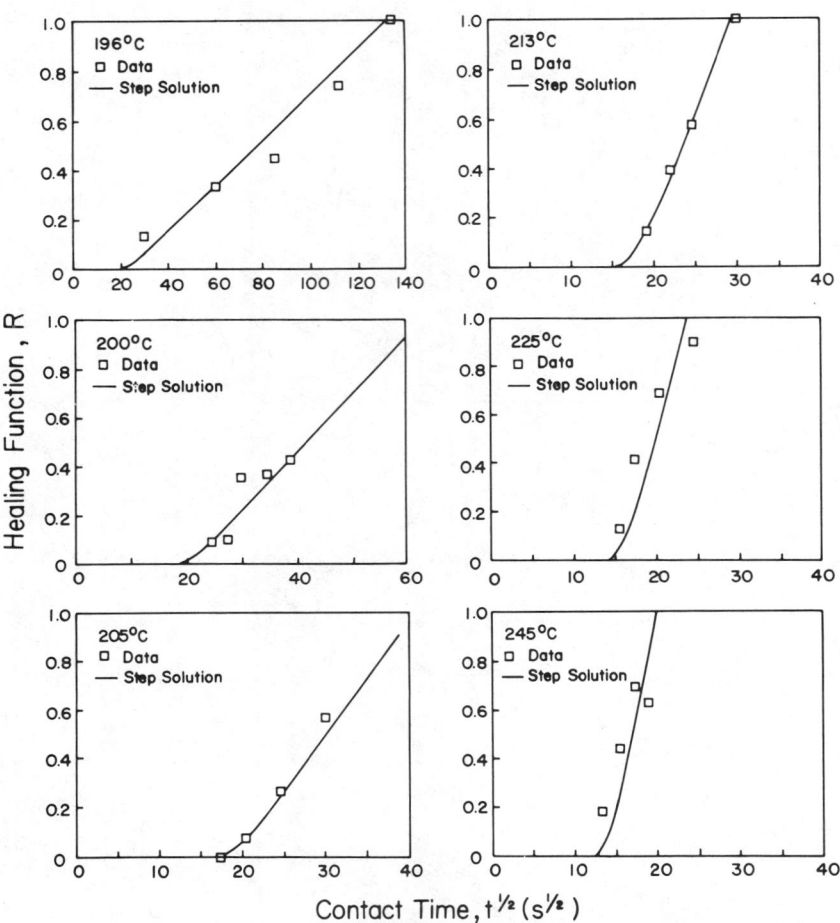

FIG. 9—*Healing function, R, versus square root of contact time. Comparison between data (symbols) and the nonisothermal healing model using a WLF temperature dependence (solid line). Note different time scales for 196 and 200°C.*

The composite data do not show the same strong time and temperature dependence that was observed in the neat resin tests. Specimens healed only once at 213°C showed a slight increase in the critical strain energy release rate with increasing contact time. A maximum of about 80% of the original interlaminar fracture toughness was recovered. Rehealing the specimens again at 213°C did not significantly improve the fracture toughness. Rehealing the DCB specimens at higher temperatures appears to increase the interlaminar fracture toughness, with several specimens achieving almost 90% of the original toughness at 245°C. However, additional tests will need to be performed before conclusions can be drawn regarding the amount of the original interlaminar fracture toughness that can be recovered upon healing.

The time required to achieve complete healing of the interlaminar fracture by autohesion can be estimated using the isothermal healing model in Eq 9. The parameter $C(T)$ was calculated using the WLF model in Eqs 10 and 11. The model predicts that an interlaminar crack will completely heal in 127 s at 213°C, 51 s at 225°C, and 22 s at 245°C. These calculations neglect

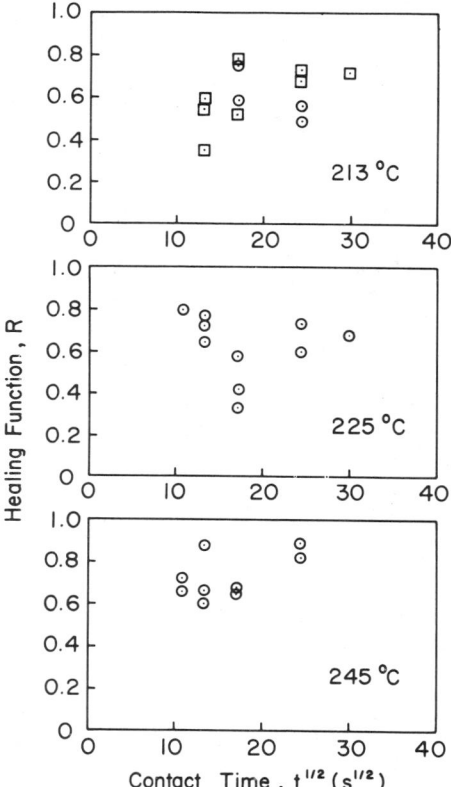

FIG. 10—*Healing function*, R, *versus square root of contact time. Each symbol represents one data point.*

wetting effects but do show that sufficient time has elapsed for complete healing of the fractured surfaces by self-diffusion. Thus, it can be concluded that wetting, intimate contact, and resin flow cause a different time dependency than observed for autohesion alone [6].

The lower toughness upon healing can be explained by different fracture mechanisms occurring in the undamaged specimens than in the healed specimens. During the first crack growth, a number of mechanisms contribute to the measured critical strain energy release rate. Among these are resin deformation, fiber/resin interfacial failure, fiber peeling, and fiber breakage. As the crack propagates, microcracks are formed on planes adjacent to the main crack plane which contribute to the amount of energy absorbed during the test. Also, the main crack may not follow a straight path but will jump between different crack planes.

Following healing of the specimen, recrack propagation will follow the path of least resistance established previously. The recrack may not cause as much fiber breakage. The absence of some energy-absorbing mechanisms results in a lower fracture energy of the healed specimens.

Scanning electron microscopy (SEM) was performed on the DCB specimens to observe the differences in the failure mechanisms between undamaged and healed specimens. The healed specimens showed a greater amount of resin at the fracture plane than the virgin specimens. This indicates that during healing some polymer flow was taking place. The flow of resin indicates that the fracture surfaces did not achieve instantaneous wetting and intimate contact during healing. A possible cause of the lack of intimate contact was due to the stray, broken fibers created during the first crack growth. These fibers did not lie parallel to the other fibers and upon healing interfered with intimate contact between the crack surfaces.

Conclusions

Mechanical tests have been investigated for characterizing autohesive bond strength development in amorphous thermoplastic resins and fiber-reinforced amorphous thermoplastic composites. Autohesive strength development in polysulfone resin was observed by measuring the refracture toughness of precracked compact tension specimens that were healed at a given temperature and contact time. Due to healing in a forced air oven, the specimens were not healed isothermally. Therefore, theories developed for isothermal healing cannot be directly applied to the data.

A nonisothermal healing model was developed by incorporating a negligible internal resistance heat transfer model into the isothermal healing model. The model was developed using a WLF temperature dependence for the parameter $C(T)$ in the healing model. Results of the nonisothermal healing model compared favorably with healed compact tension data at all temperatures. Thus, it was concluded that the CT test is suitable for characterizing autohesive strength in thermoplastic resins.

Autohesive strength in fiber-reinforced thermoplastic composites was measured using a double cantilever beam interlaminar toughness test. The data do not show a strong time or temperature dependence as observed in the neat resin tests. Furthermore, only 80 to 90% of the undamaged fracture energy can be recovered. This is accounted for by different fracture mechanisms in the virgin than in the healed (or rehealed) DCB specimens. SEM micrographs of the fracture surfaces revealed that the healed DCB specimens have more resin at the refracture plane than the virgin specimens. Resin flow and the lack of strong time and temperature dependence in the DCB data indicates that intimate contact was not established immediately. Additional data must be obtained to determine whether the DCB test is suitable for measuring autohesive bond strength at the ply interfaces of laminated thermoplastic composites. However, the test does indicate that repair of thermoplastic composites is possible although the original toughness cannot be attained.

Acknowledgments

This work was supported by NASA-Virginia Tech Composites Program, Grant NAG1-343 with NASA-Langley Research Center.

References

[1] Loos, A. C. and Dara, P. H., "Processing of Thermoplastic Matrix Composites," *Review of Progress in Quantitative Nondestructive Evaluation,* D. O. Thompson and D. E. Chimenti, Eds., Plenum Press, New York, 1987, Vol. 6B, pp. 1257–1265.

[2] Voyutskii, S. S., *Autohesion and Adhesion of High Polymers,* Vol. 4, *Polymer Reviews,* Interscience Publishers, New York, 1963.

[3] De Gennes, P. G., "Entangled Polymers," *Physics Today,* June 1983, pp. 33–39.

[4] Wool, R. P., "Relations for Healing, Fracture, Self-Diffusion and Fatigue of Random Coil Polymers," *ACS Polymer Preprints,* Vol. 23, No. 2, 1982, pp. 62–63.

[5] Wool, R. P. and O'Connor, K. M., "Craze Healing in Polymer Glasses," *Polymer Engineering and Science,* Vol. 21, No. 14, Oct. 1982, pp. 970–977.

[6] Wool, R. P. and O'Connor, K. M., "A Theory of Crack Healing in Polymers," *Journal of Applied Physics,* Vol. 52, No. 10, Oct. 1981, pp. 5953–5963.

[7] Wool, R. P. and O'Connor, K. M., "Time Dependence of Crack Healing," *Journal of Polymer Sciences: Polymer Letters Edition,* Vol. 20, 1982, pp. 7-16.

[8] Wool, R. P., "Molecular Aspects of Track," *Rubber Chemistry and Technology,* Vol. 57, 1983, pp. 307-319.

[9] Prager, S. and Tirrell, M., "The Healing Process at Polymer-Polymer Interfaces," *Journal of Chemical Physics,* Vol. 75, No. 10, Nov. 1981, pp. 5194-5198.

[10] Jud, K., Jausch, H. H., and Williams, J. G., "Fracture Mechanics Studies of Crack Healing and Welding of Polymers," *Journal of Materials Science,* Vol. 16, 1981, pp. 204-210.

[11] Skewis, J. D., "Self-Diffusion Coefficients and Tack of Some Rubbery Polymers," *Rubber Chemistry and Technology,* Vol. 39, 1966, pp. 217-225.

[12] Bothe, L. and Rehage, G., "Autohesion of Elastomers," *Rubber Chemistry and Technology,* Vol. 55, 1981, pp. 1308-1327.

[13] Bauer, R. F., "Investigation into the Mechanism of Tack of Rubbers," *Journal of Polymer Science: Part A-2,* Vol. 10, 1972, pp. 541-548.

[14] Rhee, C. K. and Andries, J. C., "Factors Which Influence Autohesion of Elastomers," *Rubber Chemistry and Technology,* Vol. 54, 1980, pp. 101-114.

[15] Hamed, G. R., "Tack and Green Strength of NR, SBR and NR/SBR Blends," *Rubber Chemistry and Technology,* Vol. 54, 1980, pp. 403-412.

[16] Boenig, H. V., Miller, C. B., and Shottafer, J. E., "Tack in Urethan Elastomers," *Rubber Chemistry and Technology,* Vol. 39, 1966, pp. 974-981.

[17] Collins, J. A., *Failure of Materials in Mechanical Design,* Wiley, New York, 1981, p. 58.

[18] Wilkins, D. J., Eisenmann, J. R., Camin, R. A., Margolis, W. S., and Bensen, R. A., "Characterizing Delamination Growth in Graphite-Epoxy," *Damage in Composite Materials, ASTM STP 775,* K. L. Reifsnider, Ed., American Society for Testing and Materials, 1982, pp. 168-183.

[19] Howes, J. C. and Loos, A. C., "Autohesive Strength Development in Polysulfone Resin and Graphite-Polysulfone Composites," *Journal of Thermoplastic Composite Materials,* Vol. 1, Jan. 1988, pp. 58-67.

[20] Chapman, A. J., *Heat Transfer,* Macmillan, New York, 1984, p. 124.

[21] Union Carbide, *Udel Polysulfone Technical Information,* Union Carbide Corporation, 270 Park Ave., New York.

[22] Harper, Charles A., *Handbook of Plastics and Elastomers,* McGraw-Hill, New York, 1976, pp. 2-50.

[23] Ferry, John D., *Viscoelastic Properties of Polymers,* Wiley, New York, 1980, p. 280.

[24] Howes, J. C., "Interfacial Strength Development in Thermoplastic Resins and Fiber-Reinforced Thermoplastic Composites," M.S. thesis, Virginia Polytechnic Institute and State University, Blacksburg, VA, 1987.

D. D. Edie,[1] B. W. Gantt,[2] G. C. Lickfield,[1] M. J. Drews,[1] and M. S. Ellison[1]

Thermoplastic Coating of Carbon Fibers

REFERENCE: Edie, D. D., Gantt, B. W., Lickfield, G. C., Drews, M. J., and Ellison, M. S., **"Thermoplastic Coating of Carbon Fibers,"** *Advances in Thermoplastic Matrix Composite Materials, ASTM STP 1044,* G. M. Newaz, Ed., American Society for Testing and Materials, Philadelphia, 1989, pp. 50–61.

ABSTRACT: A process is being developed which evenly coats individual carbon fibers with thermoplastic polymers. In this novel, continuous coating process, the fiber tow bundle is first spread over a series of convex rollers and then evenly coated with a fine powder of thermoplastic matrix polymer. Next, the fiber is heated internally by passing direct current through the powder coated fiber. The direct current is controlled to allow the carbon fiber temperature to slightly exceed the flow temperature of the matrix polymer. Analysis of the thermoplastic coated carbon fiber tows produced using this continuous process indicates that 30 to 70 vol % fiber prepregs can be obtained.

KEY WORDS: thermoplastic, carbon fiber, powder coating, composite

The demand for light-weight materials with improved strength and stiffness has led to the development of fiber-reinforced composites. The high mechanical strength, high modulus, and low specific gravity of carbon fibers have made it the predominant reinforcement fiber used in high-performance resin-matrix composites. Most of the current techniques used to manufacture carbon fiber-reinforced composites rely on the manual stacking of resin impregnated fiber tows or prepregs. One important goal in the production of high-performance composite structures is to automate the labor-intensive ply stacking step. In order for this goal to be realized, current prepreg manufacturing must be modified to meet the requirements of automated processing such as the availability of suitable prepreg materials that require no special storage or handling.

Presently, two different classes of resin systems are used in the manufacture of prepreg materials, thermoset and thermoplastic. The thermoset resins are typically applied to fiber tows in a partially reacted liquid form and, therefore, require refrigeration to inhibit curing prior to use. Thermoplastic resins, which do not require this special storage and handling, can be applied either as a pure melt or as a solution. When applied from the melt, it is difficult to produce thin, evenly coated, low-void content prepregs because of the high viscosity and poor fiber wetting characteristics of molten thermoplastics. If the individual fibers are not evenly coated, the fibers will not be evenly distributed throughout the matrix, and the strength of the final composite part will be reduced. Incomplete removal of the solvent prior to the final laminate consolidation is a major problem that is often encountered in solution coating with thermoplastic resins.

For many applications, the use of woven carbon fiber fabric structures has been identified as a viable approach to a more automated manufacturing process for high performance composites. However, several problems are encountered when uncoated fibers are woven and then impregnated with the matrix polymer. The major problem is the difficulty in obtaining a uniform

[1]Center for Advanced Engineering Fibers, Clemson University, Clemson, SC 29634-0909.
[2]BASF Fiber Corporation, Central, SC.

impregnation of all of the fiber bundles in the woven fabric. Weaving of impregnated tow materials would eliminate this problem. However, this is impractical with many of the most utilized thermosets because they are liquid and have a limited shelf-life at room temperature. While thermoplastics have an almost infinite shelf-life, present thermoplastic prepreg tows are not sufficiently flexible, presumably because of nonuniform and often thick resin distribution, to weave efficiently on a moderate speed mechanical loom and still produce acceptable fabric.

A process utilizing thermoplastic resins that produced a more uniform coating of the individual filaments in the carbon fiber tow would result in a more flexible material. Such a material would be more suitable for automated manufacturing in prepreg form and also would allow for the weaving, knitting, or braiding of improved prepreged textile structures directly. In this paper the results of an investigation into a novel, continuous process for thermoplastic coating of carbon fibers are presented. This process consists of applying a thermoplastic resin in powder form to the carbon fiber tow, followed by direct electrical heating of the carbon fiber tow to temperatures above the melt-flow temperature of the polymer. This method was developed specifically for use with thermoplastic polymers such as LARC-TPI (Langley Research Center—Thermoplastic Polyimide), which exhibit time-dependent melt viscosities. Although the exact viscous characteristics of this newly developed polymer are still being determined by Langley researchers, the material has been shown to be quite fluid when initially melted. The viscosity then increases rapidly with time. This makes it imperative that any coating technique apply and melt the polymer nearly instantaneously. The feasibility of this approach was demonstrated first using a low-melt thermoplastic polyester, Eastobond FA252. Both the LARC-TPI and Eastobond FA252 have similar particle sizes, on the order of 200 mesh. Carbon fiber tows that had been coated with Eastobond and LARC-TPI were fabricated into composite samples. These tows and composite samples then were evaluated for uniformity of coating and resin distribution using scanning electron microscopy and image analysis.

Procedures

In a preliminary study, various techniques for depositing a polymer in powder form and melting of the polymer onto a carbon fiber tow were evaluated. The coating and melting techniques were evaluated first as a simple batch process and then as a continuous coating process.

Powder Deposition

In a batch process it was demonstrated that a polymer in powder form, Eastobond FA252, could be deposited onto a carbon fiber tow using a GEMA AG (Type 708) electrostatic deposition powder coating gun. Even though this deposition technique did produce a reasonably uniform coating in the batch tests, it did not perform very well in a continuous process. Thus, a small powder extruder was constructed to deposit the powder directly onto the carbon fiber tow. A variable speed extruder drive motor was used to control powder deposition rates. A powder feed hopper was fabricated to provide sufficient agitation of the powder supply to prevent clogging of the extruder assembly.

Melting of the Polymer Coating

Three methods for heating the carbon fiber tow were evaluated: inductive heating during and after the coating step, convection heating prior to and during the coating, and direct electrical heating during and after the coating step. Upon convection heating of a tow sample that was powder coated, the polymer melted and produced a relatively nonuniform coating on the tow. The only method that showed promise was direct electrical heating of the carbon fiber tow. This was demonstrated in a batch process by passing 10 A ac through a 3K (3000 filament) carbon fiber tow coated with LARC-TPI powder and held between two alligator clips. Scanning electron microscopy revealed adequate coating of only those filaments that had good electrical contact.

Continuous Process for Coating and Melting

In the continuous process, the powder-coated tow was passed over two rollers which were connected to a d-c power supply. In order to provide good electrical contact for all the filaments in the tow and to allow for even powder coating of the tow, it was essential to spread out the tow before powder coating and heating. This was accomplished through the use of an S-wrap roller configuration containing several convex rollers. All rollers were 0.003 m in diameter, and the convex rollers had a 0.15-m radius of curvature. These low-friction rollers minimized tow tension and, thus, the contraction of the tow during polymer melting. Additionally, a spreading bar was inserted into the coating chamber to improve the powder distribution on the tow prior to heating. Figure 1 contains a schematic for the system.

Like most commercial fiber coating processes, the take-up wheel was rotated at a fixed speed and the let-off wheel was not driven. Even though each spreading roller was mounted on low-friction bearings, there was still adequate friction to keep the tow taut between all rollers and wheels.

After energizing the take-up wheel, the d-c power supply was turned on to apply a voltage between rollers H and D. A focusing infrared (IR) pyrometer was used to measure the tow temperature, and this measurement was used to manually control the applied voltage. The voltage was set at a level that resulted in a tow temperature of approximately 220°C for polyester coating trials and 280°C for LARC-TPI coating trials. Initial testing showed these to be the optimum flow temperatures for the two polymers.

Once the carbon fiber tow was passing through the apparatus and heated to the proper temperature, the polymer powder extruder was turned on and its feed hopper was filled with either polyester or LARC-TPI powder. The extruder screw was set at a rotation rate that would deposit the desired amount of dry thermoplastic powder onto the carbon fiber tow passing beneath it. The spreading bar in the coating chamber helped distribute any polymer lumps before the powder-coated tow passed between rollers H and D, where the applied voltage raised the tow temperature to the polymer melt-flow temperature.

The tow temperature was monitored throughout each coating trial, and when the desired amount of thermoplastic coated tow was obtained, the polymer powder extruder was turned off,

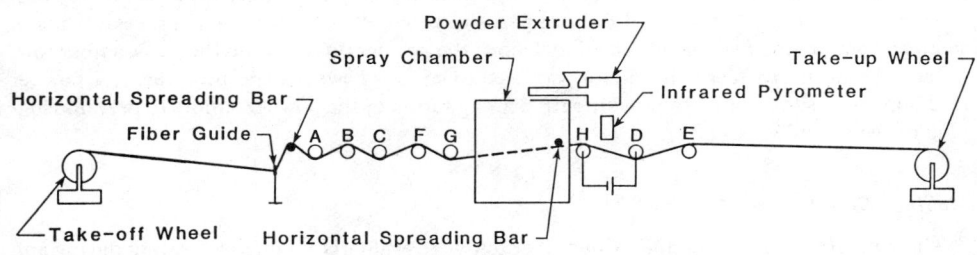

A,B,C,D and E   Flat Rollers
F,G and H   Convex Rollers

FIG. 1—*Schematic of the continuous electrical resistance heating and powder deposition apparatus using direct electrical contact and a powder extruder.*

the power bridge was deenergized, and the take-up wheel was stopped. The carbon fiber tow then was cut and the bobbin of coated carbon fiber was removed from the take-up wheel.

Product Evaluation

The tow was cut to 0.254-m lengths and analyzed to determine the uniformity of the coating. These short lengths of thermoplastic-impregnated carbon fiber tow then were molded to produce unidirectional composite samples. Finally, the composite samples were analyzed to evaluate the polymer matrix distribution.

Coating Uniformity

After each run, the continuous thermoplastic-coated tow of carbon fiber was cut into 0.254-m segments, and specimens were selected randomly for the determination of the fiber volume fraction gravimetrically. The average fiber volume fraction for this given length was calculated using the average mass of a 0.254-m length of uncoated tow, the carbon fiber density, the density of the thermoplastic polymer, and the mass of the 0.254-m specimen of thermoplastic coated tow. The variation in the average fiber volume fraction for these specimens provided an excellent indication of long-period variation in the coating process. Also, one representative 25.4-m segment was inspected using a scanning electron microscope to evaluate the microscopic evenness of the polymer coating.

Composite Evaluation

The 0.254-m segments of thermoplastic coated carbon fiber then were molded into 0.2032-m-long, 0.0127-m-wide, and approximately 0.0016-m-thick unidirectional composite samples. To form these composite samples, about 100 individual segments of thermoplastic-coated tow were trimmed and laid into the 0.2032-m-long by 0.0127-m-wide cavity of a steel mold. The mold lid was inserted and the assembly was vacuum bagged and evacuated. The evacuated mold then was placed between the platens of a Carver press that had been preheated to 160°C. The heated platens then were closed, and approximately 100 psi pressure was applied to the heated and evacuated mold. After holding at this pressure for 1 h, the mold was removed from the press and the composite sample was taken from the mold cavity.

This composite specimen was cross-sectioned, mounted in an acrylic encapsulating resin, polished, and viewed under an image analyzer to determine the uniformity of the matrix distribution in the composite specimen.

Results and Discussion

Coating Process

Continuous 30-m sections of polyester coated carbon fiber tow were produced using the procedure described under the section entitled "Product Evaluation." Preliminary runs producing 3 m of coated tow were performed to establish the proper powder deposition rate and the fiber throughput necessary to produce fiber tow coated with the desired amount of polymer. At a take-up speed of 0.041 m/s, four different powder deposition rates were used to produce 30-m sections of polyester coated tow that contained an average of 36.8, 51.3, 57.1, and 70.0 vol % carbon fiber. These 30-m lengths of polyester coated tows then were cut into 0.254-m segments to determine the long-period variability of the coating process. Table 1 shows the average and the standard deviation of these specimens. The LARC-TPI powder was extremely fine and diffi-

TABLE 1—*Process conditions for continuous coating tests.*

Run Number	Polymer Type	Measured Fiber Temperature, °C	Average Vol % Fiber	Standard Deviation
18[a]	Eastobond FA252	215	36.8	4.81
23[a]	Eastobond FA252	220	51.3	12.1
16[a]	Eastobond FA252	240	57.1	5.47
27[a]	Eastobond FA252	220	70.0	9.37
40[b]	LARC-TPI	280	78.2	2.53

[a]Carbon fiber type was Amoco (T-300), 3K tow-surface treated but unsized.
[b]Carbon fiber type was Amoco (T-300), 3K tow-epoxy sized.

cult to convey down the screw of the powder extruder. Therefore, a take-up speed of only 0.0051 m/s was used during LARC-TPI trials.

Since the take-up velocity was held constant, the variation in the amount of polymer deposited (represented by the standard deviation between specimens) can be attributed to inconsistent polymer deposition and insufficient tow spreading. The powder extruder was constructed using a motor and screw that were constructed in-house. During continuous operation, the extruder would occasionally clog due to the fine particle size of the polymer powders. This clogging was the primary cause for the long-period variability in the coating process. Since this variability in the measured coating level was caused by random clogging of the extruder, it was not a function of the extruder screw speed. This is confirmed by the standard deviations listed in Table 1. However, since neither the hopper nor the screw design were optimized for feeding fine polymer powders, the consistency of the powder deposition obtained in these experiments was quite encouraging. It is interesting to note that the LARC-TPI powder extruded with less clogging than the polyester, and this resulted in the lower standard deviation in the coating levels shown in Table 1. This is presumably because the LARC-TPI powder did not agglomerate in the powder hopper as much as the polyester powder did. However, the extremely fine size of the LARC-TPI powder made it difficult for the screw to convey the material into the coating chamber at a high enough rate to achieve greater than a 20 vol % polymer coating of the carbon fiber tow (even at the lowest fiber throughput).

The convex rolls provided adequate tow spreading, but fluctuations in the tow friction during continuous operation of the coating process caused lateral movement of the tow across these convex rolls. This gave periodic variations in the degree of spreading of the tow, which in turn caused variations in the area of tow available for polymer coating and in the uniformity of tow contact with the electrical contact rolls. While this fluctuation in spreading was not a severe problem, future plans call for the installation of a series of serrated and smooth graphite rolls to achieve a more stable tow spreading. Initial trials show this modification to be quite promising. Adequate spreading is critical if all fibers in the tow are to be evenly coated. Testing showed that the viscosity of LARC-TPI at 280°C prevented it from penetrating more than five to seven filament layers. This was true when either sized or unsized commercial carbon fiber tow was being processed. Trials indicate that the proposed graphite roll system could be used to spread the tow to a thickness well below this limit.

Coating Uniformity

Scanning electron microscopy (SEM) of the polyester-coated specimens showed that the uniformity and filament-to-filament evenness of the polymer coating improved as the amount of powder deposited decreased. Figure 2 shows an end view of a specimen that is 36.8 vol % car-

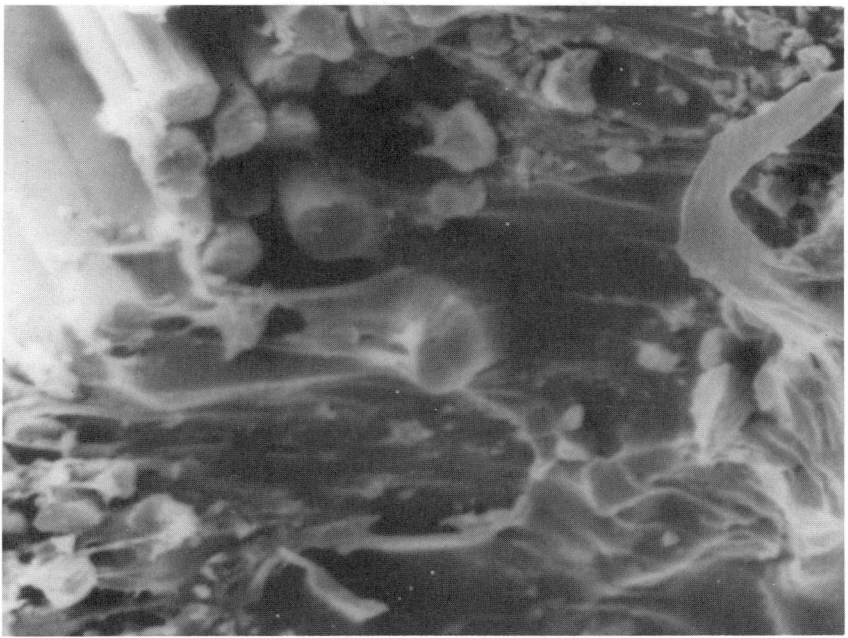

FIG. 2—*SEM photograph of the end view of a 36.8 vol % carbon fiber and 63.2 vol % polyester matrix prepreg.*

bon fiber and 63.2 vol % polyester. Note the excess of polymer and the randomly located particles of unmelted polyester. This excess of polymer appears to hinder electrical contact between the filaments and the rollers. The result is uneven heating and thus uneven coating of the individual filaments.

Figure 3 shows that this problem is reduced when the polyester coating is lowered to 48.7 vol %. Now, even though some areas have a slightly thicker coating, all filaments are coated with at least a thin layer of polymer.

Figure 4 shows that as the polymer coating continues to decrease (this specimen is 70 vol % carbon fiber and 30 vol % polyester), the coating uniformity continues to improve. This figure shows the thin coating of individual filaments which is desirable for maximum flexibility of the coated tow.

Figures 5 and 6 show that this same thin coating of individual filaments is also achievable when LARC-TPI is used. However, Fig. 7 illustrates one of the problems caused by inadequate tow spreading during process operation with the LARC-TPI. The tow was only spread to a thickness of about 14 filament layers when this specimen was coated. As a result, only the first five to seven layers of the filaments were coated with the LARC-TPI.

Polymer Distribution in Composites

As a final test to determine if the coated tow produced by this continuous process had a sufficiently even distribution of matrix polymer for use in composite materials, the 0.254-m segments of coated tow were molded into unidirectional composites. Composites containing

FIG. 3—*SEM photograph of a 51.3 vol % carbon fiber and 48.7 vol % polyester matrix prepreg.*

FIG. 4—*SEM photograph of the coated fibers in a 70.0 vol % carbon fiber and 30.0 vol % polyester matrix prepreg.*

FIG. 5—*SEM photograph of the coated surface of a carbon fiber and LARC-TPI prepreg.*

FIG. 6—*SEM photograph of the coated fibers in a 78.2 vol % carbon fiber and 21.8 vol % LARC-TPI matrix prepreg.*

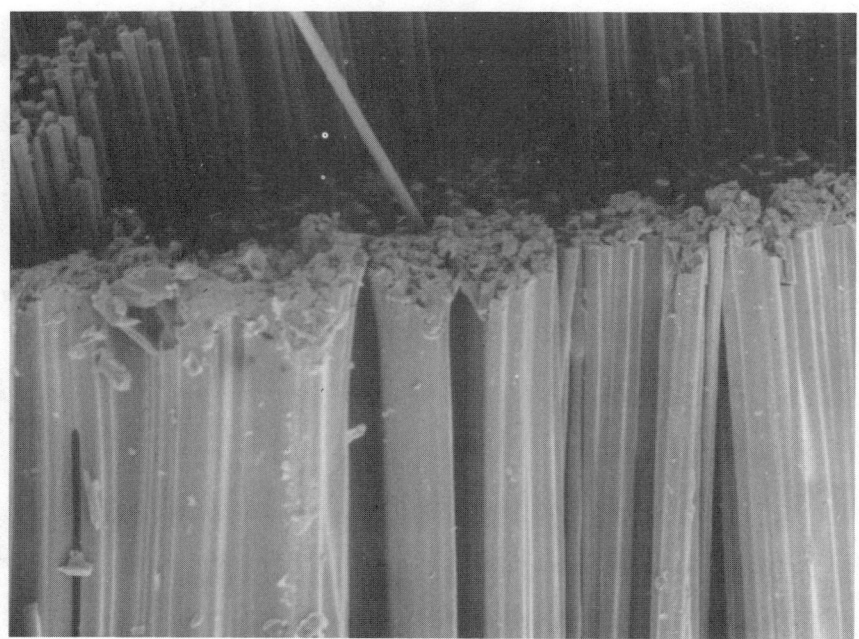

FIG. 7—*SEM photograph of the end view of a 78.2 vol % carbon fiber and 21.8 vol % LARC-TPI matrix prepreg.*

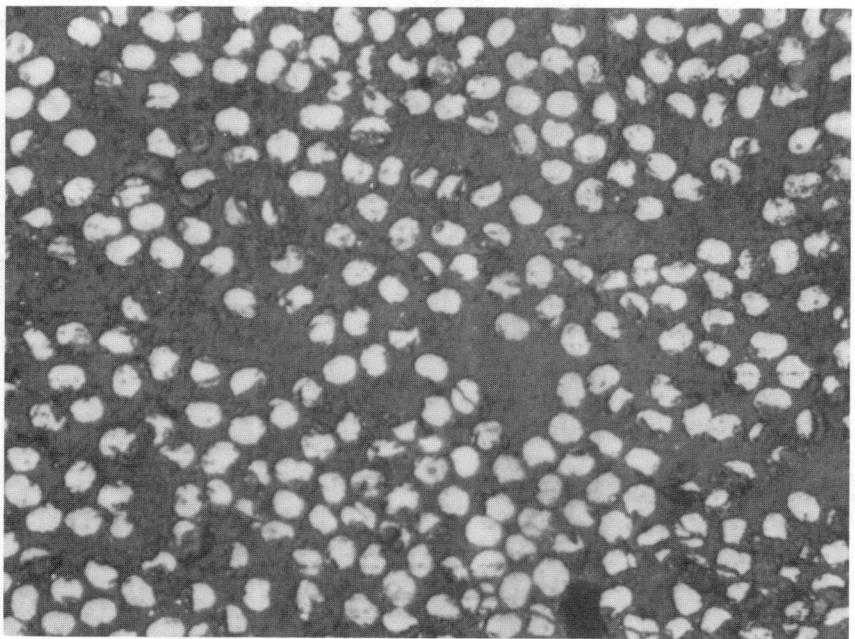

FIG. 8—*Image analyzer photograph of a cross-sectional view of a 36.8 vol % carbon fiber and 63.2 vol % polyester matrix composite.*

36.8, 51.3, 57.1, and 70 vol % carbon fiber in a polyester matrix were fabricated as described previously. These specimens then were mounted in an acrylic encapsulating resin, cut perpendicular to the fiber axis, polished, and inspected using an image analyzer to determine the evenness of the distribution of the matrix polymer within the composite.

Using the image analyzer, Fig. 8 shows a typical view of a specimen containing 36.8 vol % carbon fiber and 63.2 vol % polyester. All fibers are surrounded by a relatively even coating of matrix polymer. This is typical of all four composite specimens and indicates that the uneven coating, as observed by SEM inspection of some of the individual tow segments, tends to improve during composite molding. However, Fig. 9, a lower magnification of this same specimen, shows that there are limitations to this redistribution that can occur during molding and that areas of unreinforced matrix polymer exist within this composite.

Figure 10 shows that when the fiber volume was increased to 70 vol %, the fibers still were surrounded by a relatively even coating of the matrix polymer. However, the lower magnification in Fig. 11 shows that the distribution of matrix polymer is much more uniform throughout the specimen. This confirms the trend observed by SEM inspection of the individual tow segments: as the fiber volume percent increases to levels of 60% or more, the polymer coating becomes more uniform.

Further evidence of this additional redistribution of matrix polymer during molding is shown in Fig. 12. This is a typical cross-sectional view of the composite which was fabricated using the tow segments containing an average of 78.2 vol % carbon fiber and 21.8 vol % LARC-TPI. Even though in many areas of the individual coated segments only half of the filaments were coated, the polymer distribution in the composite was surprisingly even. This shows that the slightly uneven coating of the LARC-TPI was overcome by the temperature and pressure of the composite molding process.

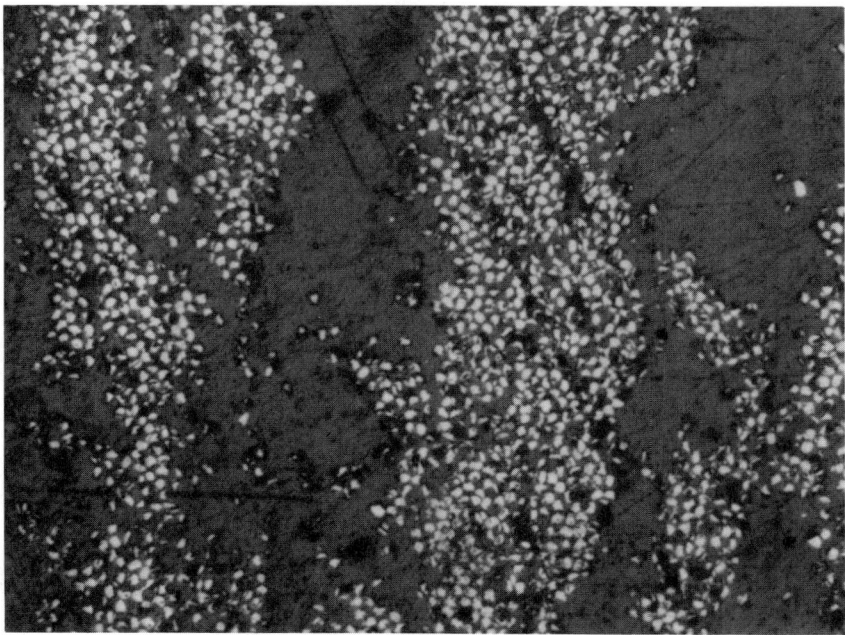

FIG. 9—*Lower magnification image analyzer photograph of a cross-sectional view of a 36.8 vol % carbon fiber and 63.2 vol % polyester matrix composite.*

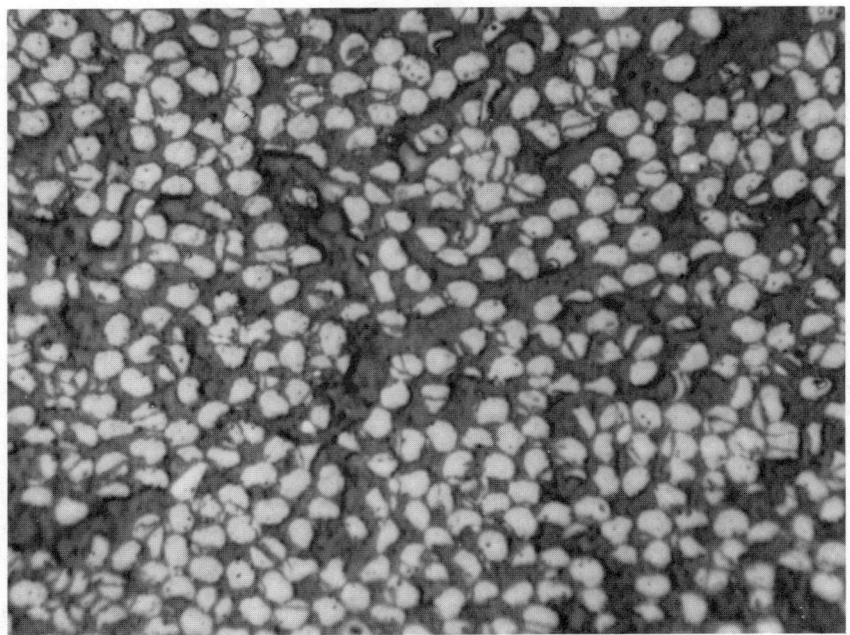

FIG. 10—*Image analyzer photograph of a cross-sectional view of a 70.0 vol % carbon fiber and 30.0 vol % polyester matrix composite.*

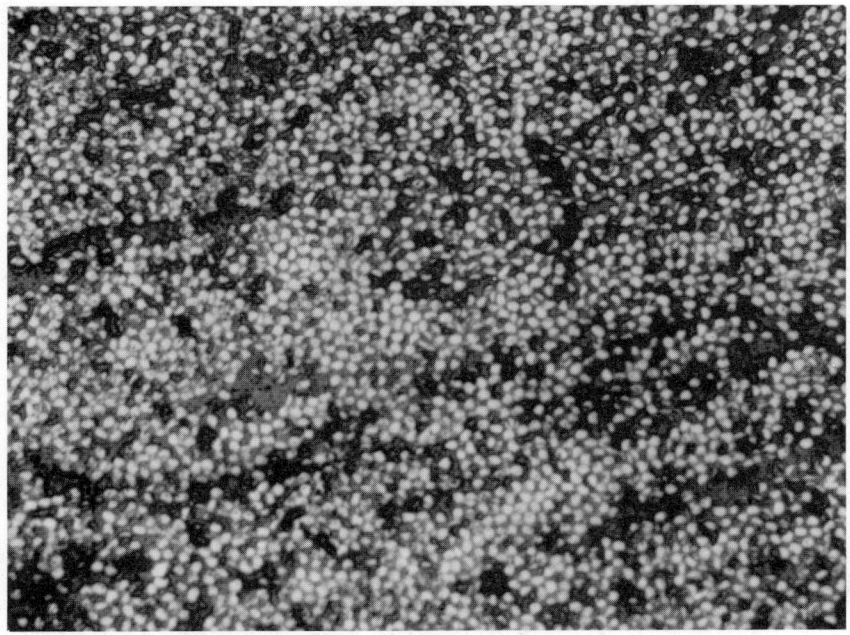

FIG. 11—*Lower magnification image analyzer photograph of a cross-sectional view of a 70.0 vol % carbon fiber and 30.0 vol % polyester matrix composite.*

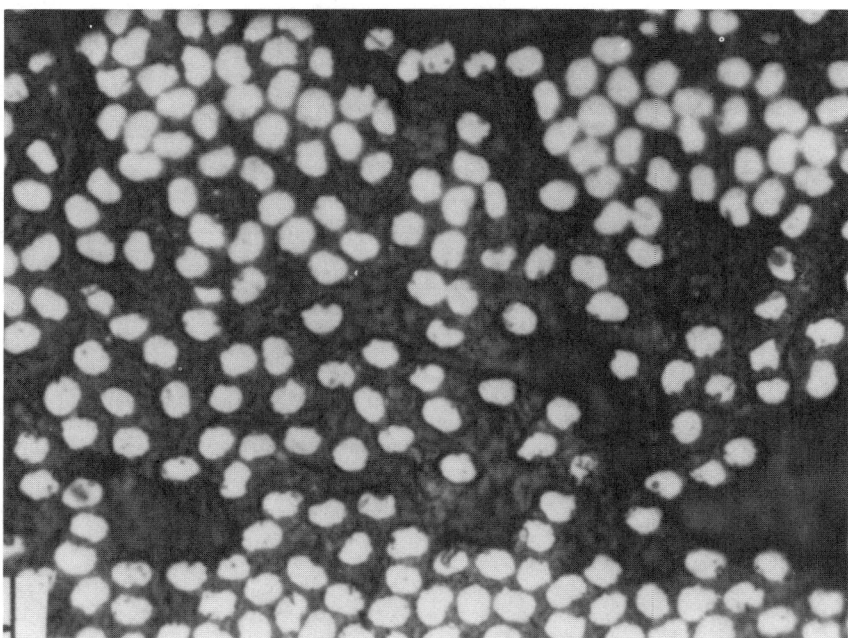

FIG. 12—*Image analyzer photograph of a cross-sectional view of a 78.2 vol % carbon fiber and 21.8 vol % LARC-TPI matrix composite.*

Conclusions

Carbon fiber tow can be coated with either polyester or LARC-TPI in the continuous process as developed. As might be expected, the spreading of the carbon fiber tow during processing affects both fiber heating and polymer distribution. An increased amount of tow spreading increases the electrical contact of the fibers and results in a more uniform temperature distribution within the tow. Also, the polymer distribution across the coated carbon fiber tow was more uniform when tow spreading was increased. Trials during which LARC-TPI was applied showed that if the tow were spread to a thickness of five to seven filament layers, the polymer will penetrate the entire tow. The long-term variation in the fiber volume percent during operation of the present continuous coating process was caused by the uneven deposition provided by the powder extruder.

Fortunately, as the fiber volume percent increases (and thus the polymer coating decreases), the coating uniformity improves because of improved electrical contact. Thus, while coating was somewhat uneven on tows that were 40 vol % fiber and 60 vol % polymer, it was quite satisfactory on tows that were 60 vol % fiber and 40 vol % polymer. However, in all cases when composites were made from both polyester and LARC-TPI coated carbon fiber tow, the distribution of the matrix polymer improved.

Acknowledgments

Support for this work was provided by the National Aeronautics and Space Administration, Langley Research Center, under NASA research grant NAG-1-680. Robert M. Baucom serves as the NASA technical officer for the project.

Robert M. Hackett[1] and Shyam N. Prasad[1]

Pultrusion Process Modeling

REFERENCE: Hackett, R. M. and Prasad, S. N., **"Pultrusion Process Modeling,"** *Advances in Thermoplastic Matrix Composite Materials, ASTM STP 1044,* G. M. Newaz, Ed., American Society for Testing and Materials, Philadelphia, 1989, pp. 62–70.

ABSTRACT: A basic one-dimensional heat transfer model of the pultrusion process for thermosetting resin composites was formulated. The model, employing a Galerkin finite-element approach, is the basic development. It can be expanded to a general characterization of the process and extended to characterize the pultrusion process for thermoplastic resin composites. The application of the basic model is demonstrated.

KEY WORDS: pultrusion, thermosetting resins, thermoplastic resins, finite-element modeling, heat transfer, composite materials processing, process simulation

Pultrusion is a process through which high-modulus, lightweight composite structural members such as beams, truss components, stiffeners, etc. can be manufactured. The operation, simple in concept but complex in detail, basically consists of pulling a number of fiber tows, or mats, through a thermosetting resin bath and then through a heated die where the wetted fiber bundle cures, thus producing a structurally sound part exiting the die, or, in the case of the pultrusion of thermoplastic composites, the material enters the die in the form of impregnated rovings.

In the pultrusion of thermosetting resin composites, the primary mechanism is the chemical reaction occurring during "cure," while in the pultrusion of thermoplastic resin composites there is no reaction during "consolidation." A key requirement for the pultrusion of thermoplastic composites is a means of applying sufficient heat and pressure to consolidate the impregnated rovings as they are formed by the die. A major advantage of thermoplastic pultrusion is the ability to reform the part after consolidation. There is strong interest in the pultrusion of thermoplastic composite material systems [1] and significant activity has been reported [2], but the technique will require further development before it is widely employed.

Details of the formulation and development of a pultrusion process simulation model presented here focus directly upon the pultrusion of thermosetting resin composites, but it is felt that the same modeling techniques can be directed toward a similar development of a simulation of the pultrusion of thermoplastic resin composites.

Process Description

In the pultrusion of thermosetting resin composites, the steady-state reaction process is initiated by the application of heat to the chemically active material. The reaction progresses while under pressure within the die. It is exothermic, and at some position within the die the relationship of heat flux is inverted and the degree of cure progresses to the point where shrinkage allows the part to release from the die wall. Reinforcing material is assembled and oriented to

[1]Professor and chairman, and professor, respectively, Department of Civil Engineering, The University of Mississippi, University, MS 38677.

enter the die in the configuration necessary for the development of the desired mechanical properties of the produced structural member. Heat is supplied to the process by electrically heated metal platens. Cooling water at the die entrance controls the transition from ambient to die temperature. Hydraulic pullers provide a continuous movement of the product.

Precise control of the thermal and chemical phenomena occurring within the die is of utmost importance. If the reaction proceeds too rapidly, the composite can bond to the die surface. This results in a loss of production, a low-quality product, and possible damage to the die. Once the process is in progress, the pulling force must be regulated to ensure that the line speed does not vary, since fluctuations thereof translate directly into variations in cure conditions.

Outwardly the process is deceptively simple in that the function is well understood. However, without an in-depth knowledge of the interaction of the various system variables, one cannot achieve efficient performance of the process. Essentially, only general qualitative information about what occurs inside the pultrusion die is presently available. In the liquid zone of the material, the temperature of the die exceeds that of the resin, with the temperature of both increasing. Within the gel zone, the peak exotherm of the resin is reached, usually being well above the temperature of the die. In effect, over the remainder of the process the die is drawing heat from the curing composite, thereby reducing thermal shock to the product upon its exit from the die. The material traveling through the process undergoes a number of dynamic changes as a result of the temperature environment within the die. The nature of these changes is manifested, to an extent, through the variation of the viscosity of the resin over the length of the die. Over the initial portion, the viscosity decreases as the temperature of the material increases through conduction. This reduction aids in the continuing wet-out of any unsaturated fibers. At the point at which the chemical reaction is initiated, the change in viscosity is reversed, and the viscosity rapidly increases through the stages of gelation and final cure. It is desirable for this reaction to occur under sufficient pressure to ensure composite integrity and to minimize internal porosity that can occur from vapor pressures within the reacting material.

The pressure at the material-die interface is a measurement of normal surface forces which, combined with the appropriate coefficients of friction, yield a measurement of frictional force, or resistance to pull. Pressures can be associated with the viscosity of the resin, the volumetric ratios of fiber and resin, the coefficient of thermal expansion of the materials, the cross-sectional geometry of the cavity, the length of die over which there is material contact, the coefficients of friction of the die with respect to the liquid, gel, and solid, and the degree of shrinkage of the solid. The efficiency of the process can be greatly enhanced through a better understanding of the pressure distribution.

The purpose of this description has been to define the pultrusion process with respect to thermosetting resin systems, highlight the most important process variables, indicate the complex interdependence among these variables which renders an intuitive grasp of the process almost impossible to attain, and, thus, emphasize the need for analytical definition of process variable interactions in order to gain insight into the process.

Process Modeling

The complexity of the pultrusion process is emphasized in the above discussion, and it goes without saying that a comprehensive analytical model of the process would be extremely difficult to develop. Previous activities in this regard, as well as with respect to the broader field of polymer process modeling, are documented and discussed in Refs 3–5. Successful modeling of these processes, on any level, is highly dependent upon the application of sophisticated numerical techniques.

The intent in this effort is to formulate a relatively straightforward one-dimensional heat transfer model of the thermoset-pultrusion process and to develop a solution for the resulting second-order governing differential equation using the finite-element method. The resulting one-dimensional model will obviously not be extensive enough to adequately characterize and thus provide a means for simulating the pultrusion process. However, it could serve as a basic model, an expansion of which might lead to the development of full-scale simulation models of the pultrusion process for both thermosetting and thermoplastic resin systems.

Heat Transfer Model

Consider a pultrusion die of length L having a thin rectangular cross section of thickness t. The steady-state heat transfer condition in the die is governed by

$$K\left(\frac{\partial^2 T}{\partial x^2} + \frac{\partial^2 T}{\partial y^2}\right) - r\frac{\partial T}{\partial x} = 0 \tag{1}$$

where

T = temperature of the material,
K = thermal conductivity of the material, and
r = rate of pultrusion.

The rate of pultrusion is given by

$$r = u\rho C_p \tag{2}$$

where

u = pultrusion line speed,
ρ = mass density of the material, and
C_p = heat capacity of the material.

If we integrate Eq 1 with respect to y and define T to be the mean temperature (thickness averaged) we obtain the following equation

$$K\frac{d^2 T}{dx^2} - r\frac{dT}{dx} + \frac{2h}{t}(T_d - T) = 0 \tag{3}$$

where

T_d = die temperature, which varies along the length of the die, and
h = die-material interface convection coefficient.

If we designate the properties and variables associated with the liquid (pregelation) and solid (postgelation) regions with the subscripts 1 and 2, respectively, we can write Eq 3 as

$$K_1\frac{d^2 T_1}{dx^2} - r_1\frac{dT_1}{dx} + \frac{2h_1}{t}(T_d - T_1) \tag{4a}$$

$$K_2\frac{d^2 T_2}{dx^2} - r_2\frac{dT_2}{dx} + \frac{2h_2}{t}(T_d - T_2) \tag{4b}$$

Solution Technique

The finite-element method is employed as the technique for obtaining a solution for Eqs 4a and 4b. The solution procedure entails developing a finite-element model for Region 1, and a similar finite-element model for Region 2. Each model consists of approximately 100 one-dimensional elements. A boundary condition for the finite-element model of the liquid region is that T_1 at $x = 0$ is set equal to the ambient temperature; a boundary condition for the finite-element model of the solid region is that T_2 at $x = L$ is set equal to T_d at the die exit. The position of the gel region (the solidification front) is defined by satisfying the continuity of flux at that location. This condition is stated by the following

$$q_1 - q_2 = \rho H u \tag{5}$$

where

H = heat of reaction,
q_1 = heat flux in the positive x direction in the liquid phase, and
q_2 = heat flux in the positive x direction in the solid phase.

These fluxes are given by the following

$$q_1 = -K_1 \frac{dT_1}{dx} \tag{6a}$$

$$q_2 = -K_2 \frac{dT_2}{dx} \tag{6b}$$

The solution procedure is thus based upon applying the "source" defined in Eq 5 to the nodal point which is common to both finite-element models but which is contained in the model of the solid region. The length of the two models is adjusted until the value of T_1 at the last nodal point in the liquid region is equal to the value of T_2 at the first nodal point in the solid region (the solidification front). The initial guess of the location of the interface of the two regions is at the midlength of the die. A search routine then locates the interface, with the total number of elements (liquid plus solid) remaining constant.

Finite-Element Formulation

The finite-element method to be applied for the solution of Eqs 4a and 4b will now be formulated. Two well-known methods exist for the solution of this type of field problem: the variational method and the method of weighted residuals. Because of limitations associated with the variational method related to the existence of particular functionals, the method of weighted residuals is widely employed. The weighting function that must be chosen in the method of weighted residuals may take one of several forms, such as Galerkin (named after its Russian inventor), least squares, collocation, as well as others. The Galerkin method employs weighting functions of the same type as the interpolation functions that approximate the variable over the element. The least squares method minimizes the square of the errors. The collocation method produces zero error at a number of specified points. Broad treatment of the method of weighted residuals is found in Refs 6 and 7, as well as in a number of other textbooks on the finite-element method.

Following the general weighted-residual method formulation for the steady-state heat conduction-convection equation found in Ref 6, and then specializing to the Galerkin one-dimensional approximation with respect to Eq 3, we define a set of equations of the form

$$[k]\{T\} + \{f\} = 0 \tag{7}$$

where

$$k_{ij} = \int_0^{L^{(e)}} \left(\frac{dN_i}{dx} K \frac{dN_j}{dx}\right) dx + \int_0^{L^{(e)}} \left(N_i\, r\, \frac{dN_j}{dx}\right) dx + \int_0^{L^{(e)}} \left(N_i\, \frac{2h}{t}\, N_j\right) dx \qquad (8)$$

and

$$f_i = -\int_0^{L^{(e)}} \left(N_i\, \frac{2h}{t}\, T_d\right) dx \qquad (9)$$

The material/system properties K, r, and h are approximated by constant values over the length of an element $L^{(e)}$, and T and T_d are nodal point values. The shape functions for the linear one-dimensional element are given by the following

$$N_i = \frac{L^{(e)} - x}{L^{(e)}} \qquad (10a)$$

$$N_j = \frac{x}{L^{(e)}} \qquad (10b)$$

Performing the integrations indicated in Eqs 8 and 9 yields the following terms

$$k_{ii} = \frac{K^{(e)}}{L^{(e)}} - \frac{r^{(e)}}{2} + \frac{2h^{(e)}L^{(e)}}{3t} \qquad (11a)$$

$$k_{ij} = -\frac{K^{(e)}}{L^{(e)}} + \frac{r^{(e)}}{2} + \frac{h^{(e)}L^{(e)}}{3t} \qquad (11b)$$

$$k_{ji} = -\frac{K^{(e)}}{L^{(e)}} - \frac{r^{(e)}}{2} + \frac{h^{(e)}L^{(e)}}{3t} \qquad (11c)$$

$$k_{jj} = \frac{K^{(e)}}{L^{(e)}} + \frac{r^{(e)}}{2} + \frac{2h^{(e)}L^{(e)}}{3t} \qquad (11d)$$

and

$$f_i = f_j = -\frac{h^{(e)}L^{(e)}T_d}{t} \qquad (12)$$

The coefficients defined in Eqs 11a, 11b, 11c, and 11d comprise the 2 × 2 conduction-convection matrix of the finite-element formulation. The vector defined in Eq 12 constitutes the element nodal point sources.

It can be noted from Eqs 11b and 11c that $k_{ij} \neq k_{ji}$ and, therefore, that the operator is not self-adjoint. Results obtained with the model that demonstrate the type of numerical oscillations inherent in the formulation are shown in Fig. 1. The "upwinding" procedure described in Ref 8 is employed in the model in order to eliminate the observed, and expected, localized oscillatory effect.

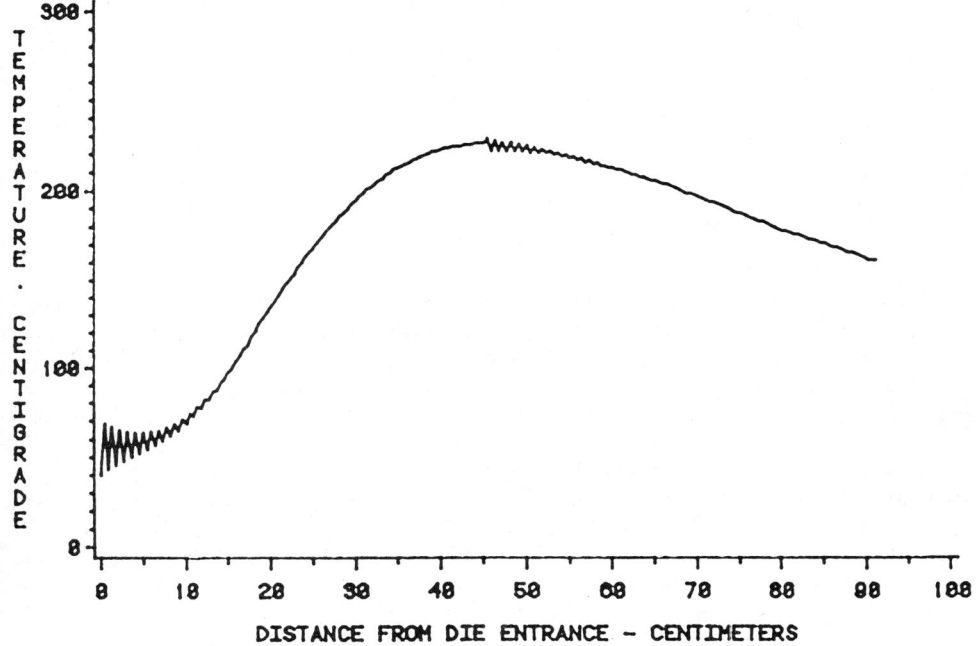

FIG. 1—*Finite-element model results showing oscillatory behavior due to the operator not being self-adjoint.*

It also may be noted that a general second-order ordinary differential equation can be converted to the desired form (for which a functional can be obtained) by a suitable transformation of coefficients. For example [7]

$$A(x)\frac{d^2T}{dx^2} + B(x)\frac{dT}{dx} + C(x)T + D(x) = 0 \tag{13}$$

may be converted to the self-adjoint form

$$K(x)\frac{d^2T}{dx^2} + P(x)T + D(x) = 0 \tag{14}$$

through the relationships

$$K(x) = \exp\left[\int \frac{B(x)}{A(x)}dx\right] \tag{15a}$$

$$P(x) = \frac{C(x)}{A(x)} \exp\left[\int \frac{B(x)}{A(x)} dx\right] \qquad (15b)$$

The relationships expressed in Eqs 15a and 15b are derived in Ref 6.

Model Application

The developed finite-element pultrusion simulation model was applied to the processing of a graphite/epoxy resin composite. The length of the modeled die was 91.44 cm. The thickness of the modeled part was 1 cm. Two solutions were generated, one for a line speed of 0.593 cm/s and the other for a line speed of 0.085 cm/s. For both cases the modeled material was graphite/epoxy. The graphite was AS4-W-12K, Hercules Type I, and the epoxy resin was Shell Epon 9310. The fiber volume fraction was 0.67.

Although the resin physical properties actually vary over the process, typical values for the given material were used in the finite-element model. Of primary importance are the values of H, the heat of reaction of the resin, which was taken to be 475.0 J/g, and h, the die-material interface convection coefficient, which varies significantly from the liquid to the solid region, but was held at a constant value of unity in the model.

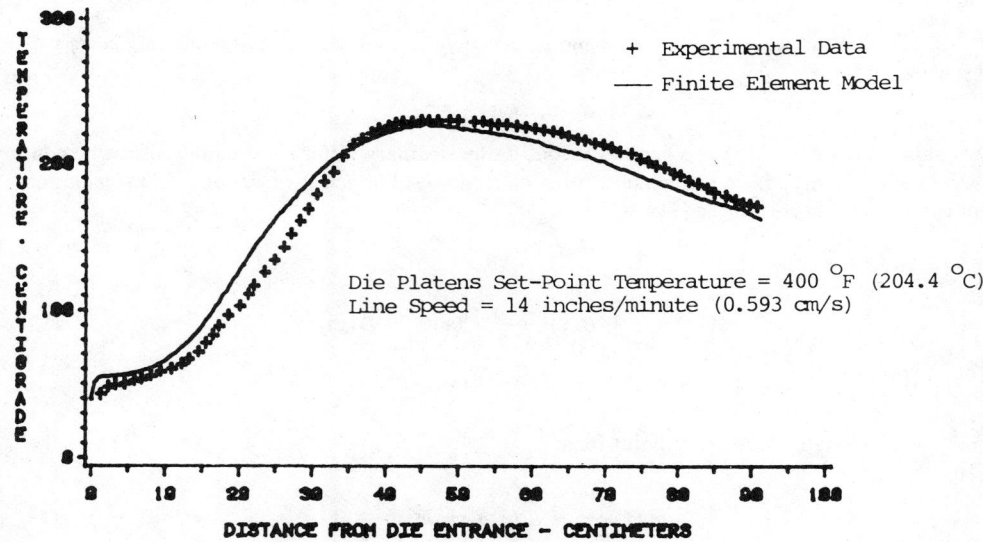

Pultrusion of a 3/8-inch Diameter Rod
Material: Graphite/EPON Resin 9310
EPON CURING AGENT 9360
EPON CURING AGENT Accelerator 537

FIG. 2—*Comparison of pultrusion data and finite-element results; line speed = 14 in./min (finite-element data is for 1-cm-thick rectangular cross section).*

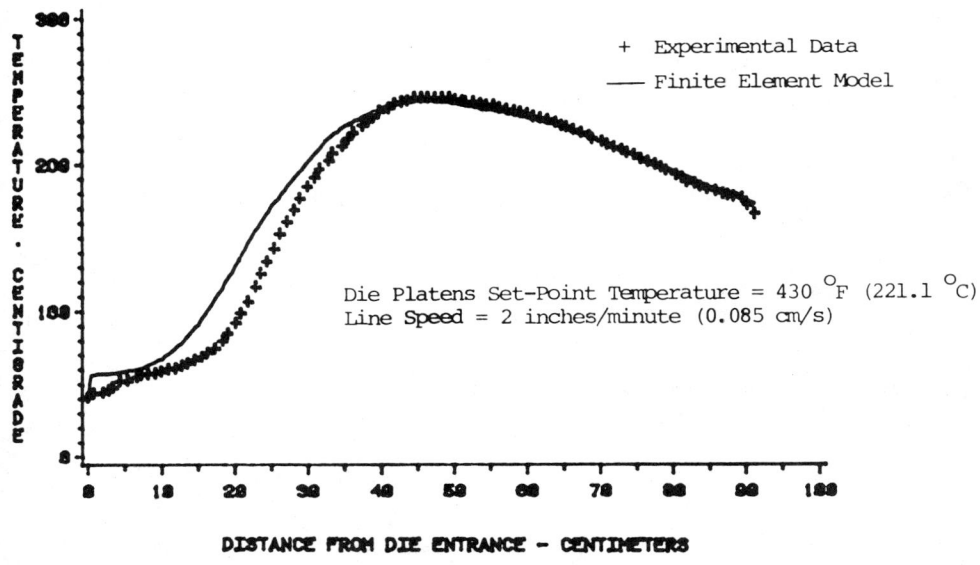

FIG. 3—*Comparison of pultrusion data and finite-element results; line speed = 2 in./min (finite-element data is for 1-cm-thick rectangular cross section).*

The results obtained from the finite-element model are shown in Figs. 2 and 3. These are compared with data obtained from the pultrusion of the same material system, although with a different cross section, at the NASA/George C. Marshall Space Flight Center [9]. All other aspects of the actual pultruding and the modeling were equivalent. The die temperature profile used in the model was identical to that of the Pulstar pultruder, manufactured by Pultrusion Technology, Inc., that was used to pultrude the graphite/epoxy composite.

Figures 2 and 3 show good agreement between the model and the actual pultrusion process, which is encouraging. Further development of the model will require a better quantitative definition of certain physical and thermal properties such as, for example, resin permeability and die-material interface convection coefficient.

Summary

A basic finite-element model of the pultrusion of thermosetting resin composites has been formulated. Comparisons of model-generated results with pultrusion data provide reason to believe that the basic model might be successfully expanded and extended to the modeling and characterization of the thermoplastic resin composite pultrusion process.

Acknowledgment

This work was supported in part by the Martin Marietta Corp. through Contract No. A71211.

References

[1] Newaz, G. M., "Advanced Thermoplastic Composites," *ASTM Standardization News,* Vol. 15, No. 10, October 1987, pp. 32–35.

[2] O'Conner, S. E. and Beever, W. H., "Polyphenylene Sulfide Pultruded Type Composite Structures," *42nd Annual Conference,* The Composites Institute of the Society of the Plastics Industry, Inc., Cincinnati, OH, 2–6 Feb. 1987.

[3] Roylance, D., "Use of 'Penalty' Finite Elements in Analysis of Polymer Melt Processing," *Polymer Engineering and Science,* Vol. 20, No. 15, October 1980, pp. 1029–1034.

[4] Alyward, L., Douglas, C. D., and Roylance, D., "Analysis of Cure in Composites Processing," *Proceedings,* 16th National SAMPE Technical Conference, 9–11 Oct. 1984, pp. 148–158.

[5] Batch, G. L. and Macosko, C. W., "A Computer Analysis of Temperature and Pressure Distributions in a Pultrusion Die," *42nd Annual Conference,* The Composites Institute of the Society of the Plastics Industry, Inc., Cincinnati, OH, 2–6 Feb. 1987.

[6] Zienkiewicz, O. C., *The Finite Element Method,* McGraw-Hill, New York, 1977, pp. 42–92.

[7] Allaire, P. E., *Basics of the Finite Element Method,* Wm. C. Brown Publishers, Dubuque, IA, 1985, pp. 167–300.

[8] Carey, G. F. and Oden, J. T., *Finite Elements—Computational Aspects,* Vol. III, Prentice-Hall, Inc., Englewood Cliffs, NJ, 1984, pp. 280–289.

[9] Hackett, R. M. and Vaughan, J. G., "Characterization and Optimization of the Pultrusion Process," Final Report to the Martin Marietta Corporation under Research Contract No. A71211, December 1987.

Thermal and Mechanical Behavior

Ray J. Bankert,[1] *Nicholas D. Lambropoulos,*[2] *Mark S. Shephard,*[2] *and Sanford S. Sternstein*[1]

Thermoplastic Matrix Composites: Finite-Element Analysis of Mode I and Mode II Failure Specimens

REFERENCE: Bankert, R. J., Lambropoulos, N. D., Shephard, M. S., and Sternstein, S. S., **"Thermoplastic Matrix Composites: Finite-Element Analysis of Mode I and Mode II Failure Specimens,"** *Advances in Thermoplastic Matrix Composite Materials, ASTM STP 1044,* G. M. Newaz, Ed., American Society for Testing and Materials, Philadelphia, 1989, pp. 73–90.

ABSTRACT: A finite-element analysis was conducted to evaluate the stress distributions within Mode I and Mode II failure specimens, assuming both isotropic and orthotropic elastic material properties. The effects of anisotropy on both the magnitude and the location of the highest stress concentration at the vicinity of the crack tip are significant. The results from modeling realistic blunt crack tip geometry and resin-rich zones imply that local variations in the microstructure strongly influence the stress state near the crack tip and therefore the measured fracture properties. In addition, the features of a viscoelastic model for thermoplastic matrices are described. This model will be used in future investigations of matrix-dominated failure phenomena.

KEY WORDS: Mode I failure, Mode II failure, anisotropy, orthotropic model, isotropic matrix material, fiber waviness, matrix viscoelasticity, resin-rich zones

A complete and accurate description of the mechanical properties of composite materials is essential to their proper application as engineering materials. The testing techniques used to determine fracture properties are well documented and widely used. For example, the slender double cantilever beam (DCB) specimen with all zero degree plies is often used to measure Mode I delamination fracture toughness, a matrix-dominated property.

In an idealized, homogeneous isotropic elastic material, fracture behavior can be treated using the well-known concepts of linear or nonlinear elastic fracture mechanics. However, such treatments give little insight into local failure processes and their relationship to material properties and structure. In addition, for materials that exhibit time-dependent behavior (for example, viscoelastic solids), the concepts of classical fracture mechanics fail to predict, without ad hoc modification, time-dependent aspects of the failure process associated with loading history. Notwithstanding the complexity associated, a detailed knowledge of the local stress field in the vicinity of a crack or other flaw is needed to fully understand failure processes in time-dependent solids. Unfortunately, the prediction of local stress field behavior as related to the sample loading history and flaw geometry requires detailed knowledge of the material's constitutive equation. A rudimentary, yet illustrative example of the local stress field-path dependent failure approach to both constant strain rate failure and creep rupture in a viscoelastic solid is given elsewhere [1].

[1]Center for Composite Materials and Structures, Rensselaer Polytechnic Institute, Troy, NY 12180-3590.
[2]Center for Interactive Computer Graphics, Rensselaer Polytechnic Institute, Troy, NY 12180-3590.

Composite materials pose additional problems because anisotropy causes considerable variation in fracture properties for flaws with different orientations. The interaction of fibers and matrix, especially close to the crack tip, can also be expected to play a major role in determining local stresses and raises the additional question of the role of the local structural features such as fiber volume fraction and waviness. Clearly, a continuum fracture mechanics approach, based on smeared global properties without taking into account the local microstructure, is not accurate for evaluating local stresses. Conversely, the detailed analysis of an infinite number of fibers with random spacing, etc., is beyond reasonable treatment and would, if tractable, lead to useless detail.

Microscopic examinations of thermoplastic composites generally indicate a significant distribution of fiber spacing and resin-rich zones. In addition to such distributions within plies, there are also resin-rich layers between plies. When one considers that a thermoplastic matrix-carbon fiber single ply with a nominal fiber volume fraction of 60% has about a 20 to 1 ratio of longitudinal to transverse modulus, it becomes clear that an analysis of the effects of fiber volume fraction and the distribution of resin-rich zones on the stress concentrating ability of a flaw involves considerations of strong local variations in anisotropy. Thus, while all solid materials have microstructural variations from point to point, composites exhibit extreme local variations, ranging from pure resin (nearly isotropic and low modulus) to densely populated fiber regions (high modulus and anisotropy). It is the size scale and degree of variation that distinguish composites from more traditional materials.

In this paper, the first of several on this subject, the effects of anisotropy on the stress-concentrating ability of a flaw in both Mode I and Mode II fracture tests are considered. The effects of an extreme variation in local structure then are considered by introducing a pure resin layer at the tip of a crack. The tension-compression yield stress asymmetry exhibited by thermoplastics is shown to affect the local stresses developed at the tip of a crack in the resin-rich layer.

A combination of both a discrete analysis in the vicinity of a crack tip together with a smoothing model for far field behavior seems appropriate, although computationally intensive. The use of an orthotropic, elastic far-field material model with interdispersed layers of isotropic matrix material described by a suitable time and stress state–dependent constitutive equation appears to be a reasonable starting point for unidirectional thermoplastic matrix composites. The finite-element method provides a useful tool which can be used for analysis of such a composite and, in particular, can be used for describing the detailed stress fields around cracks. The choice of an optimum mesh at regions with steep gradients is of major importance.

Much attention has been devoted to the use of classical fracture mechanics and the measurement of strain energy release rates [2–8]. The use of a sharp crack is common in the literature [8], but it predefines the starting point for propagation and introduces a locking mechanism that may not be realistic. Thermoplastic matrix materials, on the other hand, cannot support infinitely sharp cracks. Accordingly, blunt cracks are investigated here.

After the finite-element model has been introduced, numerical results from linear elastic finite-element analyses of both Mode I and Mode II failure tests are presented. The degree of anisotropy and the effects of the crack length, boundary conditions, and resin-rich layers are the variables considered. Next, an effort was undertaken to complement the results of a linear elastic analysis with the development of a more sophisticated viscoelastic time-dependent material model for the thermoplastic matrix. An analytical development of the constitutive model is not presented here, but some preliminary results which illustrate its behavior are given. These results form the basis for continuing research in this area.

Finite-Element Model

To determine the detailed stress distribution in specimens subject to Mode I and Mode II loading conditions, with particular reference to the stress concentration at the crack tip, finite-element analyses have been performed using the ABAQUS F.E. package.

The geometry of the specimen used in both Mode I and Mode II is shown in Fig. 1. In Mode I, only half of the specimen needs to be analyzed due to symmetry, while in Mode II, the entire sample has to be taken into account. The external load is applied to the sample in the form of distributed forces or displacements. In Mode I, the load is applied perpendicular to the crack plane at the cracked edge of the beam, while the boundary conditions are simple supports at the symmetry centerline. In Mode II, the load is applied again perpendicular to the crack plane but at the uncracked edge of the beam. Two sets of boundary conditions are used:

1. Boundary Conditions Set 1 (Clamped)—The cracked edge of the beam is clamped, that is, both in-plane nodal displacements at that edge are fixed.
2. Boundary Conditions Set 2 (Simple Support)—The nodes at the cracked edge of the beam are not allowed to move in the vertical direction, but they can move freely in the horizontal direction, so the crack lips can slide between each other.

The thickness of the crack is $1/100$ of the beam thickness and no friction is assumed between the crack lips, since in the actual test a thin film usually is inserted. The use of a blunt crack instead of a sharp one allows for an infinite number of possible points (stress-concentrations) from which the crack may continue propagating.

The finite-element modeling of these test cases was aided by the finite quadtree mesh generator [9], which provided the meshes shown in Fig. 1. The level of refinement at the areas with steep stress gradients (vicinity of the crack tip) compared to the coarseness of the mesh in the far field, is easily obtained through the specification of mesh control parameters associated with the geometric definition of the specimen. Eight-noded isoparametric quadrilateral plane-strain elements employed in ABAQUS [10] are utilized with a 2×2 Gaussian integration rule. In Mode I, the displacements were large, so a large displacement analysis was made to see if such an option was important. As suspected, it did not alter the stress fields in the critical areas. Subsequently, linear strain-displacement relations were used throughout this study.

Homogeneous orthotropic material properties were used to model uniaxial composites, and the results were compared to those obtained for a fully isotropic material. Parametric studies for various levels of anisotropy are also presented. In an effort to model the heterogeneous microstructure close to the crack tip, a thin layer with isotropic material properties was introduced, while the rest of the beam remains orthotropic (Fig. 1).

Mode I Numerical Results

Initially, an isotropic material model was used with properties representative of a random glass mat type of composite as follows

$$\text{Young's modulus} = 6.9 \text{ GPa}$$
$$\text{Poisson's ratio} = 0.3$$

Load control was simulated by applying a constant 6.25 N force at the end of the specimen for various crack lengths. The load was applied in the y-direction, the crack lies in the z-direction and plane-strain restrictions are placed in the x-direction. The results (Fig. 2a) for this case showed that the highest stress concentration occurs at the crack tip and the opening stress (y-direction) component is the dominant one. The stress levels were found to vary linearly with crack length, under load control.

To describe more accurately the stress state in the composite, an orthotropic material model was used. The specimen was also assumed to be axisymmetric with respect to the fiber direction, producing five independent elastic constants

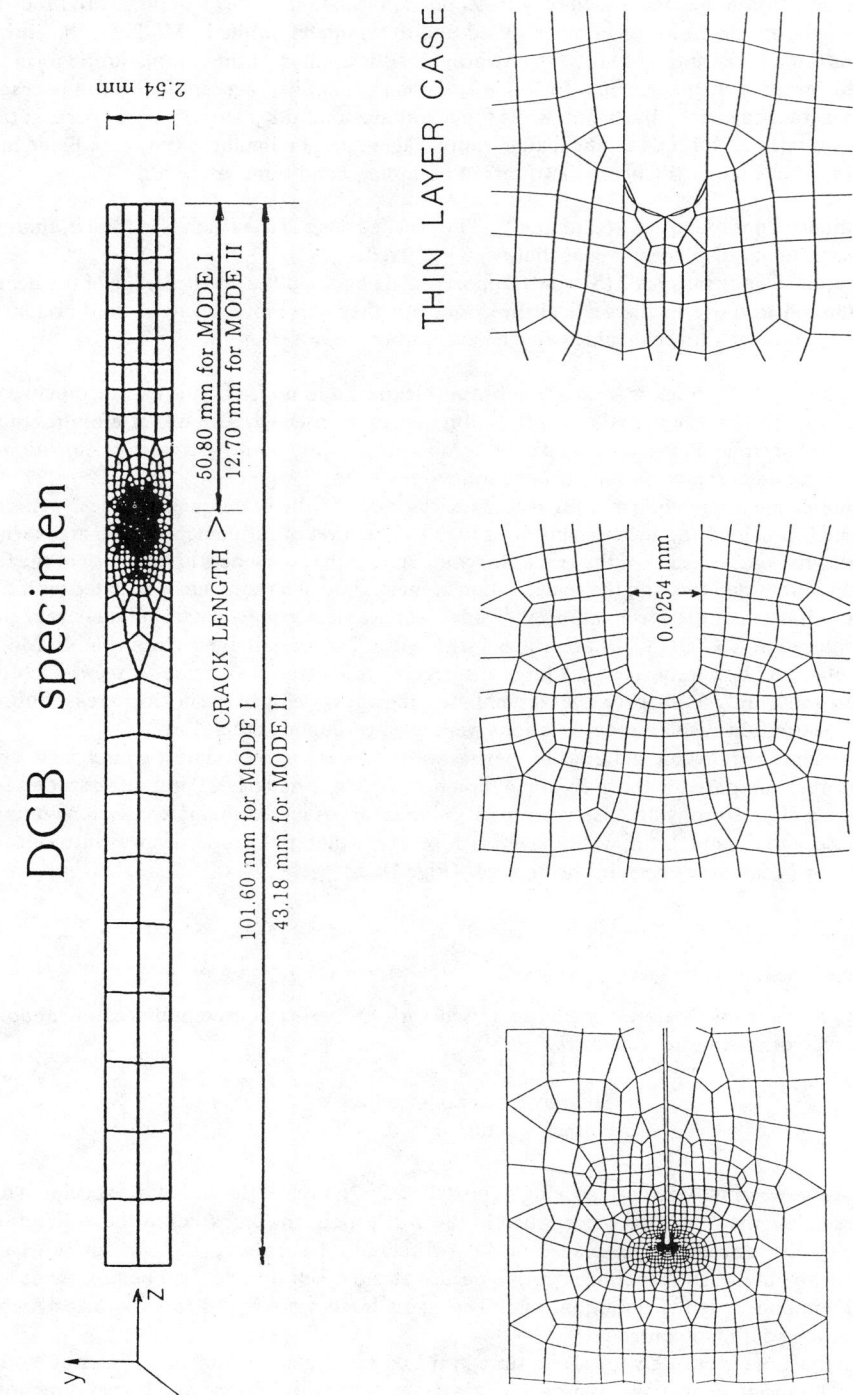

FIG. 1—*Geometry and finite-element mesh for the DCB specimen.*

DCB—Mode I

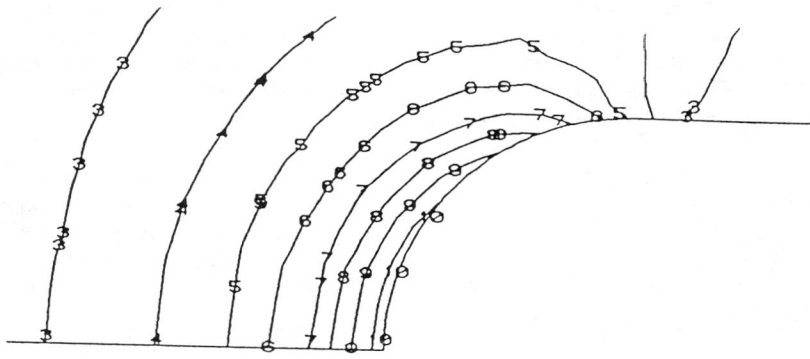

a. Isotropic

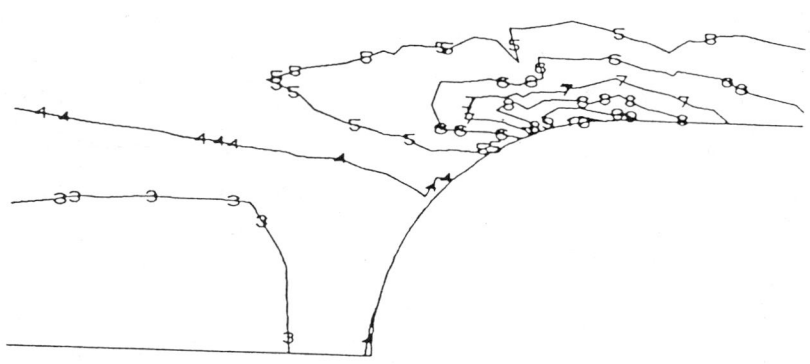

b. Orthotropic

FIG. 2—*Maximum principal stress distribution for* (a) *isotropic, and* (b) *orthotropic beam.*

Longitudinal Young's modulus	=	138.0 GPa
Transverse Young's modulus	=	6.9 GPa
In-plane shear modulus	=	3.4 GPa
Longitudinal Poisson's ratio	=	0.2
Transverse Poisson's ratio	=	0.4

The fibers and the crack both lie in the z-direction. In the Mode I orthotropic analysis, the specimen is still considered to be homogeneous. The results showed two major differences. Firstly, the highest stress concentration moved over the shoulder of the crack (Fig. 2b). Secondly, the opening mode stress is no longer dominant, rather the maximum principal stress is in the z or fiber direction. The implication is that the crack path will not remain straight on the local level. The large stress parallel to the crack (z-direction) will tend to diversify the crack advance, which might explain why a smooth, straight crack is rarely found in thermoplastic composites.

To see how large the amount of anisotropy must be to cause this dramatic shifting of the maximum principal stress from the tip to the shoulder of the crack, the ratio of the longitudinal modulus, E_1, to the transverse modulus, E_2, was varied. Initially, the $E_1:E_2$ ratio was 20:1. This was done by decreasing the value of the longitudinal modulus, keeping the other four independent elastic constants fixed. Figure 3 shows that the swing takes place with only a small amount of anisotropy.

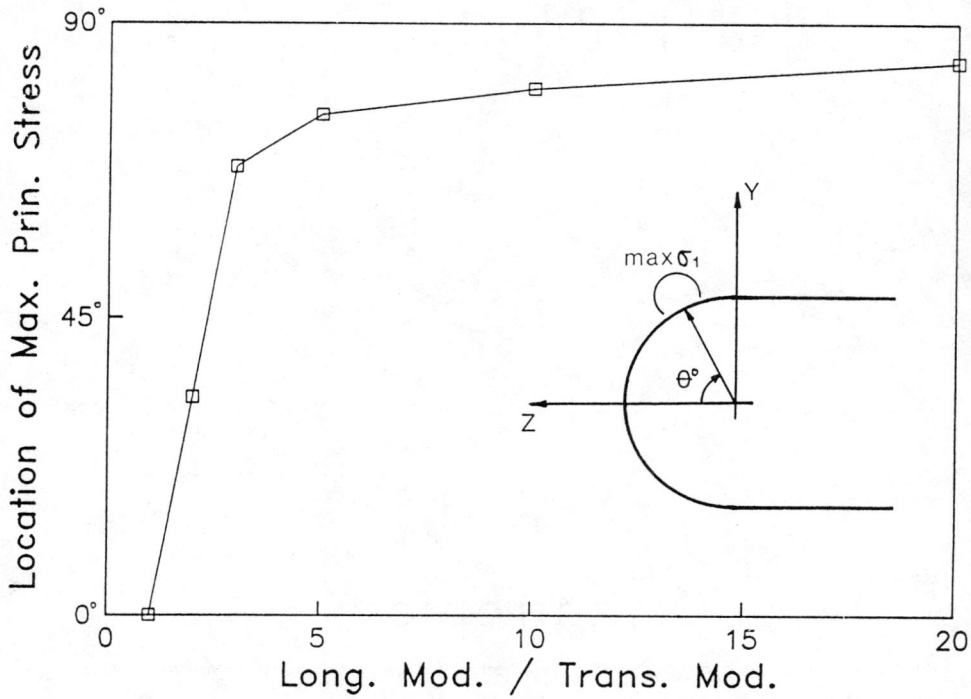

FIG. 3—*The location of the maximum principal stress as a function of the ratio of the longitudinal versus the transverse Young's modulus.*

The effect on the matrix during the move from a y to z dominated stress state is complicated by the fact that the z-component will be carried primarily by the fibers. It is interesting to notice that during this transition the matrix sees increased shearing (Fig. 4). In the isotropic case the in-plane shear is only 46% of the opening normal stress. However, at a $E_1 : E_2$ ratio of 20:1, the shear component is about 60% of the normal stress.

Mode II Numerical Results

The first step was to analyze a similar specimen using the same isotropic and orthotropic material properties as in Mode I. Load control was simulated by applying a constant 190 N force perpendicular to the uncracked end of the beam. By clamping the cracked end of the beam, the vicinity of the crack tip is under pure Mode II conditions.

Figures 5 and 6 show the in-plane shear stress contours for fully isotropic and orthotropic beams, respectively. In the isotropic case, the stresses are higher but reasonably spread out, while in the orthotropic case the highest shear stress concentration is localized over the shoulder of the crack. The next step was to introduce a resin-rich layer in front of the crack tip and consider the effects of a heterogeneous local microstructure on the shear stress distribution. The

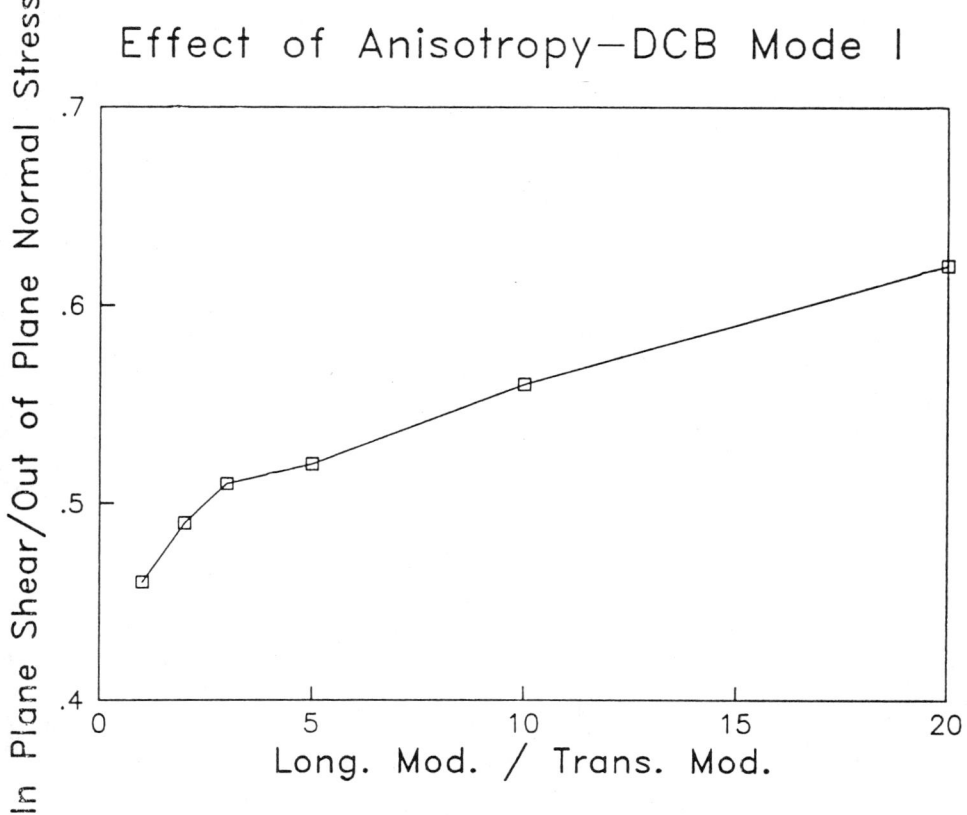

FIG. 4—*The relative magnitude of the in-plane shear stress as a function of the ratio of the longitudinal versus the transverse Young's modulus.*

Mode II

```
IN-PLANE SHEAR STRESS (MPa)
   C.L.              VALUE
    1               00.00
    2               34.50
    3               69.00
    4              103.50
    5              138.00
    6              172.50
    7              207.00
    8              241.50
    9              276.00
   10              310.50
```

FULLY ISOTROPIC

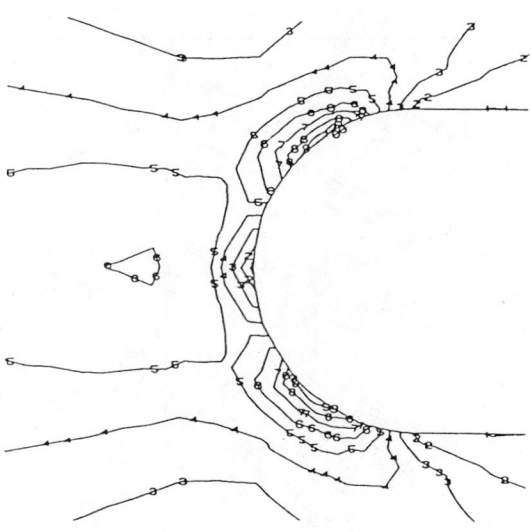

FIG. 5—*In-plane shear stress distribution for fully isotropic beam.*

magnitude of the stresses does not vary from the fully orthotropic case, but the highest shear stress concentration is now localized along the crack path, inside the layer with isotropic material properties (resin) (Fig. 7).

When varying the degree of anisotropy in the fully orthotropic but homogeneous specimen, it was observed that it also takes only a small amount of anisotropy (ratio $E_1:E_2$ equal to 5:1) to reach the results of the strongly orthotropic case (ratio 20:1) (Fig. 8).

Mode II

IN-PLANE SHEAR STRESS (MPa)

C.L.	VALUE
1	00.00
2	14.00
3	28.00
4	42.00
5	56.00
6	70.00
7	86.00
8	100.00
9	114.00
10	128.00

FULLY   ORTHOTROPIC

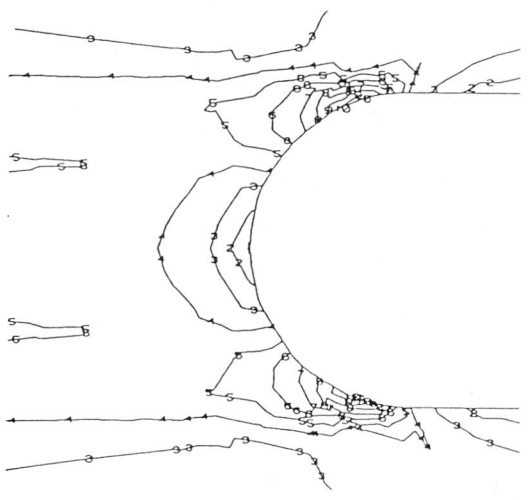

FIG. 6—*In-plane shear stress distribution for fully orthotropic beam.*

When varying the crack length, no significant difference to the shape of the shear stress distribution is observed, but the magnitude of the maximum shear stress increases with increasing crack length (Fig. 9). The effects of the boundary conditions on the test also were examined. By allowing the crack lips to slide freely but with no motion in the vertical direction, stresses, twice as high as those obtained when clamping the cracked end of the beam, are observed (Fig. 9).

The increase in beam compliance, which is directly proportional to the tip deflection obtained for an applied constant load, is nonlinear with increasing crack length. This increase is found to be more dramatic for boundary conditions set No. 2, simple support (Fig. 10).

Mode II

IN—PLANE SHEAR STRESS (MPa)

C.L.	VALUE
1	00.00
2	14.00
3	28.00
4	42.00
5	56.00
6	70.00
7	86.00
8	100.00
9	114.00
10	128.00

ORTHOTROPIC BEAM
WITH ISOTROPIC THIN LAYER

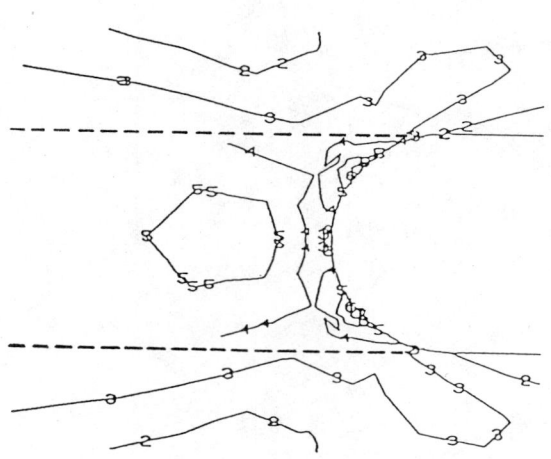

FIG. 7—*In-plane shear stress distribution for orthotropic beam with isotropic thin layer (shown in dashed lines).*

Discussion

Mode I

The DCB Mode I test is thought to characterize the matrix-controlled failure of composites due to delamination, namely the interlaminar separation resulting from the crack opening stresses. Due to the highly orthotropic nature of uniaxial composites, the crack tip lies in a complicated stress field consisting of a large stress component parallel to the crack direction.

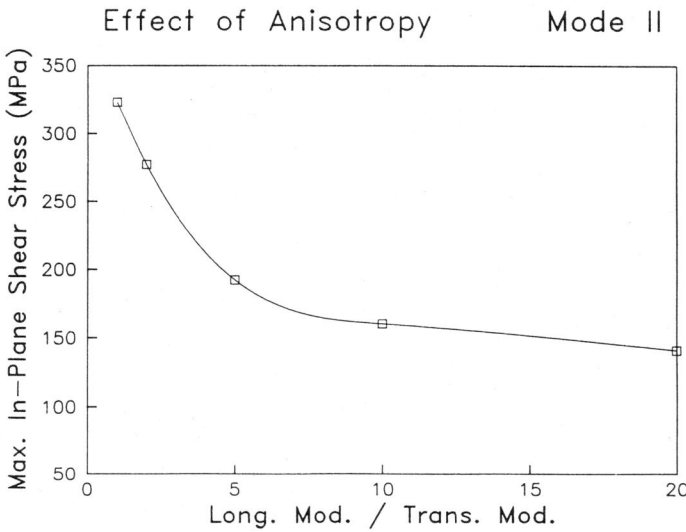

FIG. 8—*Maximum in-plane shear stress as a function of the ratio of the longitudinal versus the transverse Young's modulus.*

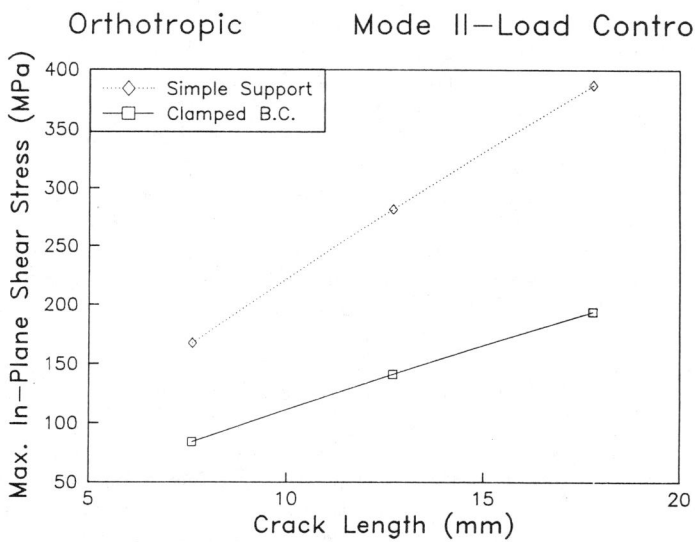

FIG. 9—*Effect of crack length on the maximum in-plane shear stress for two sets of boundary conditions.*

The off-center location of the maximum principal stress will be conducive to crack growth away from the tip. The result will be an abundance of crack jumping which is common in thermoplastic composites. Fiber bridging appears to be an integral part of the test geometry and not an unwanted side effect.

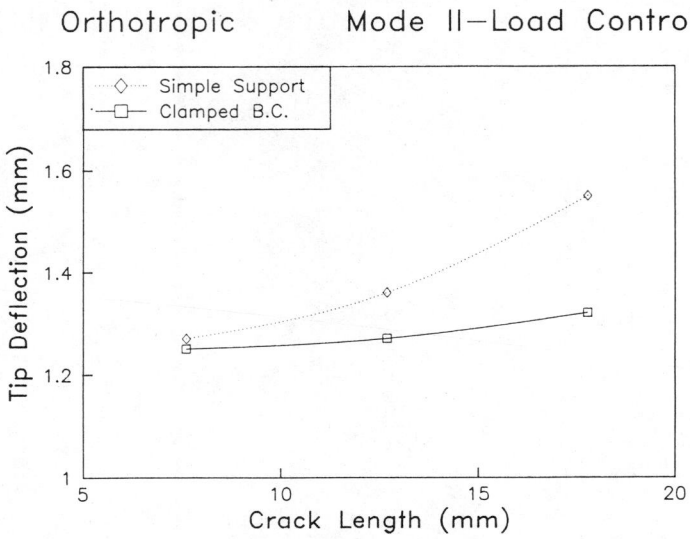

FIG. 10—*Effect of crack length on the tip deflection (beam compliance) for two sets of boundary conditions.*

When interpreting composite data, the fiber volume fraction is always a major area of concern. It has a large effect on the longitudinal modulus; a 10% drop in fiber volume fraction reduces E_1 by approximately 17%. This change in $E_1:E_2$ ratio has a profound effect on the stress field. The evenness of the fiber distribution will vary the $E_1:E_2$ ratio on a local scale. The crack will advance into an ever-changing stress field. The result will be a tortuous, twisting type of crack path that is characteristic of thermoplastic composites.

Another characteristic of thermoplastic composites is that the onset of fast fracture occurs very abruptly without being preceded by slow crack growth. It is very difficult to produce stable crack propagation. Thermoplastic composites undergo more of a stick/slip type of crack advance. These accelerating/decelerating transitions usually are explained by a process zone development argument. Our results strongly suggest that the randomness of the microstructure must be taken into account in order to find the answer.

In glassy polymers, the hydrostatic and deviatoric responses are governed by distinct volume and shear relaxation processes. A positive mean normal stress (dilatation) promotes crazing and can lead to brittle fracture. The deviatoric component is associated with a ductile shearing process which is an alternative energy dissipative mechanism. Therefore, in thermoplastics, the ratio of dilatational to deviatoric strain energy plays a large role in determining the nature of crack advance. In some glassy polymers the ductile-brittle transition is extremely sharp and the dividing line so fine that the competition between crazing and shear can occur side by side within the same specimen [11].

The ability of closely packed fibers to modify the energy transferred to the matrix by increasing the shearing component is a surprising result with numerous implications. The idea that continuous fibers simply place constraints on the thermoplastic matrix, preventing favorable energy dissipative mechanisms from fully developing, must be questioned. The opening mode DCB specimen for uniaxial composites has a complicated ratio of Mode I and Mode II effects at the crack tip. Uneven packing and variable fiber volume fraction will produce a changing stress state, not only with respect to location to the crack tip but also possibly determine the type of crack growth. Either a rapid, sharp brittle advance or a stable ductile one, depending on the nature of the microstructure, currently is encountered.

Mode II

Thermoplastics, as a rule, are quite ductile in shear. Even unmodified polystyrene that is brittle in tension can be deformed 50 to 100% in shear. The origin of this strong dependence on stress state resides in the competition between crazing (a cavitational mode) and shear band formation, as described elsewhere [12]. Uncontrolled craze growth leads to crack propagation. Thus, the ability to deform a thermoplastic without fracture is often governed by the suppression of crazing in favor of shear yielding (shear stress versus tension), or by the control of craze growth (as in the rubber dispersed phase in impact modified thermoplastics). For the reasons cited, the propagation of a crack through the matrix phase of a thermoplastic matrix composite should be strongly affected by the local stress state.

In this regard, the Mode II results are suggestive of the effects which local variations in structure can have. It is clear that the material anisotropy results in a localization of the shear stress concentration over the shoulder of the crack. It follows that variations of fiber volume fraction along the crack path will result in changes in local anisotropy and, therefore, will alter both the gradient of the shear stress distribution and the magnitude of the maximum value. The computations involving the inserted film of neat matrix show that stresses of about the same magnitude are produced at the crack tip but they are now strongly concentrated along the crack path. Thus, the local fiber volume fraction influences the amount of local anisotropy and thereby affects the local distribution of stresses. Clearly, a simple correction for global volume fraction effects on macroscopic fracture toughness seems inappropriate.

It is noted that the Mode II problem produces an intrinsically unstable crack growth because larger cracks produce higher shear stress concentration at constant beam deflection. This is to be contrasted with the stable Mode I problem (Fig. 11).

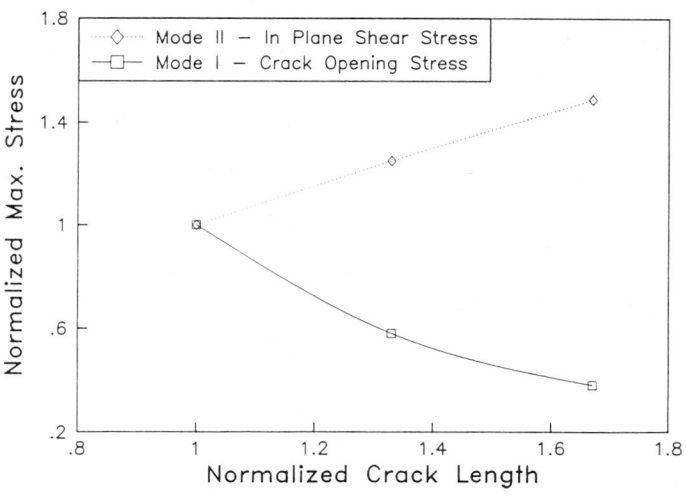

FIG. 11—*Stable (Mode I) versus unstable (Mode II) crack growth.*

Finally, the strong effect of the clamped edge boundary condition on the local stress concentration suggests that it is vital to obtain reproducible clamping conditions when performing the Mode II experiments.

Viscoelastic Constitutive Model for Thermoplastic Matrix

The conclusion from both Mode I and Mode II results is that the fracture properties of composites are very sensitive to the local microstructure (degree of anisotropy, fiber volume fraction, resin-rich zones, etc.) around a flaw. In addition, detailed knowledge of the material's constitutive equation is required in order to fully understand failure processes in time-dependent solids, such as thermoplastics.

A nonlinear viscoelastic material model has been developed. The emphasis is on capturing essential characteristics of thermoplastics, such as rate dependence, stress component interactions, and transient behavior. The basic assumption is to separate the hydrostatic from the deviatoric component of stress/strain state. The hydrostatic stress produces elastic dilatational response, while the deviatoric stress follows a specific time-dependent constitutive law. The Eyring stress-biased barrier model is generalized to 3-D such that the hydrostatic stress correctly modifies the deviatoric response.

The ABAQUS finite-element code, including the option of a "User Defined Subroutine—UMAT," was selected as the platform for implementing the constitutive equation. At each integration point of the finite-element analysis, this subroutine provided the material relation between the instantaneous increment in stresses and strains, according to the proposed constitutive law.

Figure 12(a) illustrates the behavior of the material model under tension at three different strain rates. For higher loading rate, the model can support higher stresses at the same strain level. For the same strain rates, there is a significant difference between tension and compression (Fig. 12b), due to the effect of the hydrostatic stress on the material behavior. Both tests are within reasonable correlation with available experimental data for thermoplastics. The zero-tension cycling loading test (Fig. 12c), exhibits the ratcheting effect, that is, the area of advancing hysteresis loops decreases.

In an effort to show the significance of the detailed knowledge of the material's constitutive relation, the previously presented material model is incorporated into the Mode II analysis. A thin, resin-rich layer with isotropic but nonlinear viscoelastic behavior is introduced in front of the crack tip and constant rate loading is applied up to a stress level far inside the post-yield range of the material's response.

A qualitative more than quantitative selection of results is presented in Fig. 13. In-plane shear stress contours at three different steps of the loading history illustrate the progressive redistribution of stresses around the crack tip. Load factor 1 corresponds to the elastic range of the material's behavior, therefore the response is similar to the isotropic linear elastic thin-layer case (Fig. 7). As the stresses at both the highest stress concentrations increase and reach the onset of nonlinear response (Load Factor 6), significant variations are observed. The upper half of the crack tip which is under tension can no longer support stresses as high as the lower half which is under compression. This can be explained by the different behavior of the model in tension versus compression (Fig. 12b). For load factor 6, the compressive part is still in the linear elastic range, while the tensile part is already in the post-yield range. When the compressive part enters its post-yield range (Load Factor 10), the variations in stress distribution are found to be even more dramatic.

It is obvious now that, for small scale modeling (for example, the vicinity of a crack tip) where the effects of the microstructure are significant, the use of a smoothing procedure through a mixing model is inappropriate. On the other hand, when global noncritical properties (for example, stiffness) of a composite system are under consideration, a good mixing model is appro-

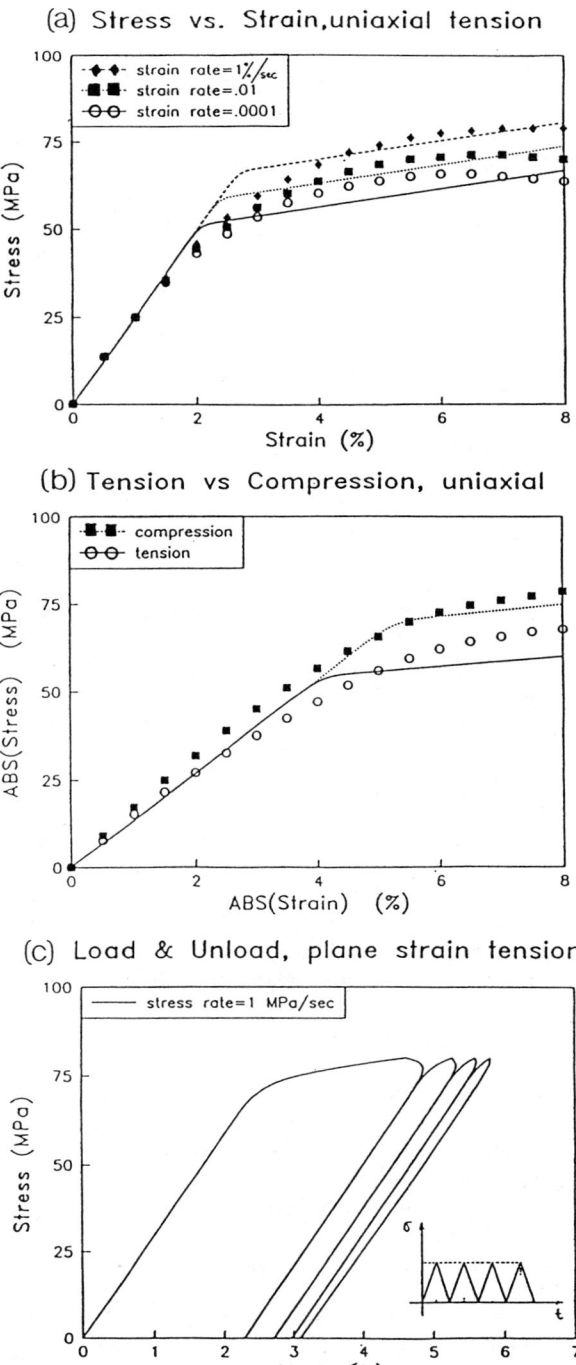

FIG. 12—*Basic features of the nonlinear time-dependent constitutive relation for thermoplastics (symbols in figures* (a) *and* (b) *indicate experimental data and lines represent the predictions of the model).*

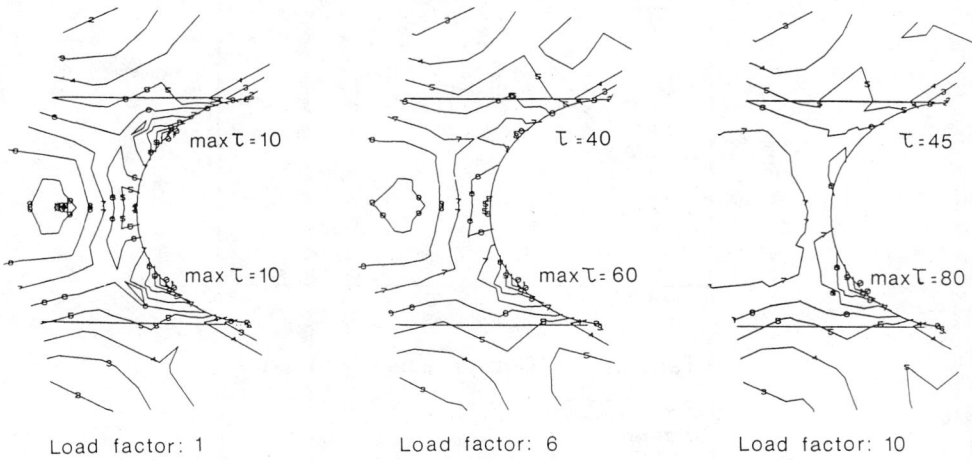

Mode II

ORTHOTROPIC BEAM WITH ISOTROPIC - VISCOELASTIC THIN LAYER

Load factor: 1   Load factor: 6   Load factor: 10

FIG. 13—*In-plane shear stress distribution at three different levels of applied load.*

priate. The combination of both scale levels in the same analysis is a difficult task which requires special attention in the effort to capture those phenomena that affect the overall behavior.

However, the incorporation of the proposed constitutive relation for the matrix into a reliable mixing model is a goal worthy to undertake. The Periodic Hexagonal Array (PHA) model developed by Dvorak and Teply [13] for the elastoplastic behavior of metal matrix composites was selected, where the nonlinear matrix behavior will be predicted by the time-dependent viscoelastic constitutive relation. The structure of the PHA is suitable for finite-element analysis because it can be very easily used as a user-defined subroutine providing material behavior information. The main objective is to evaluate the overall instantaneous stiffness of the composite medium in terms of local moduli of the constituents (linear elastic fiber and nonlinear viscoelastic matrix).

Conclusions

The frequently held view that fibers limit plastic zone growth in the matrix and, therefore, reduce the fracture toughness relative to the neat matrix material, is too simplistic. It is our belief that a local stress field computation, taking into account the local microstructure and matrix constitutive equation, must be path averaged to obtain the macroscopically observed fracture toughness, rather than computation based on global properties. In other words, the question of when to average is the key issue. To make further progress, a more accurate material model to numerically account for the randomness of the microstructure is under development.

This paper shows that anisotropy plays a major role in determining the stress concentration and stress state for cracks in both Mode I and Mode II specimens. Of particular note is the location of the maximum tensile stress on the surface of the crack tip as a function of the degree of anisotropy. This result suggests that a crack propagating in a composite with strong variations in local fiber volume fraction will jump, or alter its course, as it encounters regions of different anisotropy. Thus, the correction for fiber volume fraction among samples of different values is not simple and must also account for each sample's distribution of fiber volume fraction (or alternatively the resin distribution or fiber spacing distribution). Additionally, our results suggest that interply cracks will follow tortuous paths and ultimately link up to form a delamination.

Specifically, with regard to the well-known observations of fiber bridging, the wide scatter, and lack of reproducibility which appear to accompany double cantilever beam Mode I fracture tests, we suggest the following:

1. Fiber bridging is a manifestation of nonplanar crack growth, or crack jumping associated with the fiber volume distribution, that is, with the extreme local variations in anisotropy and its effects on stress concentration value and location relative to the crack tip. Experiments on DCB samples of polycarbonate-carbon fiber composites and microscopic studies of the crack path and fiber volume distribution (or sample heterogeneity) support this conclusion [14].

2. Comparisons of DCB fracture toughness parameters for purposes of in-situ performance are to be viewed with extreme caution since the variations in observed fracture toughness due to fiber volume fraction and its distribution may be as strong as the variations due to intrinsic resin behavior. Comparative studies of different matrices should be conducted on samples of equivalent heterogeneity.

Following the same logic, it would appear that variations in local anisotropy resulting from fiber waviness would also affect the stress field at a crack tip. In this case, it is the orientation of the orthotropic axes (and not local fiber volume fraction) with respect to the crack axis that is relevant. Calculations on the effects of fiber waviness amplitude and period on the stress field around a crack in a DCB specimen have been performed [15]. We find that a waviness of just 5° in amplitude and a period of 300 fiber diameters results in a 300% variation of stress concentration with position in the waviness cycle. We view this as further computational verification of the role which local anisotropy plays in determining the fracture properties of composites. The combined effects of variations in both fiber volume fraction and local anisotropy direction represent a computational challenge yet to be overcome.

We believe that the computations presented here (and our recent studies yet to be published) provide guidance for the fracture characterization of composites as well as guidance for the manufacturing of composites insofar as the control of microstructure is concerned.

Acknowledgments

The work described in this paper was funded by the National Aeronautics and Space Administration and the Air Force Office of Scientific Research under Grant No. NGL-33-018-003 and by a supplemental Grant from NASA-LANGLEY (NAG-1-253). One of the authors (R. J. B.) is an employee of the GE Research and Development Center, which has provided time and facilities for his graduate work. The authors would also like to thank Hibbitt, Karlsson, and Sorensen, Inc. for use of the ABAQUS F.E. package.

References

[1] Cessna, L. C., Jr. and Sternstein, S. S., "Viscoelasticity and Plasticity Considerations in the Fracture of Glasslike High Polymers," *Fundamental Phenomena in the Material Sciences,* Vol. 4, *Fracture of Metals, Polymers and Glasses,* L. J. Bonis, J. J. Duga, and J. J. Gilman, Eds., Plenum Press, NY, 1967, p. 45.

[2] Keary, P. E., Ilcewicz, L. B., and Trostle, J., "Mode I Interlaminar Fracture of Composites Using Slender Double Cantilever Beam Specimens," *Journal of Composite Materials,* Vol. 19, 1985, p. 154.

[3] Devitt, D. F., Schapery, R. A., and Bradley, W. L., "A Method for Determining the Mode I Delami-

nation Fracture Toughness of Elastic and Viscoelastic Composite Materials," *Journal of Composite Materials,* Vol. 14, 1980, p. 270.

[4] Gillespie, J. W., Jr., Carlsson, L. A., and Smiley, A. J., "Rate-Dependent Mode I Interlaminar Crack Growth Mechanisms in Graphite/Epoxy and Graphite/PEEK," *Composites Science and Technology,* Vol. 28, 1987, p. 1.

[5] Whitney, J. M., Browning, C. E., and Hoogsteden, W., "A Double Cantilever Beam Test for Characterizing Mode I Delamination of Composite Materials," *Journal of Reinforced Plastics and Composites,* Vol. 1, 1982, p. 297.

[6] Carlsson, L. A., Gillespie, J. W., Jr., and Tretheway, B. R., "Mode II Interlaminar Fracture of Graphite/Epoxy and Graphite/PEEK," *Journal of Reinforced Plastics and Composites,* Vol. 5, 1986, p. 170.

[7] Lee, S. H., "A Comparison of Fracture Toughness of Matrix Controlled Failure Modes: Delamination and Transverse Cracking," *Journal of Composite Materials,* Vol. 20, 1986, p. 185.

[8] Gillespie, J. W., Jr., Carlsson, L. A., and Pipes, R. B., "Finite Element Analysis of the End Notched Flexure Specimen for Measuring Mode II Fracture Toughness," *Composites Science and Technology,* Vol. 27, 1986, p. 177.

[9] Baehmann, P. L., Wittchen, S. L., Shephard, M. S., Grice, K. R., and Yerry, M. A., "Robust, Geometrically Based, Automatic Two-Dimensional Mesh Generation," *International Journal for Numerical Methods in Engineering,* Vol. 24, 1987, p. 1043.

[10] ABAQUS, "User's Manual," Hibbitt, Karlsson, and Sorensen Inc., Version 4, 1982.

[11] Takemori, M. T. and Kambour, R. P., "Discontinuous Fatigue Crack Growth in Polycarbonate," *Journal of Materials Science-Letters,* Vol. 16, 1981, p. 1108.

[12] Sternstein, S. S., "Mechanical Properties of Glassy Polymers," *Treatise on Materials Science and Technology,* Vol. 10, part B, *Properties of Solid Polymeric Materials,* J. M. Schulty, Ed., Academic Press, NY, 1977, p. 541.

[13] Dvorak, G. J. and Teply, J. L., "Periodic Hexagonal Array Models for Plasticity Analysis of Composite Materials," *Plasticity Today: Modeling, Methods and Application,* A. Sawczak, Ed., Elsevier, New York, p. 625.

[14] Buhrmaster, C. and Sternstein, S. S., Rensselaer Polytechnic Institute, to be published.

[15] Lambropoulos, N. D., Shephard, M. S., and Sternstein, S. S., Rensselaer Polytechnic Institute, to be published.

J. Ramey[1] and Anthony Palazotto[2]

A Study of Graphite/PEEK Under High Temperatures

REFERENCE: Ramey, J. and Palazotto, A., "**A Study of Graphite/PEEK Under High Temperatures,**" *Advances in Thermoplastic Matrix Composite Materials, ASTM STP 1044,* G. M. Newaz, Ed., American Society for Testing and Materials, Philadelphia, 1989, pp. 91–112.

ABSTRACT: An experimental evaluation of graphite/PEEK, APC-2, was carried out on specimens with various size holes acting under tension and compression loading. Quasiisotropic symmetrical laminates were tested at RT, 121, 135, and 149°C, and the Pipes three-parameter failure criterion was correlated with experimental results. Scanning electron microscopy photographs were used to determine specific notch effects on failure modes.

KEY WORDS: thermoplastic, strength, high temperature, compression, tension

Thermoplastics have been developed to improve fracture toughness in comparison to widely used thermosetting polymers such as epoxy systems. The lower fracture toughness and brittleness of the thermosets limit their notched capability to a fraction of their unnotched strength.

If a composite material is to be used in a given situation, certain aspects of fracture toughness must be considered. Designers must be able to choose correctly the best composite material for a particular application. They must be able to compare toughness characteristics from material to material so design constraints can be met. Designers must also know the effect of holes or cutouts on the strength of the material. Notch effects vary from material to material and can influence failure mechanisms, thus producing undesired results if the material properties were not well understood during the design phase. Temperature variations also affect the characteristics of materials, and these variations must be known.

A semicrystalline thermoplastic matrix poly (ether-ether-ketone) (PEEK) has been developed by ICI (Imperial Chemical Industries) with a reported improvement in toughness over thermosets and an increased resistance to typical aircraft solvents [1–4]. This paper reports on an experimental study performed on an Aromatic Polymer Composite-2 (APC-2), a 60% by volume of graphite fiber/PEEK composite. Quasiisotropic $(0/45/-45/90)_{2s}$ symmetrical laminates were tested monotonically in tension and compression to failure at room temperature and temperatures just below (121 and 135°C) and slightly above (149°C) the glass transition temperature ($T_g = 145°C$) for APC-2.

Experimental data were used to determine functions for a three-parameter failure criterion, and the results were compared to previous findings for APC-1 and graphite/epoxy [5]. Scanning electron microscopy (SEM) photographs were used to determine specific notch effects on failure modes.

[1]Graduate student, Air Force Institute of Technology, Wright-Patterson Air Force Base, OH 45433.
[2]Professor, Aeronautics and Astronautics Department, Air Force Institute of Technology, Wright-Patterson Air Force Base, OH 45433.

Experiment

All specimens were flat, rectangular plates cut from 406-mm square panels using fabrication methods for thermoplastic suggested by ICI, and were inspected by X-ray techniques to identify any defects. The typical tension specimen is shown in Fig. 1 with lengths of approximately 170 to 220 mm and with a width greater than three times the notch size. The end tab material was 1.6-mm thick glass epoxy bonded to the APC-2 specimen using a general purpose epoxy. The specimens were heat treated for 1 h at 93°C to ensure complete bonding. A Bridgeport milling machine was used to drill 2.54, 5.0, 10.0, and 15.2-mm diameter holes. Dog bone specimens for compression testing of unnotched specimens were machined on a router, equipped with an abrasive cylinder.

The tension tests were performed using an Instron Model 1115 at a constant cross head displacement rate of 1.25 mm/min. In order to prevent end tab adhesive problems (slipping) at the elevated temperatures considered (121, 135, and 149°C), specimens were heated in a cylindrical chamber using an adjustable hot air gun, avoiding the need for enclosing the tab area within the heat chamber. Specimens were equipped with thermocouples and allowed to thermally stabilize for 10 min prior to testing at the elevated temperatures. The temperature test equipment is shown in Fig. 2. Table 1 includes material properties at the above temperatures for specimens without holes.

The compression tests were performed using an MTS model 312.31 with a 156 kN capacity and a constant cross head displacement rate of 0.76 mm/min. The specimens were placed in an antibuckling fixture, consisting of two aluminum plates supporting the entire gage length as shown in Fig. 3. The compression specimens were tested in the same manner as the tension specimens previously addressed. A total of 120 specimens were tested, three for each hole radius at each elevated temperature and at room temperature. Figures 4 and 5 show the typical failed tension and compression specimens, respectively. The SEM characteristics of the compression failure will be discussed subsequently.

Discussion

Figure 6 shows the effects of temperature on the strength of quasiisotropic laminates for notched and unnotched APC-2 specimens to APC-1 considering (notched strength), σ_n^∞. One can notice the rate of strength reduction with temperature for both the APC-2 and APC-1 when

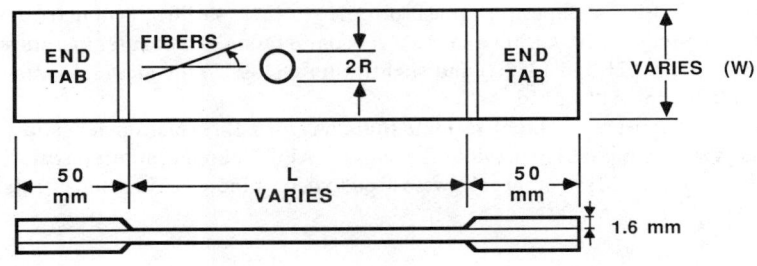

W>3 x DIAMETER
L>170 mm

FIG. 1—*Typical geometry of notched specimens.*

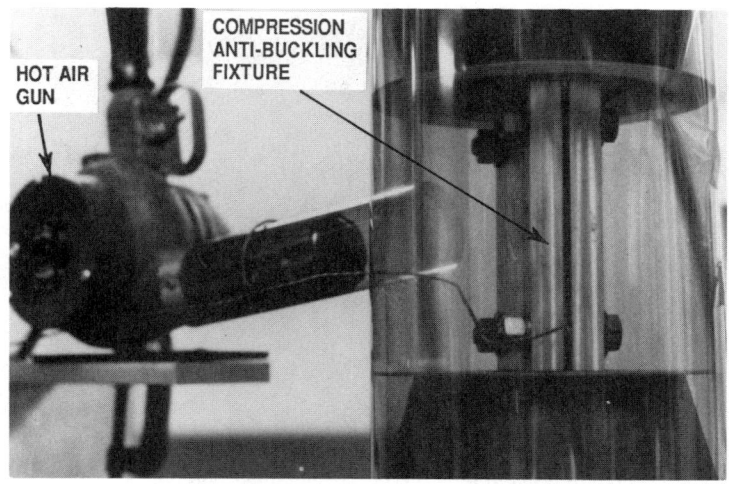

FIG. 2—*Heat system setup.*

TABLE 1—*Typical properties of the unidirectional.*

GRAPHITE-PEEK SYSTEM (APC-2)			
Material Property	Room Temperature	121°C	149°C
Longitudinal tensile modulus, GPa	129	118	119
Transverse, tensile modulus, GPa	10.1	5.3	4.0
Shear modulus, GPa	4.8	. . .	. . .
Poisson's ratio (major)	0.37	0.375	0.351
Fiber volume fraction	0.6	0.6	0.6
Average thickness of ply, mm	0.13	0.13	0.13

compression loading is considered in comparison with tension loading. For compression loading, an average reduction of about 40% can be seen for each hole diameter from room temperature to beyond the glass transition temperature, T_g. The tensile failure stress, however, does not indicate any strength reduction until very near T_g. Tension strength reductions are not as large as compression reductions. The tensile and compressive strength of APC-2 is generally greater than APC-1.

Figure 7 compares the effect of temperature on the strength of a specimen for APC-2 and graphite epoxy. The graphite/epoxy laminate demonstrates greater strength than the APC-2 for tension and compression. Note from Table 2, however, that the epoxy laminate has a 7% fiber volume fraction increase over APC-2. Examining the graphite epoxy system, one can see that the hole diameter affects the amount of the tensile strength near T_g. For 5.0 and 7.6 mm radius holes, the tension strength shows a downward trend when compared with the smaller radius holes. It must be pointed out that the σ_n^∞ values used in the previous discussion are nothing

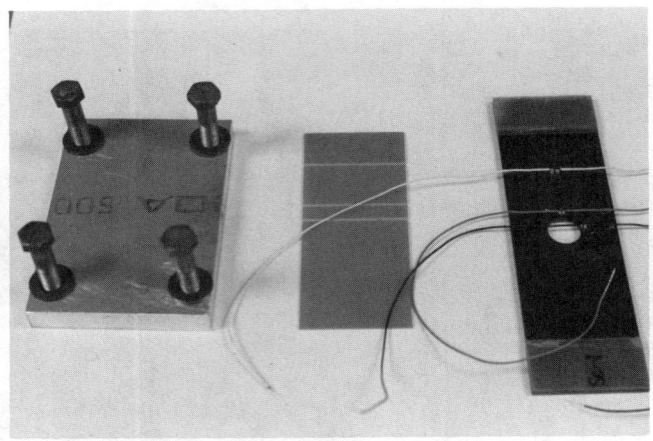

FIG. 3—*Compression specimen and antibuckling support.*

FIG. 4—*Failed tension specimens.*

more than the gross failure stress of the notched specimen adjusted to the finite width correction factor presented in Ref 6.

A great deal of investigation has been carried out to predict the notch strength of a composite. The article by Awerbuck and Madhukar [7] contains over 300 references related to composite notch strength and gives a review of the numerical and experimental results. It is evident that, because of the complex details within the damage zone occurring during the failure mode, an

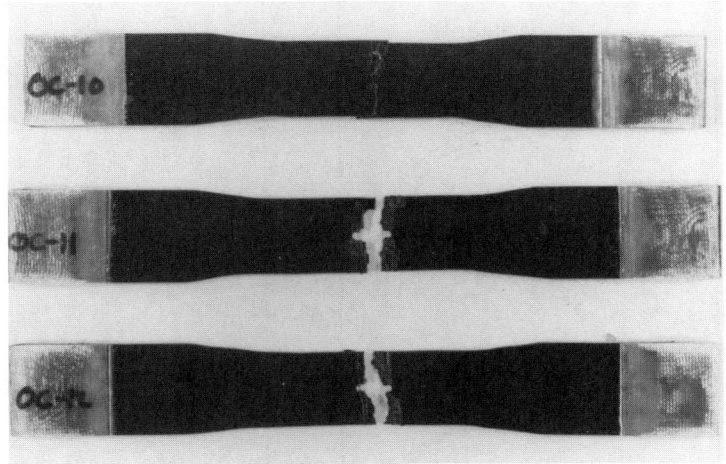

FIG. 5—*Failed compression specimens.*

empirical approach to failure predictions within a laminate presently is more practical. However, Ref 7 does not include any findings related to the APC-2 material.

In this paper, the authors used for the strength prediction the three-parameter method developed by Pipes et al. [8]. The results are shown in Fig. 8. The three-parameter method takes into consideration the effect of hole size on the value of d_0. The value, d_0, is a discrete point away from the hole in which the laminate axial stress at that point is equal to or greater than the strength of the unnotched material [6]. In order to account for the hole size dependency of d_0, Ref 8 introduced an expression representing this dependency in the so-called point criterion as follows:

$$\sigma_n^\infty/\sigma_0 = 2\{2 + f(R)^{-2} + 3f(R)^{-4} - (K_T - 3)[5f(R)^{-6} - 7f(R)^{-8}]\}^{-1} \qquad (1)$$

where

σ_n^∞ = notch strength corrected for finite width,
σ_0 = unnotched strength,
$f(R) = [1 + R^{m-1} R_0^{-m} C^{-1}]$,
K_T = stress concentration factor for the laminate,
R_0 = reference radius, usually chosen as 1 mm for simplicity, and
R = notch radius.

The undetermined parameters are the notch sensitivity factor, C, and the exponential parameter, m. Increases in the notch sensitivity factor for a given radius result in a reduction of laminate strength for the given radius. The exponential parameter, m, acts to change the slope of the notch sensitivity curves. Intermediate values of $m = 0.5$ and $C = 10$ are expected for notch radii less than 2.5 mm.

The tension strength predictions from the three-parameter model are very good. The curve indicates that use of this method would give slightly conservative failure predictions for the

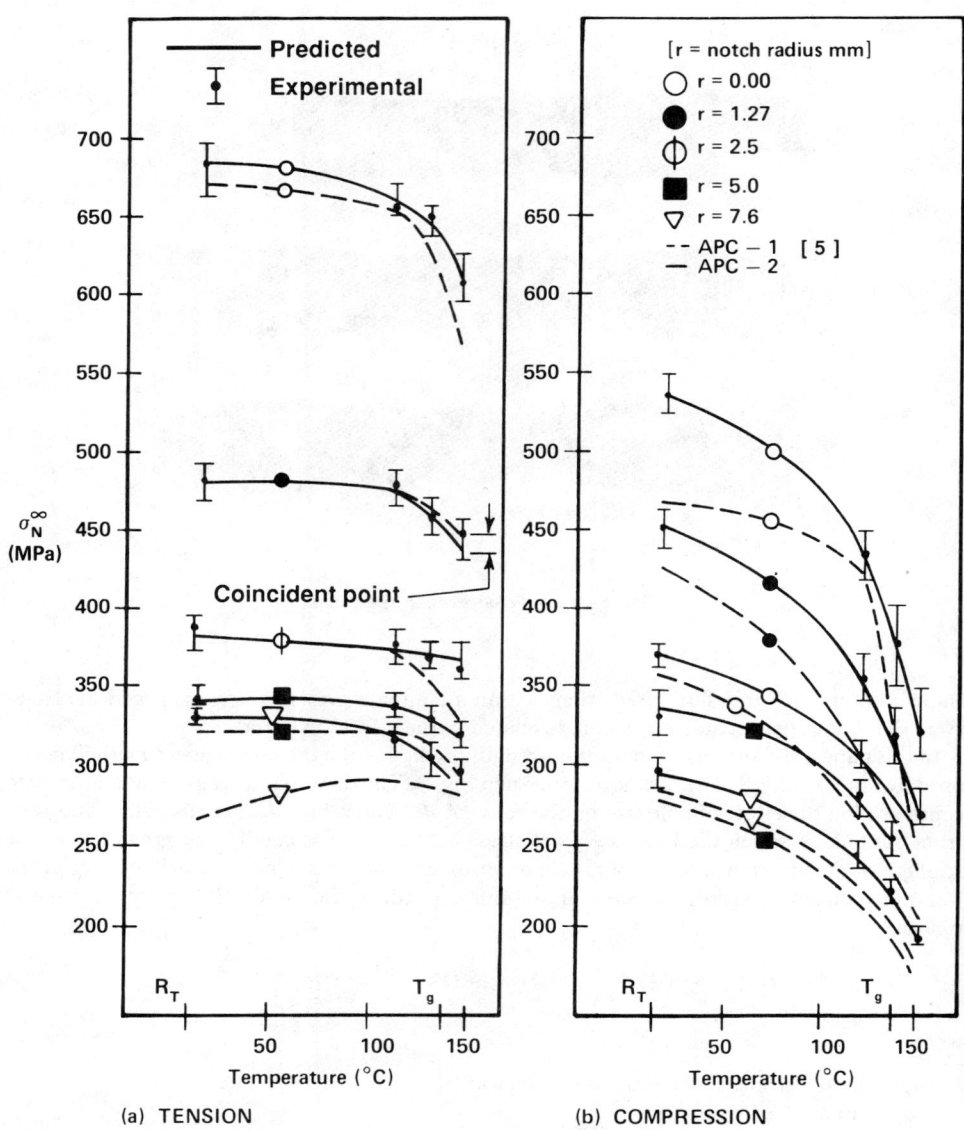

FIG. 6—*Strength for notched and unnotched quasiisotropic Specimen APC-2 and APC-1 specimens.*

given temperature range. The compression predictions indicate that judgement in the use of the Pipe's failure equation should be exercised when using the model since the specimens tested at 121°C, for all radius holes, were consistently below the prediction. Also, the room temperature specimen for 2.5, 5.0, and 7.6 mm radius holes were below the predicted value.

The notch effects were studied further by investigating the micromechanical failure modes. The compression study only is reported here. Graphite/Epoxy and APC-1 were examined previously with the scanning electron microscope (SEM) [5,9], and thus only the APC-2 data show-

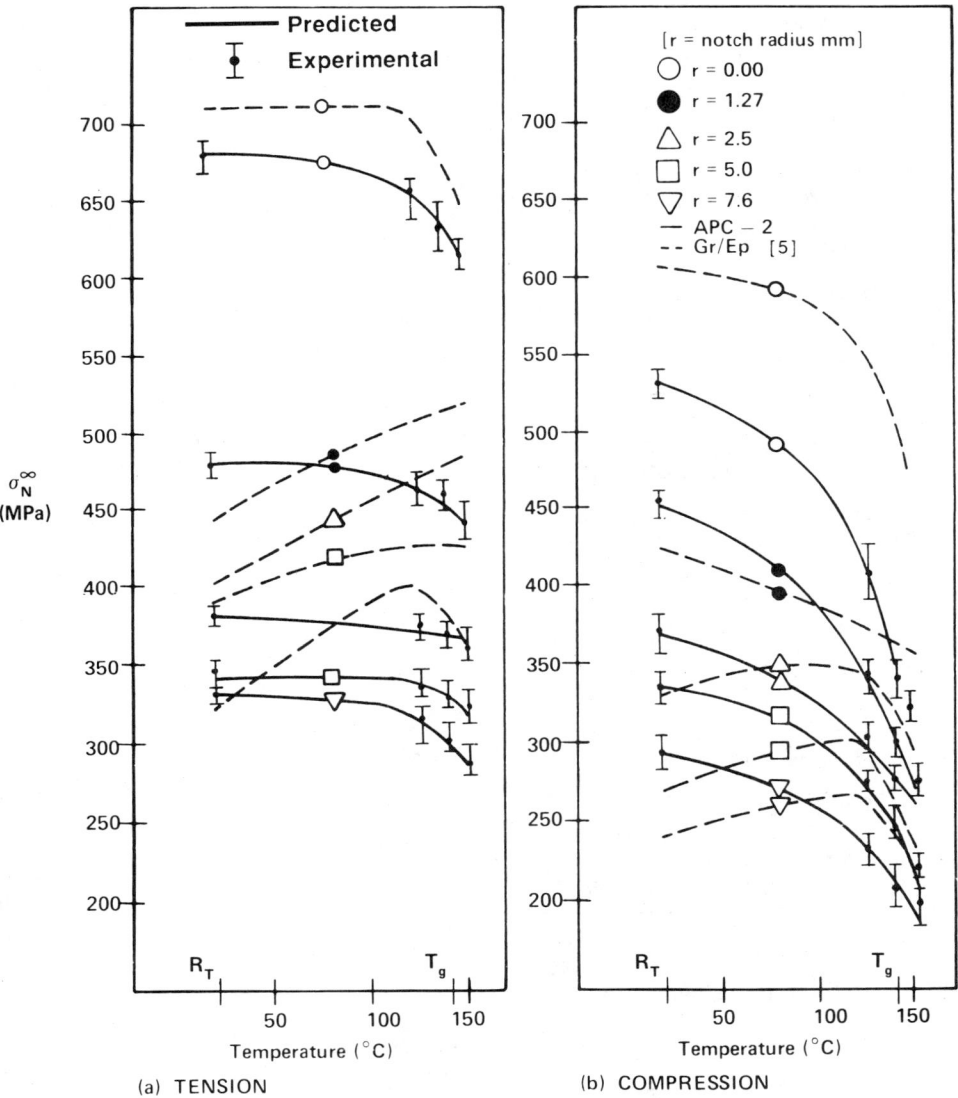

FIG. 7—*Strength for notched and unnotched APC-2 and Gr/Ep specimens.*

TABLE 2—*Comparison of laminate systems.*

Material	Fiber	Matrix	% Fiber Volume
APC-2	AS-4	PEEK	60
APC-1	XAS	PEEK	56
Gr/Ep	AS-4	3502	67

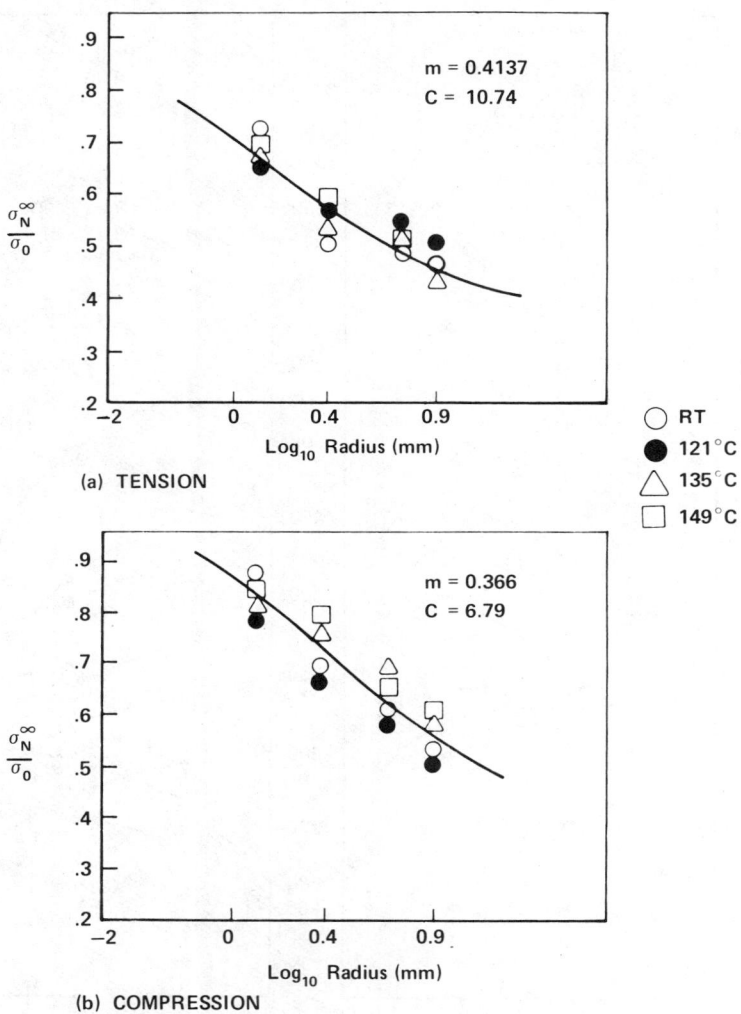

FIG. 8—*Comparison of predicted and experimental failure stresses.*

ing the catastrophic compression failure are reported in this paper. For this investigation, a panel with a hole radius of 2.5 mm was arbitrarily chosen.

When tested to failure, the compression specimens demonstrated a unique failure mode. Compression specimens tested at room temperature broke into two separate pieces. Specimens tested at 121, 135, and 149°C did not fail in this manner. Each elevated-temperature specimen remained in an interconnected form. This failure pattern prevented examination through the thickness of the compression specimen. Instead, only the top and in-hole views were available for observation without disturbing the actual fracture failure mechanisms themselves. Figure 9 indicates the patterns for viewing the given specimens. This failure mode seems to support the idea that the room temperature matrix reinforces the fibers more efficiently than the elevated temperature matrix. Similar comments have been made in Refs *10* and *11*.

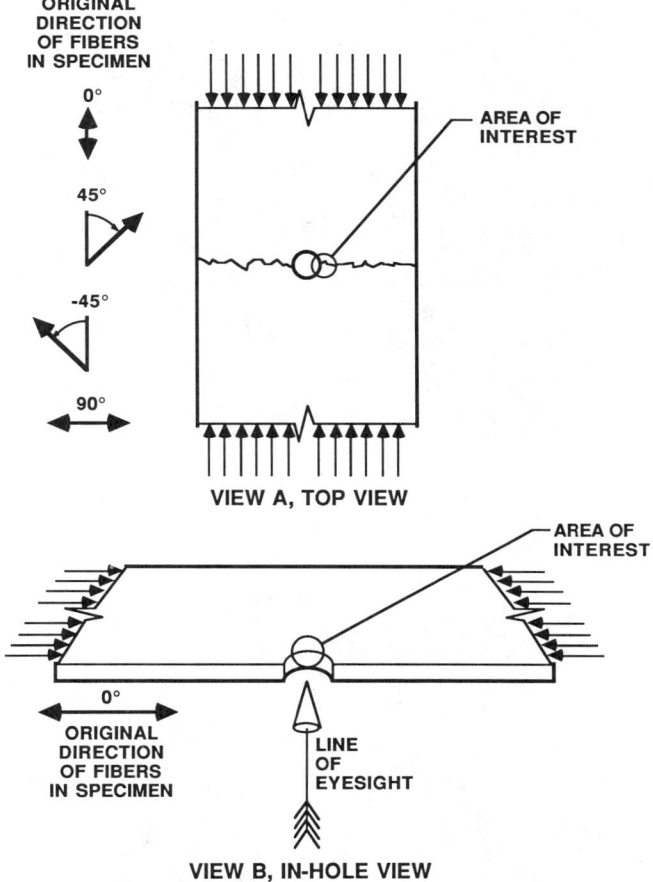

FIG. 9—*Area of examination for compressive specimen.*

The in-hole or free edge of the hole view for compression failures is presented in Fig. 10. The crippling effect (a fiber deforming under shear and bending action) at failure (also shown in Ref *10*) easily can be observed for the 0° fibers. The room temperature failure shown in Fig. 10(*a*) indicates that the 0° fibers have failed to such an extent that they have turned perpendicular to the load direction (also the original 0° direction). This figure also demonstrates the intralaminar separation of the first and sixteenth plies. These plies demonstrate a separation of the laminate into separate directions. Part of the 0° oriented ply (the fifth and twelfth plies) fails into the cutout while the uppermost 0° ply fails out of the plane. Interlaminar stresses and intralaminar stresses are distributed in various manners throughout the specimen. One may observe the independent response of the plies in Fig. 10(*b*). In Fig. 11, it appears that interlaminar stresses cause the matrix deformation between the 0° crippled fibers and the adjacent 90° and 45° fiber plies.

Closer examinations of the crippled 0° fibers are presented in Figs. 12 to 15 for the various experimental temperatures. It can be seen that 0° plies have failed from crippling modes. Figure 12 demonstrates excellent adhesion; the matrix has failed in a partially cohesive mecha-

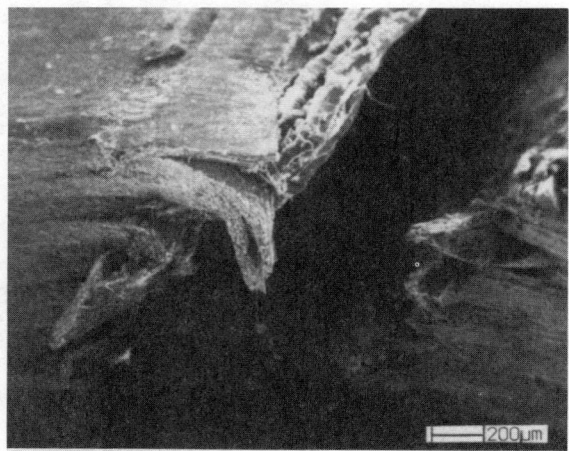

(a) RT

(b) 121°C

FIG. 10—*Compressions failure specimen, in-hole view, RT and 121°C (×32).*

nism. The matrix at room temperature is ductile, as observed between the fibers in the lower microphotograph of Fig. 12.

Figure 13 shows that the matrix is not as adhesive to the fiber at 121°C when compared with the room temperature specimen. However, a complete adhesion failure mechanism is not seen at 121°C, as some matrix material still adheres to the fibers. The matrix at 121°C appears more ductile and more plastic than the matrix at room temperature, as implied by the deformed nature of the existing matrix in Fig. 13.

The 135°C specimen in Fig. 14 depicts more cohesive failure than adhesive failure. The matrix can be seen adhering to the fiber, but the toughness of the matrix seems to be diminishing

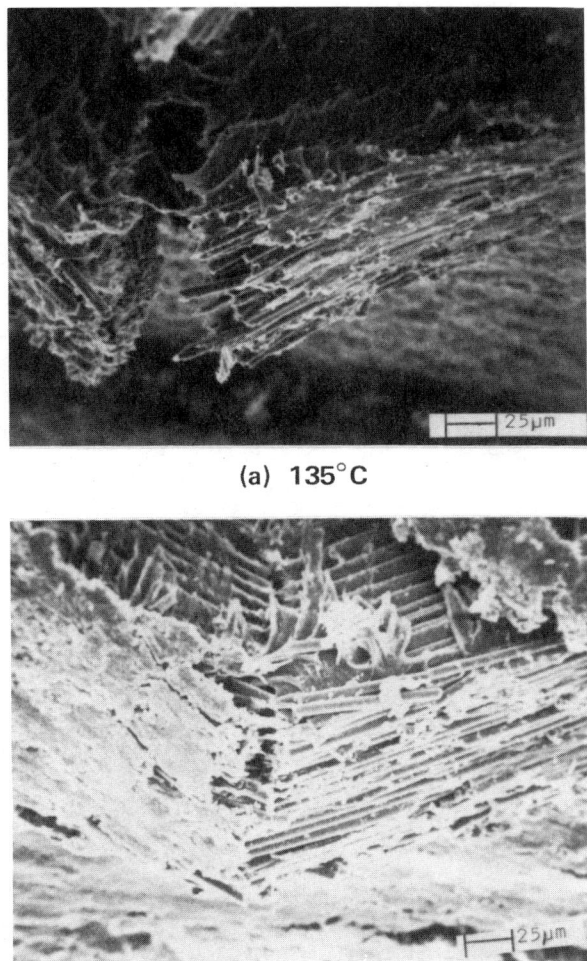

(a) 135°C

(b) 149°C

FIG. 11—*Compression failure specimen, in-hole view, 0° oriented fibers with interlaminar matrix defor-mation (×260).*

(as shown in Fig. 6). The resistance to shearing deformation within the matrix appears to be much less at 135°C when compared with the room temperature microphotographs.

The 149°C specimen in Fig. 15 demonstrates cohesive and adhesive failure. The matrix can be seen as a very thin film on some fibers, while on other fibers the matrix appears to have greater deformation.

Very large magnification of the compression specimens are presented in Figs. 16 and 17. These figures illustrate more graphically matrix-to-fiber bonding characteristics and are presented to amplify the previous discussion of matrix-to-fiber bond.

The room temperature specimen matrix at a 6500 magnification factor (Fig. 16(*a*)) shows

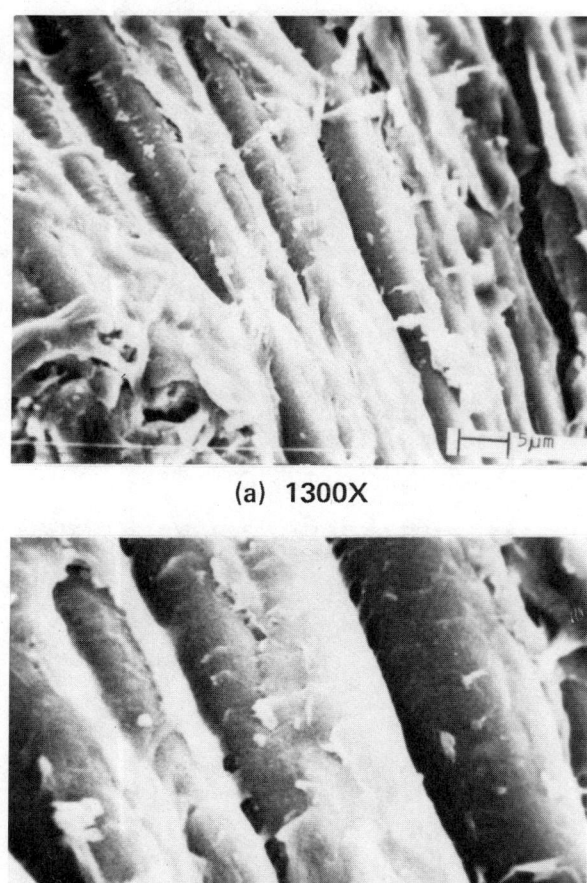

(a) 1300X

(b) 3250x

FIG. 12—*Compression failure specimen, in-hole view, 0° oriented fibers with interlaminar surface, room temperature.*

shear failure (hackles). The matrix is ductile, yet capable of distributing and resisting stress. The separation of the fibers from the matrix is also shown. Contrast the large deformation of the matrix in the 121°C specimen of Fig. 16(*b*) with the room temperature. Although the matrix is more plastic, it is less cohesive than it appeared at room temperature. In the 121°C specimen, the fiber fracture surface should be noted also. The bending failure of the fiber can be observed with the lower portion of the photograph being the tension loaded side, while the upper portion is compressively loaded. Figure 17 illustrates similar characteristics at the high temperatures.

Figures 18 and 19 are high magnification photographs that illustrate the interlaminar matrix deformation for the 0° fiber-oriented plies at elevated temperatures. If one compares Figs. 18

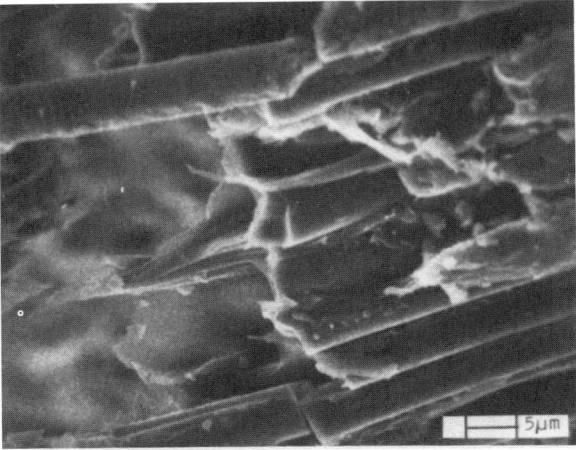

(a) 1300X

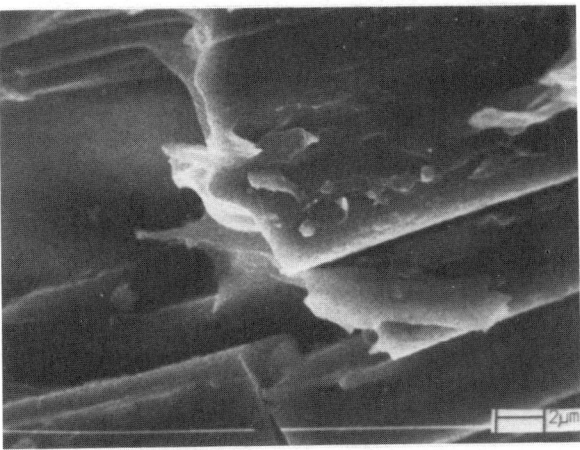

(b) 3250X

FIG. 13—*Compression failure specimen, in-hole view, crippled surface of 0° oriented fibers, 121°C.*

and 19 with Fig. 12 at room temperature, the interlaminar matrix deformation between the crippled 0° ply and the 90° lamina directly adjacent shows the degradation of the matrix at elevated temperatures. It may be speculated that at elevated temperatures, the matrix flows more and provides less capability to transfer load or to provide fiber support as it becomes more flexible. Not only does the matrix provide less support for the fibers at elevated temperatures, the matrix also provides less interlaminar strength. This loss of support for the fibers and reduction of interlaminar strength easily account for the overall reduction of strength for compression loading at elevated temperatures for APC-2 (Fig. 6).

Changing the viewing perspective from the free edge of the hole to the top view, compression

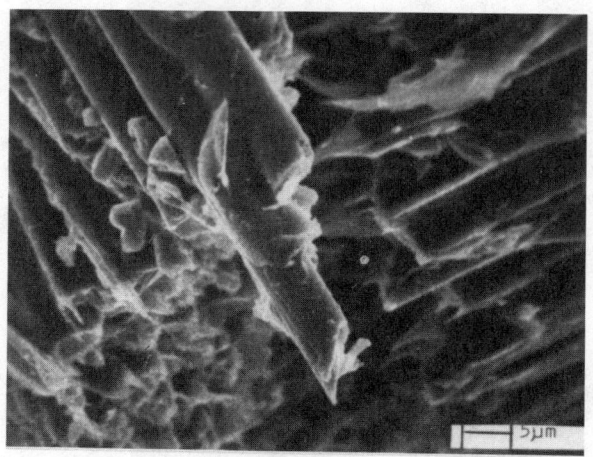

(a) 1300X

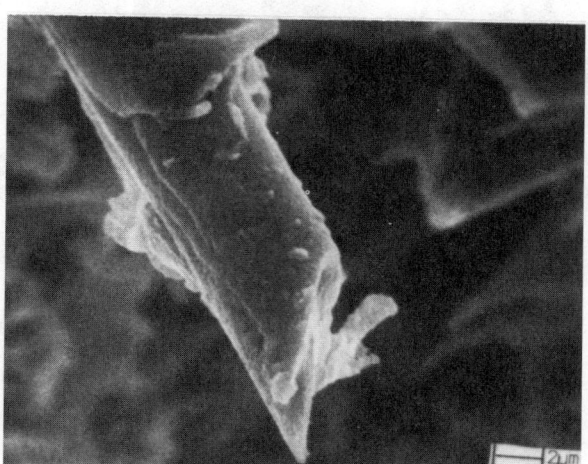

(b) 3250X

FIG. 14—*Compression failure specimen, in-hole view, 0° fibers, 135°C.*

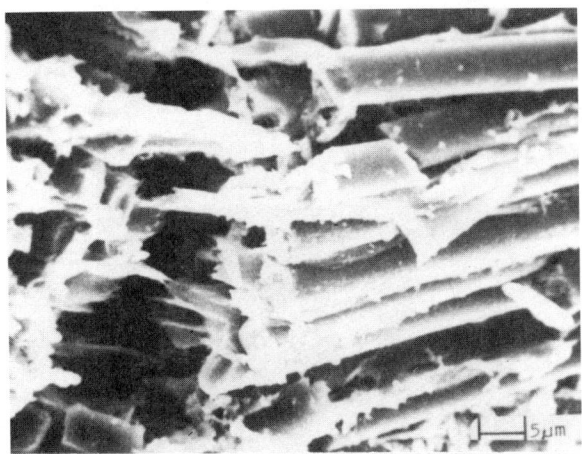

(a) 1300X

(b) 3250X

FIG. 15—*Compression failure specimen, in-hole view, crippled 0° oriented fibers, 149°C (×1300 and ×3250).*

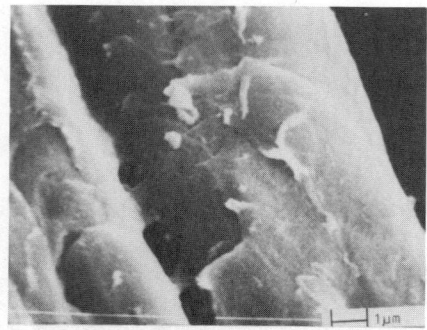

(a) RT

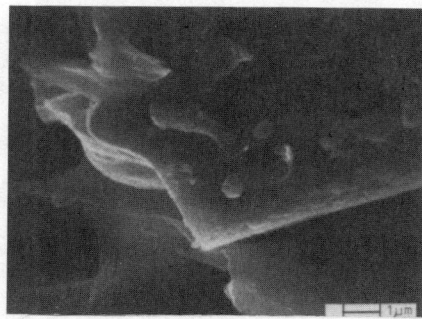

(b) 121°C

FIG. 16—*Compression failure specimen, in-hole view, crippled 0° oriented fibers, RT and 121°C, (×6500).*

failure specimens are shown in Figs. 20(*a*) and 20(*b*). In these figures it is noticed that as temperature increases, fewer fibers failed toward the hole, indicating that the fibers experienced less severe loading at elevated temperatures. Reviewing the SEM photographs, the authors have seen that an increase in specimen test temperature reduces the amount of out-of-plane crippling and intralaminar separation of the first 0° fiber ply.

An interesting verification of the intralaminar fracture of the first 0° ply is shown in Fig. 21. The original 0° fiber direction is parallel to the length of the page. The angles of the failed fibers beneath the top fractured fibers indicate that the fibers have buckled out of plane toward the viewer and remain fairly well aligned to the original direction. The lower microphotograph in Fig. 21(*b*) is an enlargement of the upper microphotograph. One should observe the matrix failure mechanisms. The matrix shows that the specimen was tested at room temperature. Due to the relative brittleness of the matrix at that temperature, the photo demonstrates matrix cracking between fibers and some ductile shearing phenomena.

The overall compression study indicates that initial failure for a notched specimen occurs at the notch where the largest stress concentrations are predicted. The 0° fiber plies initially bend toward the cutout but do this very locally. Increasing the applied load, the fibers are subjected to greater stresses and strains. Eventually sufficient deformation occurs and fracture results, originating at the hole and extending outward to the specimen edge.

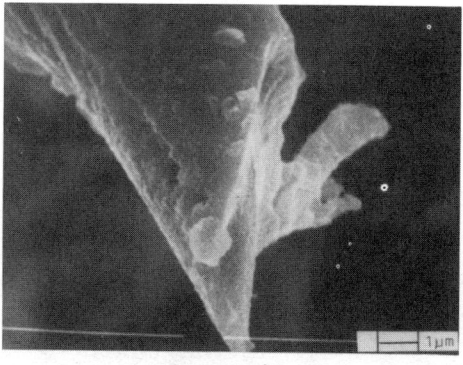

(a) 135°C

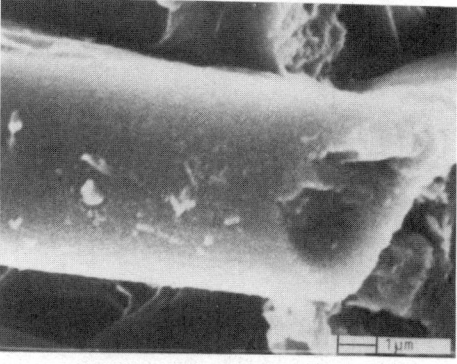

(b) 149°C

FIG. 17—*Compression failure specimen, in-hole view, crippled 0° oriented fibers, 135 and 149°C, (×6500).*

The progression of crack growth from the cutout to the edge of the specimen is also a phenomenon of the tension failure (not shown herein). Specimens fail locally around the hole in a manner very dissimilar to failure at a distance away from the hole. The matrix properties play an important role in this occurrence.

Conclusions

It is possible to make the following conclusions based upon the particular laminate considered:

1. The strength of APC-2 composite tested in tensions is only slightly affected by temperature up to the glass transition temperature.

2. The APC-2 composite demonstrated consistent strength reductions with increased temperature under compression loading. APC-1 compares in a similar fashion.

3. It is possible to model with fairly consistent results the failure strength of an APC-2 laminate with a circular hole cut out using the Pipe's three-parameter model.

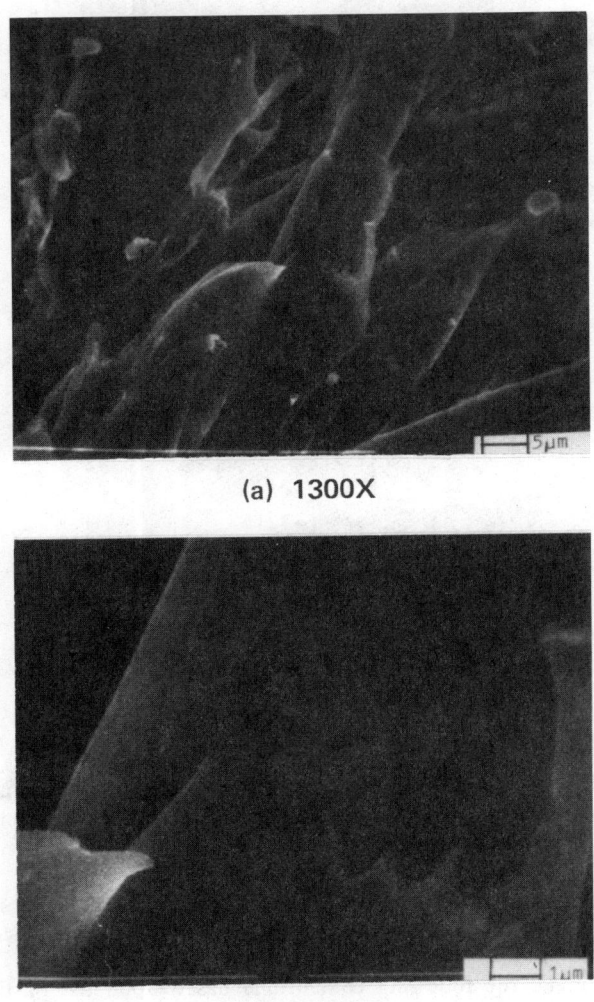

(a) 1300X

(b) 6500X

FIG. 18—*Compression failure specimen, in-hole view, crippled 0° oriented fibers, interlaminar failure, 135°C.*

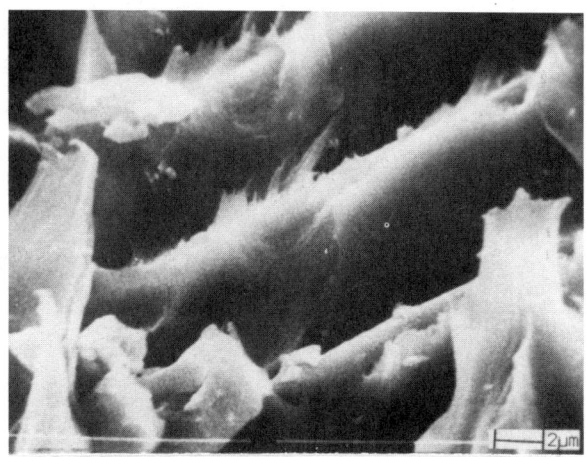

(a) 3250X

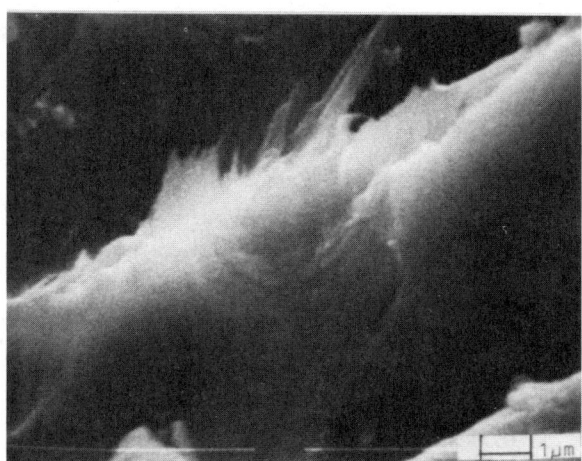

(b) 6500X

FIG. 19—*Compression failure specimen, in-hole view, crippled 0° oriented fibers, interlaminar failure, 149°C.*

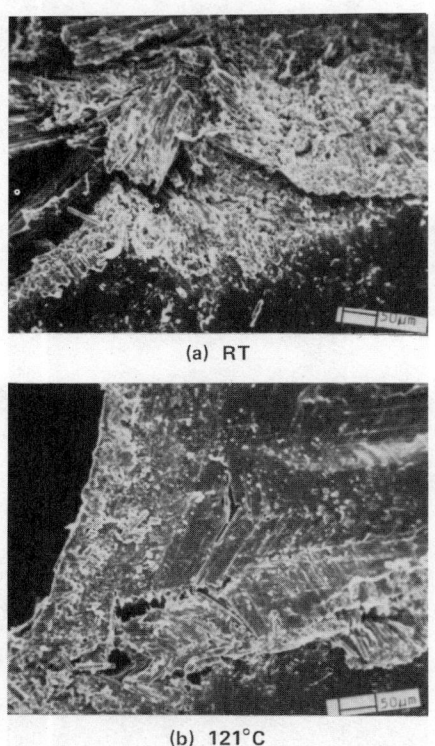

(a) RT

(b) 121°C

FIG. 20—*Compression failure specimen, top view, RT and 121°C, (×130).*

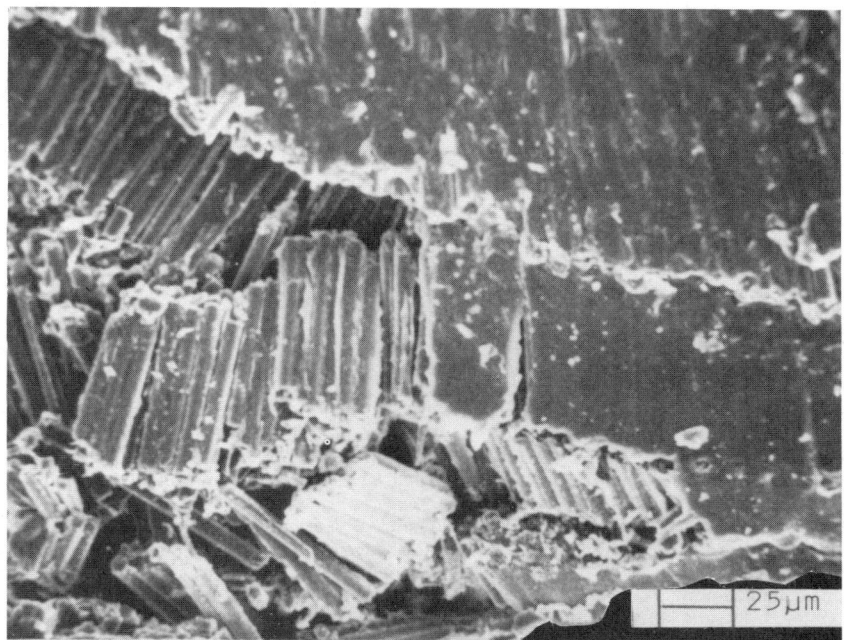

(a) 260X

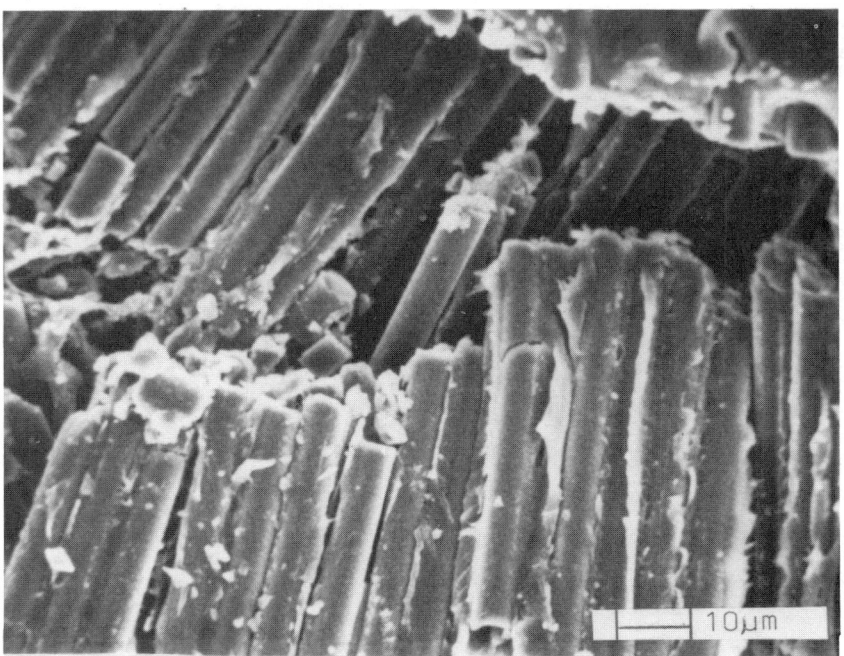

(b) 650X

FIG. 21—*Compression failure specimen, top view, interlaminar failure, RT.*

4. The observed failure modes for APC-2 were shear crippling of the fibers and delamination under compression loading. These mechanisms are similar for APC-1 and graphite/epoxy.

Acknowledgment

The authors would like to express their appreciation to J. Whitney of the Air Force Materials Laboratory for his assistance throughout this study.

References

[1] Hartness, J. T., "Polyether-ether-ketone Matrix Composite," *SAMPE Quarterly,* January 8, 1983, pp. 33–37.

[2] Cebe, P., Hong, S. D., Chung, S., and Gupta, A., "Mechanical Properties and Morphology of Poly (etheretherketone)," *Toughened Composites, ASTM STP 937,* Norman J. Johnston, Ed., American Society for Testing and Materials, 1987, pp. 342–357.

[3] Leach, D. C., Curtis, D. C., and Tamblen, D. R., "Delamination Behavior of Carbon Fiber/Poly (etheretherketone) (PEEK) Composites," *Toughened Composites, ASTM STP 937,* Norman J. Johnston, Ed., American Society for Testing and Materials, 1987, pp. 358–380.

[4] Harness, J. T., "An Evaluating of Poly-Ether-Ether-Ketone Matrix Composite Fabricated from Unidirectional Prepreg Tape," *Proceedings of the 29th National SAMPE Symposium,* Vol. 29, Society for the Advancement of Materials and Process Engineering, Covina, CA, 1984, pp. 459–474.

[5] Malik, B., Palazotto, A., and Whitney, J., "Notch Strength of Gr/PEEK Composite Material at Elevated Temperatures," AIAA paper 85-0648, presented at the 1985 AIAA Structures, Structural Dynamics, and Materials Conference, Orlando, FL, April 1985.

[6] Whitney, J. M. and Nuismer, R. J., "Stress Fracture Criteria for Laminated Composites Containing Stress Concentrations," *Journal of Composites,* Vol. 8, July, 1974, pp. 253–275.

[7] Awerbuck, J. and Madhukar, M. S., "Notched Strength of Composite Laminate: Predictions and Experiments, A Review," *Journal of Reinforced Plastics and Composites,* Vol. 4, January, 1985, pp. 3–159.

[8] Pipes, R. B., Wetherhold, R. C., and Gillispie, J. W., "Notched Strength of Gr/PEEK Composite Materials," *Journal of Composite Materials,* Vol. 13, April, 1979, pp. 148–160.

[9] Donaldson, S. L., "Fractography of Mixed Mode I-II Failure in Graphite/Epoxy and Graphite/Thermoplastic Unidirectional Composites," AFWAL Technical Report 84-4186, Materials Laboratory, Wright-Patterson Air Force Base, Ohio, 1984.

[10] Starnes, J. H. Jr., and Williams, J. G., "Failure Characteristics of Graphite—Epoxy Structural Components Loaded in Compression," *Mechanics of Composite Materials—Recent Advances, Proceedings of the IUTAM Symposium on Mechanics of Composite Materials,* Virginia Polytechnic Institute and State University, Blacksburg Virginia, 16–19 Aug. 1982, Zui Hashin and Carl T. Herakovich Eds., Pergamon Press, Elmsford, NY, pp. 283–321.

[11] Hahn, H. T., Sobi, M., and Moon, S., "Compression Failure Mechanisms of Composite Structures," NASA CR 3988, June 1986.

E. Dan-Jumbo,[1] *S. G. Zhou,*[2] *and C. T. Sun*[2]

Load-Frequency Effect on Fatigue Life of IMP6/APC-2 Thermoplastic Composite Laminates

REFERENCE: Dan-Jumbo, E., Zhou, S. G., and Sun, C. T., "**Load-Frequency Effect on Fatigue Life of IMP6/APC-2 Thermoplastic Composite Laminates,**" *Advances in Thermoplastic Matrix Composite Materials, ASTM STP 1044,* G. M. Newaz, Ed., American Society for Testing and Materials, Philadelphia, 1989, pp. 113–132.

ABSTRACT: Load-frequency effect on fatigue life of a thermoplastic and toughened BMI composite was investigated. Experimental results were obtained using three frequencies at two load levels with three types of laminates, each containing a center hole. Besides fatigue life, the temperature history near the hole and the specimen stiffness were monitored. In addition, the hysteresis loop was recorded at various stages of the fatigue test to observe the variation in viscoelastic property of the specimen. From the fatigue test results, it was found that fatigue life of matrix-dominated thermoplastic composite laminates decreased as load frequency increased. This behavior is opposite to that of epoxy-based fiber composites. For fiber-dominated thermoplastic laminates under tension-compression load, very low load-frequencies (less than 0.4 Hz) could significantly lower their fatigue life. Analytical models were developed to account for the load-frequency effect on fatigue life of thermoplastic composites. Temperature rise near the hole was included in the model to account for temperature effect.

KEY WORDS: fatigue, load frequency, temperature, thermoplastic composite

Because of the viscoelastic property of epoxy, the fatigue life of graphite-epoxy composite laminates has been found to be highly frequency-dependent [1–3]. Moreover, if temperature rise during fatigue loading is not significant, fatigue life (in terms of cycles to failure) of graphite-epoxy composites increases as the load frequency increases. At high load levels together with high loading frequencies, the temperature rise could adversely affect fatigue life [1].

In this paper, load-frequency effects on fatigue of an IM6/APC-2 thermoplastic composite were investigated. Three layups were considered: a $[\pm 45]_{2s}$ laminate, a matrix-dominated $[\pm 45/0/\pm 45/90/\pm 45/90/\pm 45/\pm 45]_s$ laminate [the so-called (4/80/16) laminate], and a fiber-dominated $[\pm 45/0/0/45/90/-45/0/0/45/0/0/-45]_s$ laminate [or (48/48/4) laminate]. An epoxy-based IM7/5250-2-toughened BMI $[\pm 45]_{2s}$ composite laminate was also tested for comparison. All fatigue specimens contained a center hole.

In general, three load frequencies and two load levels were used in the fatigue test. Besides fatigue life, the temperature history near the hole and the specimen stiffness were also monitored. In addition, the hysteresis loop was recorded at various stages of the fatigue test to observe the variation in viscoelastic property of the specimen.

Experimental results show that the fatigue behavior of the BMI composite follows the trend of other graphite-epoxy systems. However, the tension-tension fatigue life of the $[\pm 45]_{2s}$ thermoplastic laminate decreases as the load-frequency increases. This behavior is opposite to the

[1]Structures and Design Department, General Dynamics Corp., Fort Worth, TX 76101.
[2]School of Aeronautics and Astronautics, Purdue University, West Lafayette, IN 47907.

reported behavior of graphite-epoxy composites [1,2]. For the other two lay-up laminates, that is, the matrix-dominated (4/80/16) laminate and the fiber-dominated (48/48/4) laminate, fatigue tests were conducted under tension-compression loads. Results show that these two laminates are also highly frequency-dependent.

Mathematical models were developed to account for the load-frequency effect on fatigue life of the thermoplastic composite tested. Temperature rise near the hole also was included to model the temperature effect.

Specimen Description and Test Procedure

The baseline [±45]$_{2s}$ laminate properties were tested using unnotched coupon specimens with nominal dimensions of 22.9 cm (9 in.) length and 2.54 cm (1 in.) width. For the fatigue tests, all [±45]$_{2s}$ specimens were 3.8 cm (1.5 in.) wide and 25.4 cm (10 in.) long, containing a drilled hole of 0.635 cm (0.25 in.) in diameter at midspan. Average thickness of the specimen was 0.11 ± 0.005 cm (0.043 ± 0.002 in.) for the IM6/APC-2 composite and 0.11 ± 0.005 cm (0.045 ± 0.002 in.) for the IM7/5250-2-toughened BMI composite. Glass/epoxy end tabs 3.8 cm (1.5 in.) long were used. For each material system, two specimens were selected to test the ultimate strength and the apparent Young's modulus in the loading direction.

For the matrix-dominated (4/80/16) laminate and the fiber-dominated (48/48/4) laminate, a gage length of 5.1 cm (2 in.) was chosen in order to avoid global buckling during the test. These specimens were 12.7 cm (5 in.) long and 3.8 cm (1.5 in.) wide with a 0.635-cm (0.25-in.) center hole. Glass-epoxy end tabs 3.8 cm (1.5 in.) long were used. The (48/48/4) and the (4/80/16) laminates were 0.36 ± 0.13 cm (0.14 ± 0.005 in.) thick.

Fatigue tests were conducted using load-controlled constant-amplitude tension-tension (or tension-compression) sinusoidal load cycles. The stress ratio for [±45]$_{2s}$ laminate was $R = 1/15$, and $R = -1$ for all the rest.

Specimens were tested on an MTS machine (Model 810) and were selected randomly. For each load level, specimens were tested at three different frequencies, 0.4, 2, and 10 Hz. During the test, temperature history near the hole (the maximum stress concentration site) was recorded using a thermocouple. Specimen (initial) stiffness was calculated from the load-displacement data which were recorded during the test. An oscilloscope was used to monitor the load-displacement curves and hysteresis loops at various stages of the fatigue test to observe the variation in viscoelastic property of the specimen.

Experimental Results

Static Tests

The unnotched [±45]$_{2s}$ laminates were tested, and the results are shown in Table 1. The in-plane shear modulus, G_{12}, of the unidirectional composite was obtained by using the relation

$$G_{12} = \frac{E_x}{2(1 + \nu_{xy})} \tag{1}$$

Note that both BMI and APC-2 have similar elastic properties.

The stress-strain curve for BMI composite is essentially linear up to 55 MPa (8.0 ksi), but only to 24 to 28 MPa (3.5 to 4.0 ksi) for thermoplastic laminate. Due to the toughness of thermoplastic matrix, the thermoplastic composite can sustain much greater elongation than graphite-epoxy laminate (1.15% ultimate strain for epoxy laminate and 13% for thermoplastic laminate). Although thermoplastic laminate has a much greater ultimate strength than the

TABLE 1—*Basic material properties of [±45]₂ₛ laminates.*

Material	Ultimate Load	E_x, GPa (msi)	ν_{xy}	G_{12}, GPa (msi)
IM7/5250-2	5006.25 (1125)	22.9 (3.33)	0.75	6.55 (0.95)
IM6/APC-2	9064.65 (2037)	21.5 (3.12)	0.82	5.91 (0.868)

BMI laminate, at any strain less than 1.15%, the latter has a significantly greater secant modulus.

The ultimate loads for fatigue test specimens (specimens with a center hole) were obtained and are listed below:

BMI [±45]₂ₛ : 6274 N (1410 lb) tension.
APC-2 [±45]₂ₛ : 10950 N (2460 lb) tension.
APC-2 (4/80/16) : 48060 N (10800 lb) tension, 31510 N (7080 lb) compression.
APC-2 (48/48/4) : 113030 N (25400 lb) tension, 48060 N (10800 lb) compression.

Fatigue Test Results for [±45]₂ₛ Laminates

For the IM7/5250-2 composite, only a high load level (75% of ultimate load) was considered. Three different frequencies were used in the fatigue tests. Test results are presented in Table 2. The above fatigue data are plotted in fatigue life versus load frequency in Fig. 1. The range of fatigue life and its trend (average values) are clearly shown in the figure. It is evident that the BMI composite follows the trend of other graphite-epoxy laminates; that is, fatigue life (number of cycles) increases with load frequency if temperature rise is not severe.

For the IM6/APC-2 composite, three different frequencies (0.4, 2, and 10 Hz) and two different load levels (57.9 and 75% of ultimate load) were chosen for the fatigue tests. For each condition, three specimens were tested. The test results are listed in Table 3.

Data in Table 3 are also presented graphically in Fig. 2 in which the mean and the scatter range of the fatigue life are plotted versus load frequency. It is evident that, for the thermoplastic composite, the fatigue life decreases as load frequency increases. This behavior is opposite to the behavior of conventional graphite-epoxy composites [1,2].

Figure 3 shows the histories of temperature increase for different load levels and frequencies

TABLE 2—*Fatigue life of BMI [±45]₂ₛ laminate at different frequencies.*[a]

Specimen	Frequency, Hz	Number of Cycles	Average
S3	0.4	1050	
S4	0.4	1231	1140
S5	2.0	1704	
S6	2.0	4824	3260
S7	10.0	3381	
S8	10.0	3936	3660

[a]For the 75.0% load level, P_{max} = 4708.10 N (1058 lb); P_{min} = 315.95 N (71 lb).

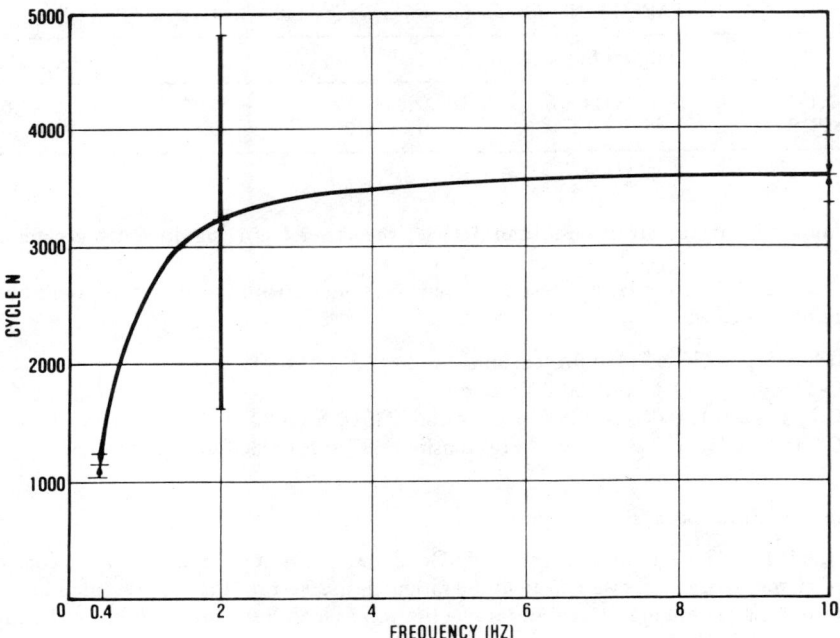

FIG. 1—*Fatigue life of* [±45]$_{2s}$ *toughened BMI (IM7/5250-2) composite under tension-tension fatigue loading.*

in the [±45]$_{2s}$ thermoplastic composite near the hole. As expected, high loads and high frequencies produced higher temperatures near the hole. For lower frequencies (0.4 and 2 Hz), the near hole temperature was seen to rise at the beginning and reach a steady level after several thousand cycles. The temperature increased rapidly shortly before the final failure of the specimen. At 10 Hz, a substantial temperature rise was noted, and the temperature in the test coupon rose so high that the smell of burning material was present.

Comparing the results shown in Fig. 3 with the results (graphite-epoxy composite) presented in Ref *1*, we find that the temperature rise in the [±45]$_{2s}$ thermoplastic composite is significantly greater than that in the [±45]$_{2s}$ graphite-epoxy composite. This may be due to the more pronounced inelastic property of the thermoplastic composite.

Variation in stiffness during cyclic loading was measured from the load-displacement data and is presented in Fig. 4. It is of interest to note that, for lower frequencies, stiffness of the specimen increases at the beginning of the test and then decreases gradually. This phenomenon could have stemmed from the cyclic hardening property of the thermoplastic laminate.

Figure 5 shows the hysteresis loops at various stages of the fatigue test for the APC-2 laminate. From these results, we note the viscoelastic behavior in this material. Further, continuous creep took place as evidenced by the shifting of the hysteresis loops.

Discussion

For the [±45]$_{2s}$ laminates, the toughened BMI specimen behaves like a brittle material with little plasticity. However, for the APC-2 thermoplastic composite, pronounced plasticity and viscoelasticity are noted.

TABLE 3—*Fatigue life of IM6/APC-2 laminate at different load levels and frequencies.*

Specimen	Frequency, Hz	Load Level, %	Number of Cycles	Average
T8	0.4	57.9[a]	20900	
T12	0.4	57.9	14864	18533
F63	0.4	57.9	19836	
T6	2.0	57.9	15900	
T9	2.0	57.9	16270	17560
F55	2.0	57.9	20510	
T4	10.0	57.9	15790	
T7	10.0	57.9	12420	10542
F52	10.0	57.9	3416	
T5	0.4	75.0[b]	12410	
T10	0.4	75.0	17170	13222
F64	0.4	75.0	10086	
T2	2.0	75.0	9570	
F53	2.0	75.0	12703	12676
F62	2.0	75.0	15755	
T3	10.0	75.0	1680	
T11	10.0	75.0	1162	1064
F56	10.0	75.0	455	
F65	10.0	75.0	957	

[a]For the 57.9% load level, $P_{max} = 6336.80$ N (1424 lb); $P_{min} = 422.75$ N (95 lb).

[b]For the 75.0% load level, $P_{max} = 8210.25$ N (1845 lb); $P_{min} = 551.80$ N (124 lb).

Figure 6 shows the failure modes of the two $[\pm 45]_{2s}$ laminates cycled to failure at 2 Hz. Both specimens failed in in-plane shear and delamination failure modes. For the BMI laminate, except for the separation of the specimen, the specimen geometry remains unchanged. However, for thermoplastic specimens, permanent deformation is obvious. For example, the width of the specimen decreased to 3.43 cm (1.35 in.) after failure, and the original circular hole became elliptical. The fiber orientation was also altered.

From the test results of $[\pm 45]_{2s}$ laminates at 75% load level, it is concluded that the APC-2 composite has a better fatigue property than the BMI composite at low load frequencies. However, at high frequencies, the temperature rise in the APC-2 laminate was much more severe than that in the BMI laminate, and consequently APC-2 laminate behaved poorly in fatigue.

Fatigue Test Results for (4/80/16) Laminate

Twenty-four specimens were used for tension-compression fatigue tests. Three different frequencies (0.4, 2, and 10 Hz) and two load levels (57.9 and 70% of ultimate compression load) were used. The fatigue data are presented in Table 4 and also in Figs. 7 and 8. At the lower load level (57.9%), the fatigue life of the laminate decreases as frequency increases. The rate of decrease is significantly greater in the high frequency region. This resembles the behavior exhibited by the $[\pm 45]_{2s}$ APC-2 composite laminate. At the higher load level (70%), fatigue life at 0.4 Hz is slightly shorter than that at 2 Hz.

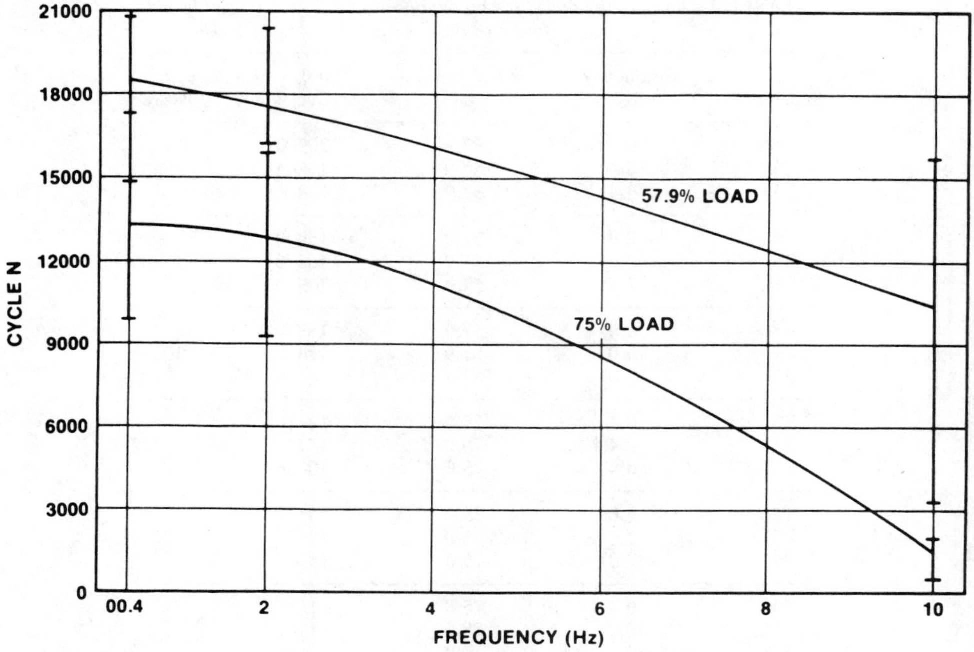

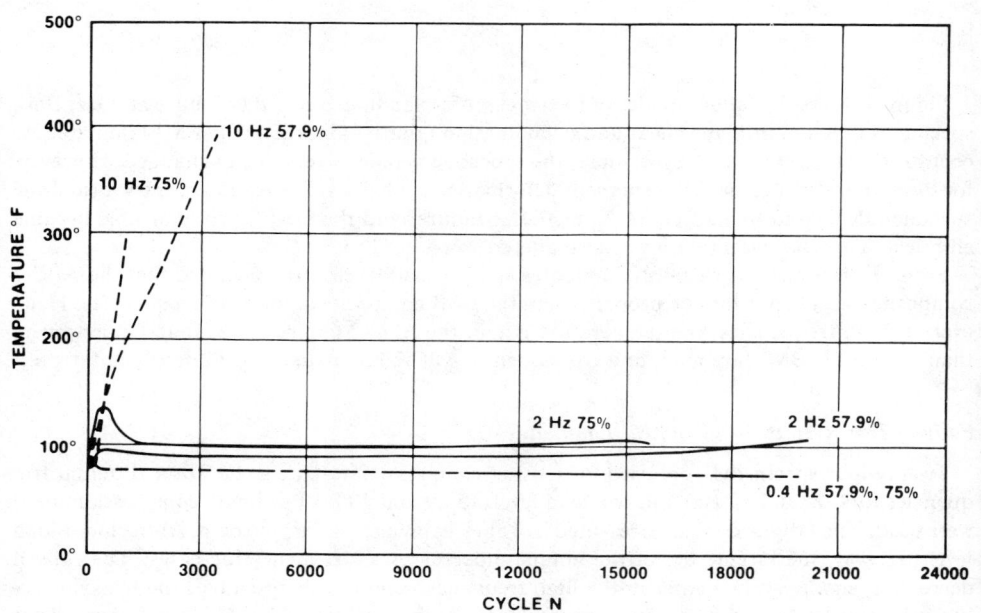

FIG. 2—*Fatigue life of [±45]₂ₛ thermoplastic APC-2 composite under tension-tension fatigue loading.*

FIG. 3—*History of temperature rise near the hole of [±45]₂ₛ IM6/APC-2 laminate.*

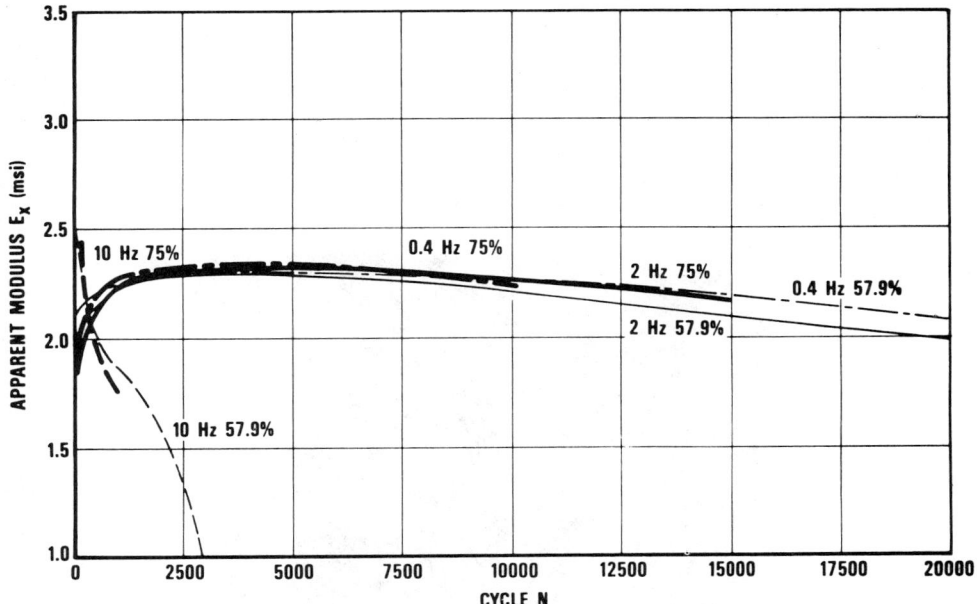

FIG. 4—*History of stiffness change during the fatigue life of* $[\pm 45]_{2s}$ *IM6/APC-2 laminate.*

Figures 9 and 10 show samples of temperature rise history of several composite specimens. The effect of load frequency is displayed clearly.

The variation of specimen stiffness, E_x, shows that the specimen usually experienced rapid decrease at two stages; at the beginning and near the end of the fatigue test. From the hysteresis loops recorded, we note that viscoelastic behavior is not significant. This is due to the existence of 0° laminae in the laminate.

In tension-compression fatigue tests, the final mode of failure usually was dominated by buckling in the compression half-cycle. Buckling usually became evident toward the end of the fatigue test.

The failure surfaces corresponding to the three load frequencies appear to be different. From the radiographs and photographs of the failed specimens as shown in Figs. 11 and 12, respectively, we see that, at low frequencies (0.4 and 2 Hz), extensive delamination occurred in the specimens. At these frequencies, failure seems to have involved local buckling of the delaminated region. At high load frequency (10 Hz), global buckling seems to dominate the failure mode.

Fatigue Test Results for (48/48/4) Laminate

For the fiber-dominated (48/48/4) thermoplastic laminate, fatigue life is very sensitive to the load level. A small variation in load may yield a very different fatigue life. In this study, twenty-two specimens were used for the tension-compression fatigue test. Three different frequencies (0.4, 2, and 10 Hz) and three different load levels (85.0, 87.5, and 88.9% of ultimate compression load) were chosen for the tests. The test results are presented in Tables 5 through 7. The most striking feature of these fatigue data is the extremely low life at 0.4 Hz. One possible explanation of this behavior is that the specimen was subjected to a load close to its buckling

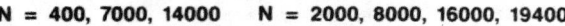

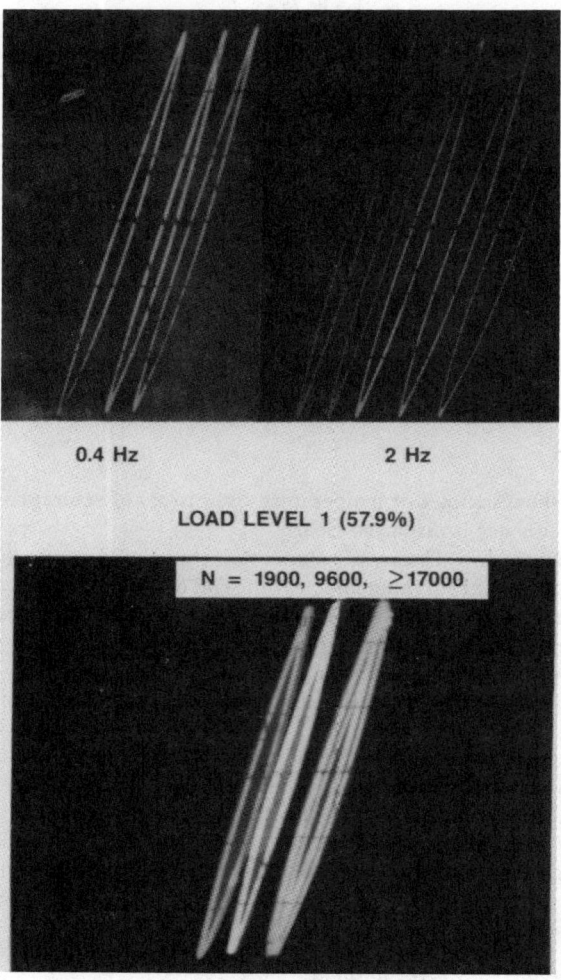

FIG. 5—*Creep and hysteresis loop under tension-tension fatigue load ([±45]₂ₛ laminate).*

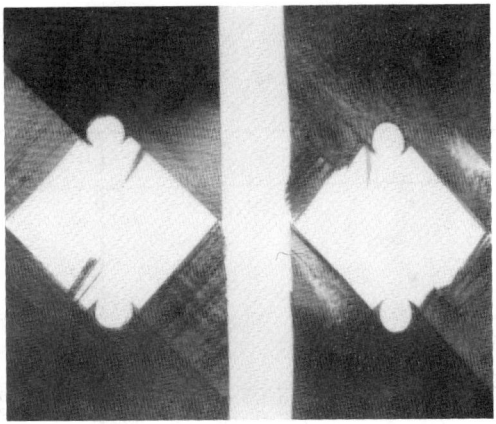

BMI   **IM6 / APC-II**

FIG. 6—*Failure surface of different [±45]$_{2s}$ laminates after fatigue test.*

TABLE 4—*Fatigue life at different load levels and frequencies for IM6/APC-2 (4/80/16) laminate.*

Specimen	Frequency, Hz	Load Level, %	Number of Cycles	Average
A4	0.4	57.9[a]	12958	
A5	0.4	57.9	11393	
A21	0.4	57.9	80795	37452
A22	0.4	57.9	44660	
A1	2.0	57.9	28060	
A2	2.0	57.9	45000	
A13	2.0	57.9	20136	30290
A13	2.0	57.9	27963	
A3	10.0	57.9	11760	
A6	10.0	57.9	8560	
A23	10.0	57.9	3600	4164
A24	10.0	57.9	2737	
A11	0.4	70.0[b]	1060	
A12	0.4	70.0	957	
A19	0.4	70.0	5092	3035
A20	0.4	70.0	5030	
A7	2.0	70.0	1395	
A9	2.0	70.0	1205	
A17	2.0	70.0	4510	3643
A18	2.0	70.0	7461	
A8	10.0	70.0	1277	
A10	10.0	70.0	1926	
A15	10.0	70.0	2381	2066
A16	10.0	70.0	2681	

[a]For the 57.9% load level, $P = \pm 18245$ N (4100 lb).
[b]For the 70.0% load level, $P = \pm 22072$ N (4960 lb).

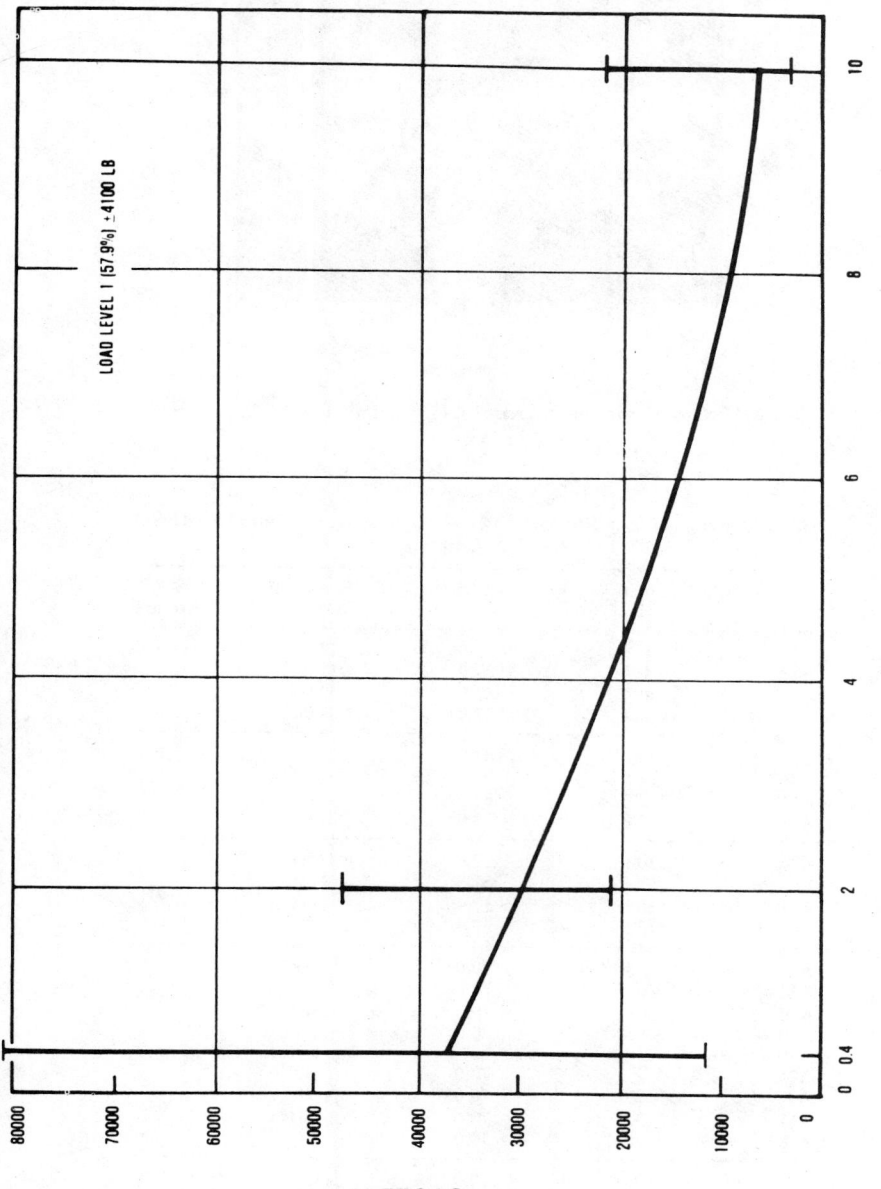

FIG. 7—*Fatigue life of (4/80/16) laminate under tension-compression fatigue loading (lower level case).*

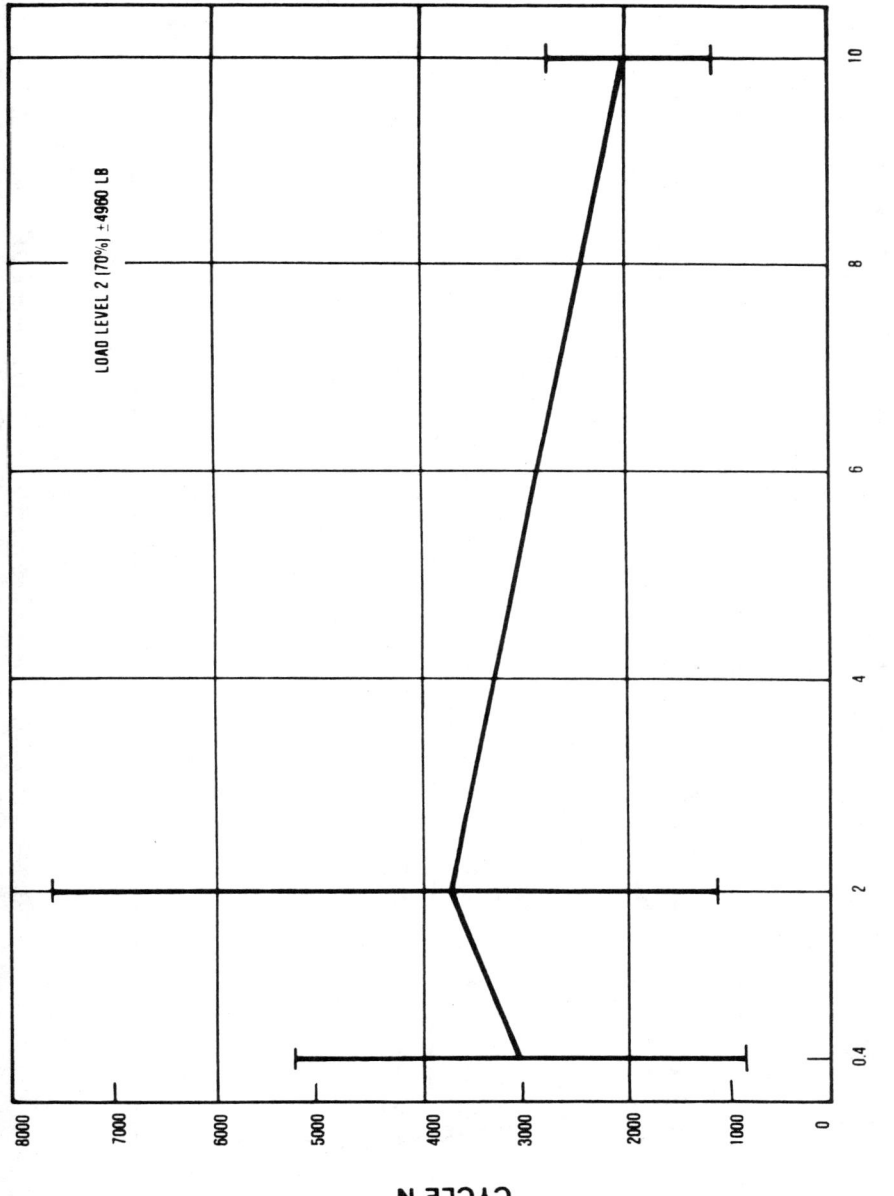

FIG. 8—*Fatigue life of (4/80/16) laminate under tension-compression fatigue loading (higher level case).*

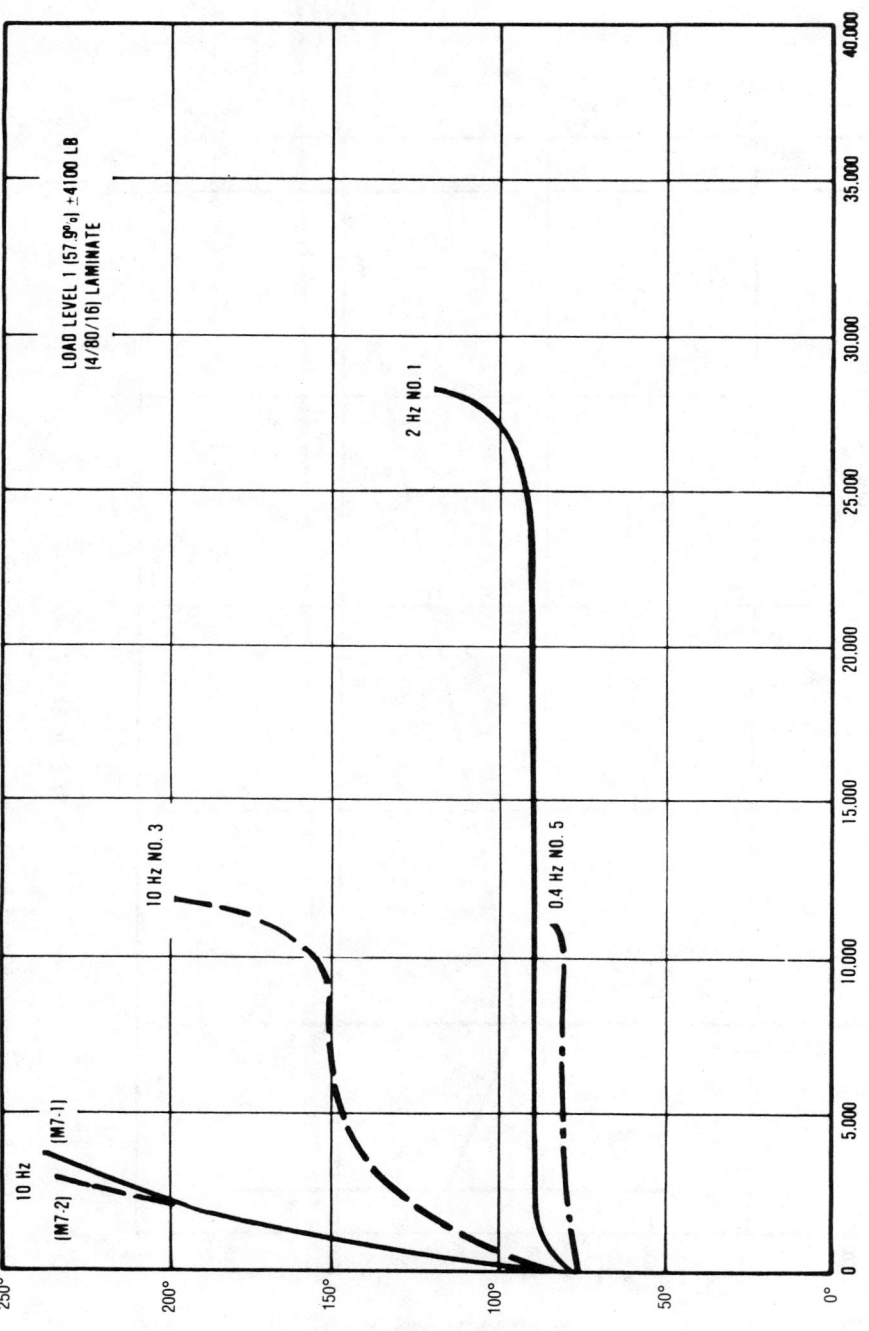

FIG. 9—*History of temperature rise near the hole of (4/80/16) laminate at 57.9% load level.*

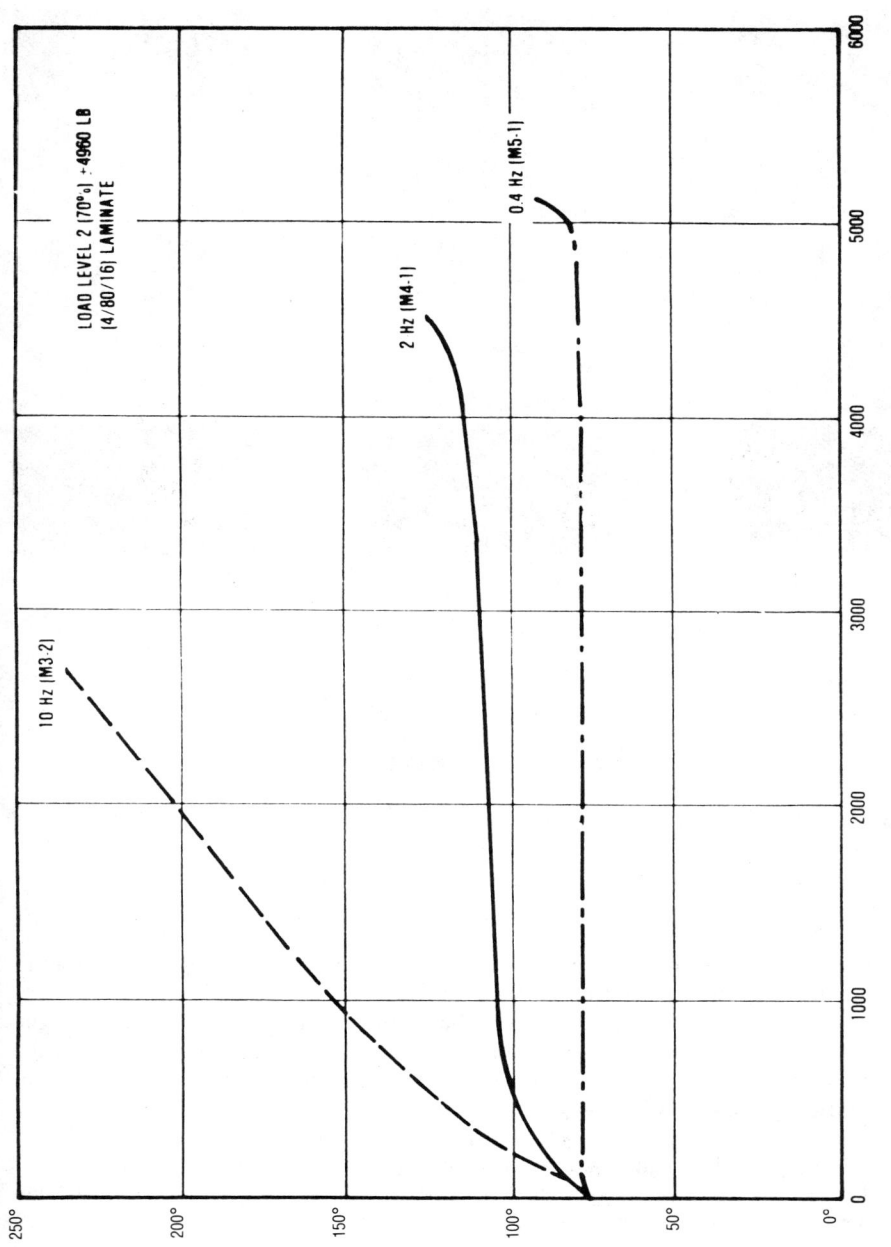

CYCLE N

FIG. 10—*History of temperature rise near the hole of (4/80/16) laminate at 70% load level.*

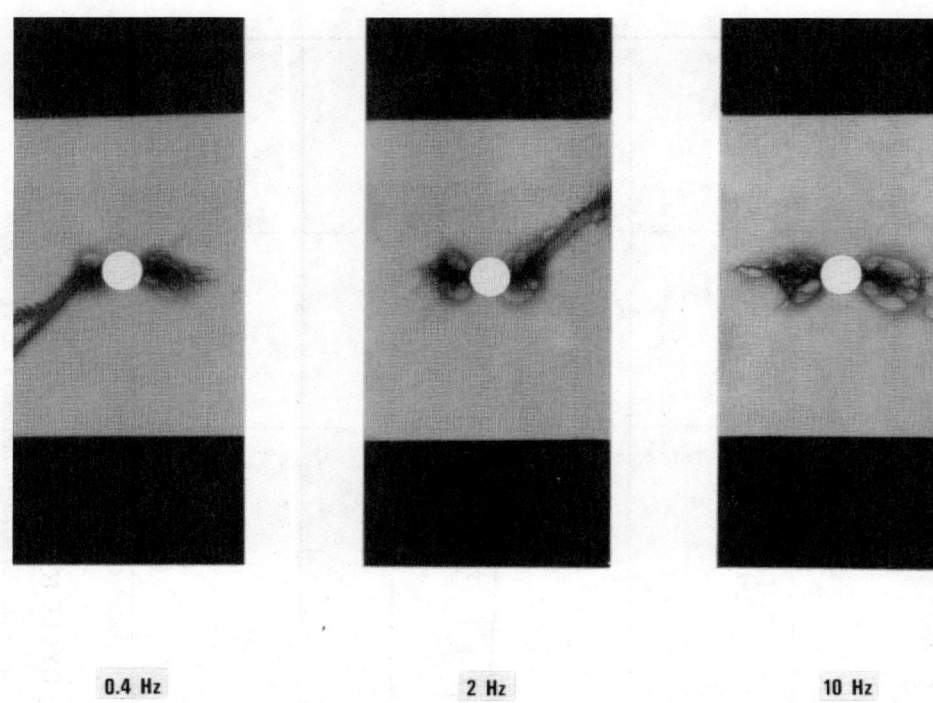

0.4 Hz   2 Hz   10 Hz

LOAD LEVEL 2 ±4960 LB

FIG. 11—*Radiographs of (4/80/16) laminate for different frequencies near fatigue failure.*

load and a longer time of load application could more easily produce buckling failure. Beyond 2 Hz, fatigue life decreases as load frequency increases.

Figure 13 shows the near hole temperature in the (48/47/4) laminate under 87.5% fatigue load. It is seen that the temperature rises steadily to reach an equilibrium level. At 10 Hz, the temperature rise is quite significant. The presence of high temperatures at high load frequencies may contribute to lowering the fatigue life.

A typical radiograph showing fatigue damage growth in the (48/48/4) laminate is presented in Fig. 14. Both delamination and transverse matrix cracking are evident.

For all load frequencies considered, stiffness experienced a sharp drop initially and then maintained a steady decrease during the fatigue test.

Frequency-Dependent Fatigue Models

In Ref *1*, a theoretical model for life prediction was given as follows:

$$N(\omega) = N(\omega_1) \frac{\omega}{\omega_1} e^{\eta[(\Delta T_1 - \Delta T)/T_0]} \tag{2}$$

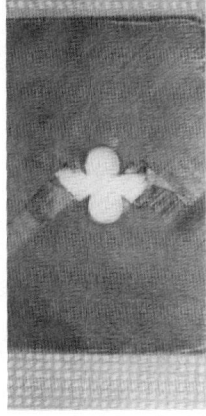
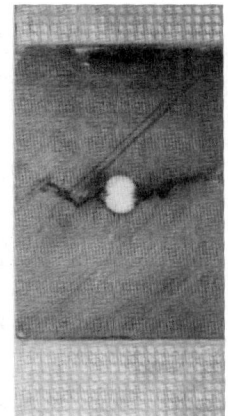

2 Hz   **10 Hz**

FLAT VIEW

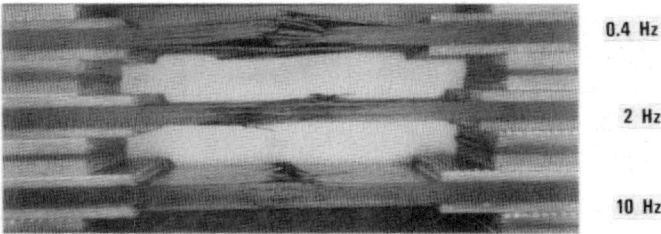

EDGE VIEW

FIG. 12—*Typical failure modes after fatigue failure for (4/80/16) laminate.*

TABLE 5—*Fatigue life at different frequencies (85% load level)[a] for IM6/APC-2 (48/48/4) laminate.*

Specimen	Frequency, Hz	Load, ±N (±lb)	Number of Cycles	Average
3C12	2.0	40850 (9180)	1469491	1114979
3C01	2.0	40850 (9180)	760467	
3C07	10.0	40850 (9180)	1228229	803753
3C09	10.0	40850 (9180)	379277	

[a]For the 85.0% load level, $P = \pm 40851$ N (± 9180 lb).

TABLE 6—*Fatigue life at different frequencies (87.5% load level)*[a] *for IM6/APC-2 (48/48/4) laminate.*

Specimen	Frequency, Hz	Load, ±N (±lb)	Number of Cycles	Average
3B71	0.4	42453 (9450)	2970	
3B72	0.4	42453 (9450)	13972	
3B61	0.4	42453 (9450)	1795	4807
3B62	0.4	42453 (9450)	489	
3B31	2.0	42453 (9450)	3979	
3B32	2.0	42453 (9450)	892778	
3B21	2.0	42453 (9450)	1328374	1156346
3B22	2.0	42453 (9450)	2400255	
3B41	10.0	42453 (9450)	4375	
3B42	10.0	42453 (9450)	1116	
3B11	10.0	42453 (9450)	1184415	419659
3B12	10.0	42453 (9450)	488730	

[a]For the 87.5% load level, $P = \pm42453$ N (±9450 lb).

TABLE 7—*Fatigue life at different frequencies (88.9% load level)*[a] *for IM6/AFPC-2 (48/48/4) laminate.*

Specimen	Frequency, Hz	Load, ±N (±lb)	Number of Cycles	Average
3A51	0.4	42720 (9600)	101	
3A52	0.4	42720 (9600)	211	156
3C06	2.0	42720 (9600)	7742	
3C01	2.0	42720 (9600)	2934	5338
3A30	10.0	42720 (9600)	2298	
3A40	10.0	42720 (9600)	437	1368

[a]For the 88.9% load level, $P = \pm42720$ N (±9600 lb).

where

ω = load frequency,
N = fatigue life,
T_0 = room temperature,
ΔT = near-hole temperature increase over T_0 at half-life, and
η = a parameter.

In Eq 2, $N(\omega_1)$ is the fatigue life at load-frequency ω_1, and ΔT is the associated temperature rise. This model predicts that fatigue life is directly proportional to load frequency if the temperature effect is negligible. A similar model was discussed by Schapery for viscoelastic media [4].

In Ref 1, the half-life temperature was used if the temperature remained constant most of the time during the fatigue test. If the temperature exhibited a steady increase, ΔT was regarded as the averaged temperature increase throughout the life.

In this study, the test data do not indicate that fatigue life is proportional to the frequency applied. Thus, we modify Eq 2 into the following form:

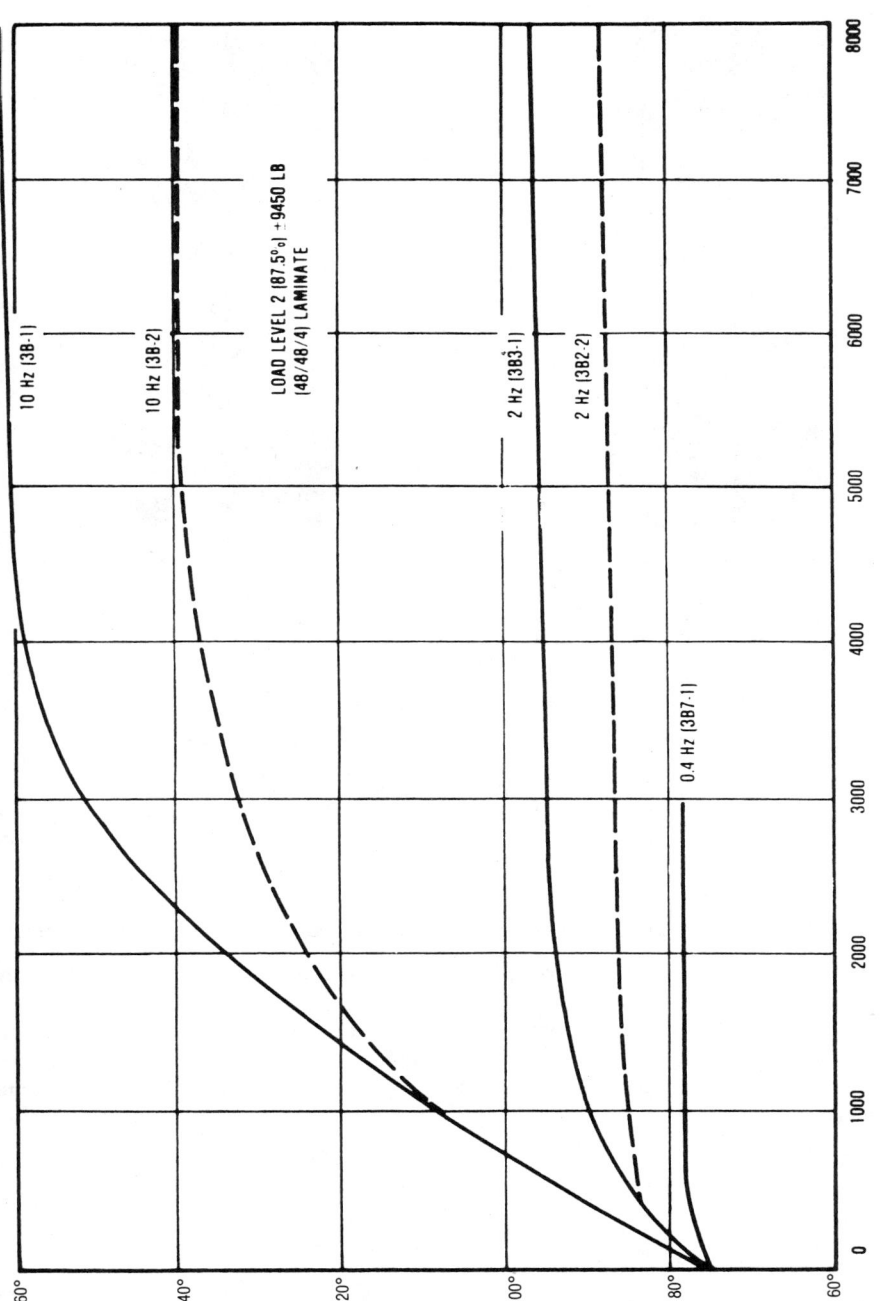

FIG. 13—*History of temperature rise near the hole of (48/48/4) laminate under 87.5% load level.*

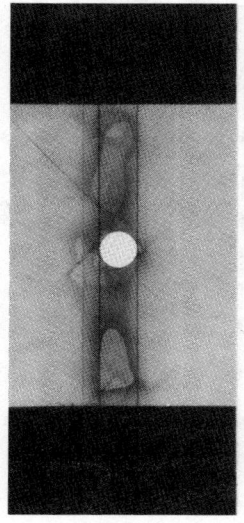 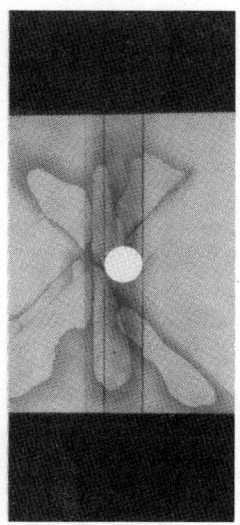

N = 860000
N_f = 892778

N = 500000
N_f = 1328374

2 Hz LOAD LEVEL 2 ±9450 LB

FIG. 14—*Radiographs of (48/48/4) laminate before failure.*

$$N(\omega) = N_0 \left(\frac{1}{p} - 1 \right)^{\alpha} \omega^{\beta} e^{-\eta p \omega} \qquad (3)$$

where

p = percentage of the maximum fatigue load level versus the ultimate load of the specimen.

The parameters N_0, α, β, and η are determined from the fatigue data. The effect of temperature is included in the exponential term.

For the $[\pm 45]_{2s}$ thermoplastic laminate, the summary of fatigue life and temperature increase near the hole at half-life is listed in Table 8. For the 26-plied laminates, summaries of data are presented in Table 9 for the (4/80/16) laminate, and Table 10 for the (48/48/4) laminate, respectively.

For the $[\pm 45]_{2s}$ laminate, values for these parameters are obtained as follows:

$N_0 = 28673$
$\beta = 0.2876$
$\alpha = 0.2995$
$\eta = 0.4834$

It is noted that, except for one case (57.9% at 10 Hz), the model predicts the fatigue life of the $[\pm 45]_{2s}$ laminate very well, as shown by the results listed in Table 11.

For the (4/80/16) and (48/48/4) laminates, the model given by Eq 3 does not work well, especially for the (48/48/4) laminate, which is highly fiber-dominated. It is well known that the fatigue life of fiber-dominated laminates is very sensitive to load level and scatters widely.

TABLE 8—*Summary of fatigue life and temperature increase for $[\pm 45]_{2s}$ IM6/APC-2 laminate.*

Frequency, Hz	Load Level 1 (57.9%)		Load Level 2 (75.0%)	
	ΔT, °F	N	ΔT, °F	N
0.4	1	18533	7	13222
2.0	17	17560	29	12676
10.0	95	10542	75	1064

TABLE 9—*Summary of fatigue life and temperature increase for (4/80/16) laminate.*

Frequency, Hz	Load Level 1 (57.9%)		Load Level 2 (70.0%)	
	ΔT, °F	N	ΔT, °F	N
0.4	4	37452	6	3035
2.0	15	30290	29	3643
10.0	69	4164	91	2066

TABLE 10—*Summary of fatigue life and temperature increase for (48/48/4) laminate.*

Frequency, Hz	Load Level 1 (85.0%)		Load Level 2 (87.5%)		Load Level 3 (88.9%)	
	ΔT, °F	N	ΔT, °F	N	ΔT, °F	N
0.4	...	...	2	4807	3	156
2.0	14	1114979	19	1156346	16	5338
10.0	68	803753	82	419659	62	1368

TABLE 11—*Predicted fatigue life of $[\pm 45]_{2s}$ IM6/APC-2 laminate by Eq 3.*

Frequency, Hz	Load Level 1 (57.9%)		Load Level 2 (75.0%)	
	N, Predicted	N, Test	N, Predicted	N, Test
0.4	17904	18533	13713	13222
2.0	18177	17560	12198	12676
10.0	3078	10542	1066	1064

TABLE 12—*Parameters N_0, β, and η for (4/80/16) IM6/APC-2 laminate.*

Laminate	Load Level 1 (57.9%)			Load Level 2 (70%)		
	N_0	β	η	N_0	β	η
(4/80/16)	47713	0.1434	0.4782	3926	0.2299	0.1674

TABLE 13—*Parameters N_0, β, and η for (48/48/4) IM6/APC-2 laminate.*

	Load Level 2 (87.5%)			Load Level 3 (88.9%)		
Laminate	N_0	β	η	N_0	β	η
(48/48/4)	412541	4.4159	1.1601	3177	2.9553	0.8602

For these two laminates, we used the following equation for each load level:

$$N(\omega) = N_0 \, \omega^\beta \, e^{-\eta p \omega} \tag{4}$$

The parameters N_0, β, and η are determined from the fatigue data. The results are listed in Tables 12 and 13.

Conclusions

Based upon the experimental results, the following conclusions have been reached:

1. The fatigue life of polymer-based composites is highly frequency-dependent. For $[\pm45]_{2s}$ laminate, evidence shows that the BMI composite follows the trend of other graphite-epoxy systems; that is, at low frequencies, fatigue life increases as frequency increases. However, the fatigue life of thermoplastic APC-2 laminate decreases as the load-frequency increases. This behavior is opposite to the behavior of graphite-epoxy composites at low frequencies.

2. Fatigue property of the APC-2 thermoplastic $[\pm45]_{2s}$ composite is far better than that of the BMI composite at low frequencies. At high frequencies, the thermoplastic $[\pm45]_{2s}$ composite may generate too much heat that reduces fatigue life.

3. For the (48/48/4) and (4/80/16) APC-2 thermoplastic laminates, the frequency effect on fatigue life of these laminates depends greatly on the load level. At high loads, low frequencies (such as 0.4 Hz) drastically could reduce the life, especially for the fiber-dominated laminate. This could be the result of creep buckling.

4. For the (48/48/4) and the (4/80/16) APC-2 thermoplastic laminates, at load frequencies above 2 Hz, fatigue life decreases as frequency increases.

Acknowledgments

This work was supported by an IRAD program with General Dynamics/Fort Worth Division. The first author extends his appreciations to S. M. Speaker and W. C. Jackson of General Dynamics for their encouragement in this work.

References

[1] Sun, C. T. and Chan, W. S., "Frequency Effect on the Fatigue Life of a Laminated Composite," *Composite Materials: Testing and Design, ASTM STP 674,* S. W. Tsai, Ed., American Society for Testing and Materials, Philadelphia, 1979, pp. 418–430.

[2] Saff, C. R., "Effects of Layup and Loading Frequency on Fatigue Life of Graphite/Epoxy," Technical Report No. NADC-81017-60, Naval Air Development Center, Warminster, PA, October 1982.

[3] Reifsnider, K. L., Stinchcomb, W. W., and O'Brien, T. K., "Fatigue of Filamentary Composite Materials," *Fatigue of Filamentary Composite Materials, ASTM STP 636,* R. Evans, Ed., American Society for Testing and Materials, Philadelphia, 1977, pp. 171–184.

[4] Schapery, R. A., "Deformation and Failure of Viscoelastic Composite Materials," *Inelastic Behavior of Composite Materials,* AMD-Vol. 13, C. T. Herakovich, Ed., American Society of Mechanical Engineers, New York, 1975, pp. 127–156.

Robert A. Simonds[1] and Wayne W. Stinchcomb[1]

Response of Notched AS4/PEEK Laminates to Tension/Compression Loading

REFERENCE: Simonds, R. A. and Stinchcomb, W. W., **"Response of Notched AS4/PEEK Laminates to Tension/Compression Loading,"** *Advances in Thermoplastic Matrix Composite Materials, ASTM STP 1044*, G. M. Newaz, Ed., American Society for Testing and Materials, Philadelphia, 1989, pp. 133–145.

ABSTRACT: The response of notched AS4/PEEK specimens to fully reversed, tension/compression loading has been investigated by examining their fatigue lives, damage initiation and propagation, and their residual strength. AS4/PEEK specimens were subjected to $R = -1$ tension/compression cyclic loading at 52.4 and 64.3% of their monotonic compression strength. The results indicate a significant difference in the response of the material at the two cyclic stress levels. At the lower fatigue stresses, the predominant damage, as determined by X-ray radiography and by deplying, is characterized by matrix cracking and delamination that initiate at the notch and grow both perpendicular and parallel to the load direction. Stiffness measurements taken during the low-level fatigue history show that compression stiffness and tension stiffness degrade throughout the fatigue lifetime. Further, specimens fatigued at the lower fatigue stresses lost compressive strength as the damage developed while they gained tensile strength. Fatigue life was defined by reduction of compression strength.

Damage to specimens fatigued at higher cyclic stresses developed much more predominantly in the direction perpendicular to the loading and much less in the direction parallel to the loading. Stiffness measurements made on these specimens showed a more rapid degradation of tension stiffness than of compression stiffness throughout the fatigue life. As with specimens fatigued at the lower stress levels, residual compressive strength decreased with damage development. However, the residual tensile strength of specimens fatigued at higher stresses decreased with damage development and the fatigue failure modes were tensile.

The difference in the response of the graphite-PEEK laminates at the two cyclic stress levels suggests that damage initiation and damage propagation play different roles in defining fatigue response.

KEY WORDS: composite material, graphite/PEEK, thermoplastic, fatigue, damage, strength, stiffness, life

Damage tolerance and structural durability are of great interest as more demanding performance requirements are established for composite materials and structures. In general terms, damage tolerance may be defined as the ability of a damaged material to maintain properties, such as strength, stiffness, and life, greater than a specified minimum level. Damage tolerance for composite materials is often quantified as interlaminar fracture toughness (measured by critical strain energy release rate), open hole compressive strength, or compressive strength after impact. Damage tolerance data for a wide range of graphite fiber-reinforced composites show that thermoplastic matrix composites are generally tougher than thermoset, non-toughened, epoxy composites. For example, the interlaminar fracture toughness of graphite

---

[1]Research engineer and professor, respectively, Department of Engineering Science and Mechanics, Polytechnic Institute and State University, Blacksburg, VA 24061.

133

polyetheretherketone (PEEK) composites is on the order of ten times greater than that for graphite epoxy composites [1].

The improved toughness, processing efficiency, and potential life-cycle cost savings of thermoplastic composites have made them attractive materials for many applications, including aircraft, automobiles, and medical implant devices [2,3]. However, much of our present knowledge about the performance of thermoplastic composite materials is based on short-term test results. Relatively little information is available on the long-term performance of thermoplastic composites, such as fatigue and creep. Results from several early studies [1,4–7] indicate that the greater short-term toughness of thermoplastic composites may not translate directly into improved performance under long-term loading conditions. Specifically, threshold strain energy release rates for delamination growth are much less than interlaminar fracture toughness values and are nearly the same as threshold values for thermosetting epoxies [1,4,7]. Secondly, there are differences in the initiation and propagation of damage in the interaction between damage modes and in the influence of fiber fracture on stiffness, residual tensile and compressive strength, and failure mode during long-term loading [5,6].

Objectives

In an earlier study, the fatigue response of two graphite fiber-reinforced, notched laminates subjected to completely reversed ($R = -1$) cyclic loads were compared [5]. The matrix materials were 5208 thermosetting epoxy and PEEK semicrystalline thermoplastic to provide a contrast in behavior based on matrix toughness. At reversed cyclic stress levels less than 60% of the monotonic compressive strength of each notched (center hole) laminate, the PEEK specimens had fatigue lives greater than those of the 5208 specimens. At low cyclic stress levels, both material systems failed in compression due to localized delamination at the hole and the attendant loss of stiffness. At high stress levels, fatigue lives of the epoxy matrix laminates were greater than those of the PEEK laminates. This reversal was attributed, in part, to a transition in failure mode of the PEEK laminates from compression at low cyclic stresses (corresponding to long life) to tension at high cyclic stresses (corresponding to short life). The failure mode of the epoxy matrix specimens was compression in all cases investigated [8].

The results presented in Ref 5 showed the need for a better understanding of the damage process and its effect on performance during long-term loading of thermoplastic composite laminates. The objectives of the study reported in this paper are the following:

1. To describe the stress-dependent damage processes during cyclic loading of AS4/PEEK notched laminates.

2. To determine the effects the damage modes and damage states have on stiffness, residual strength, life, and failure mode.

Material and Specimens

Thirty-two ply, nominally 4-mm (0.162-in.)-thick panels of AS4/PEEK (APC-2) were manufactured at Imperial Chemical Industries using a 380°C (716°F) cure process. The panels were delivered to NASA Langley Research Center where they were machined into specimens 121 mm (4.75 in.) long by 38 mm (1.5 in.) wide having a $[(0/+45/90/-45)_s]_4$ stacking sequence. A 9.5-mm (0.375-in.) hole was machined into the center of each specimen using a diamond core drill (Fig. 1). For both monotonic tensile and compressive loading and completely reversed cyclic loading, the specimens were hydraulically gripped in MTS fixtures having alignment plates to facilitate location and alignment of the specimen as indicated in Fig. 2. The unsupported length of the specimen between the grips was specified to be 61.0 mm (2.4 in.) as described in Ref 9.

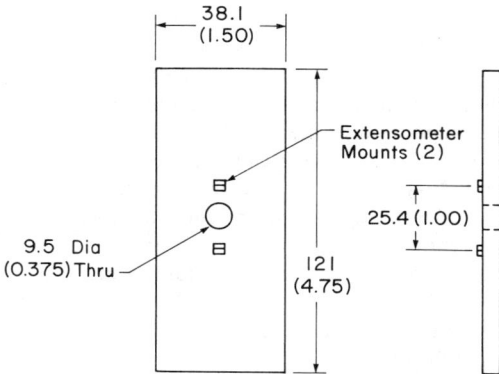

FIG. 1—*Specimen geometry, dimensions in mm (in.).*

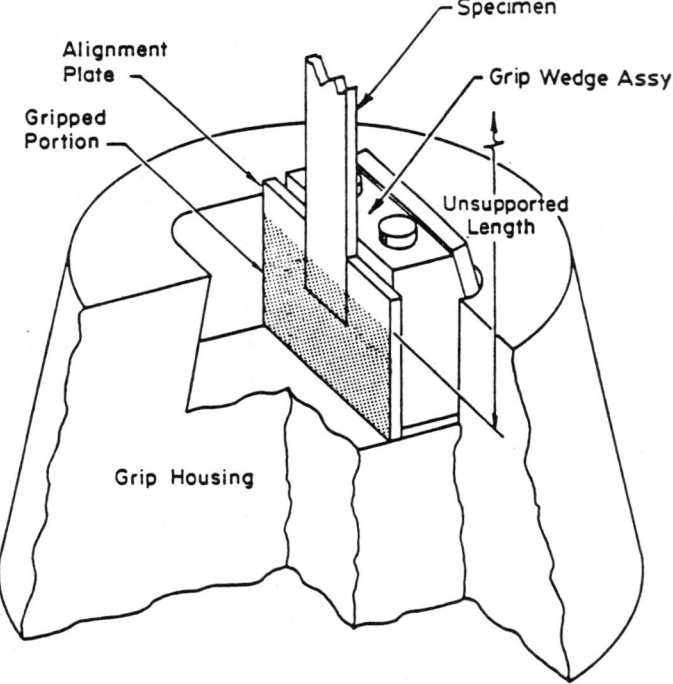

FIG. 2—*Diagram of grips.*

This length permits specimens to respond freely to monotonic and cyclic loads and to fail in a natural mode due to the damage and applied loading.

Thin aluminum tabs with engraved V-notches were bonded to the specimen with compliant silicone adhesive to locate and seat a 25.4-mm (1.0-in.) extensometer used to measure monotonic (tensile and compressive) stiffness and monitor stiffness change during cyclic loading.

Experimental Procedure

Monotonic mechanical tests were conducted using the test method described in Ref *9* and the procedure described in Ref *5*. Briefly, notched specimens were loaded monotonically in tension and in compression to determine initial stiffness and strength. Compressively loaded specimens with a 61-mm (2.4-in.) unsupported length failed by crushing. Cyclic tests were conducted in load control at a frequency of 10 Hz. Some cyclic tests were continued until the specimen failed in order to establish an *S-N* curve, some tests were terminated at selected stages of life for residual tensile or compressive strength tests, and some tests were interrupted for nondestructive inspections (penetrant enhanced X-ray radiography and ultrasonic C-scans) and destructive examinations (sectioning and deply). Prior to all sectioning and deply examinations, the specimens were inspected nondestructively.

During fatigue loading, stiffness measurements were periodically taken by halting the load cycling, manually ramping the load to the maximum tension cyclic load, taking a strain reading, manually ramping through zero load to the absolute maximum cyclic compressive load, and taking another strain reading before returning to zero load. The load was held at the maximum tensile and compressive levels only for the short amount of time necessary to read the strain values so as to minimize any time-dependent effects. In this manner, a tension-compression (*T-C*) secant stiffness could be calculated by dividing twice the stress magnitude by the tensile strain plus the absolute maximum compressive strain (Fig. 3).

Results

Monotonic and Fatigue Test Results

Initial strength, fatigue life, and residual strength data for AS4/PEEK notched specimens are given in Ref *5*. A brief summary of results is included herein for ease of reference:

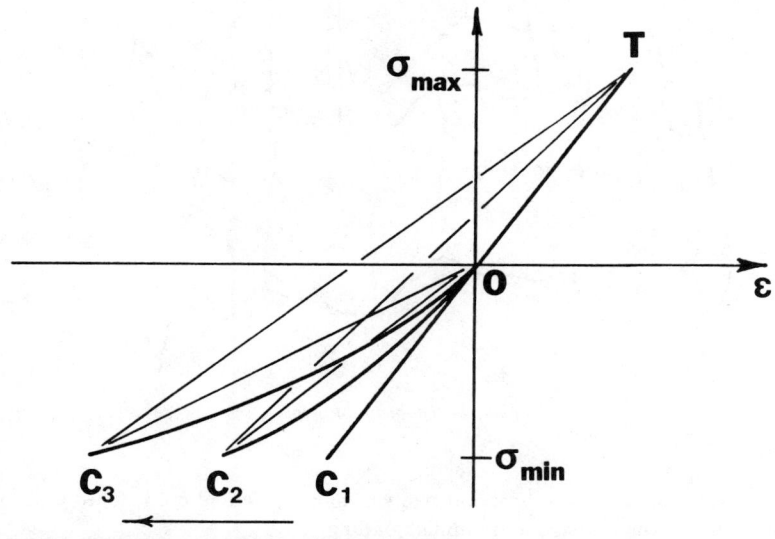

FIG. 3—*Determination of T-C stiffness, 0-T stiffness, and 0-C stiffness during fatigue testing with 0-C stiffness decreasing with increasing fatigue cycles.*

1. Monotonic tensile strength: 351 MPa (50.9 ksi).
2. Monotonic compressive strength: 290 MPa (42.0 ksi).
3. S-N data: Table 1 and Fig. 4.

Figure 5 shows the result of plotting the T-C stiffness of several specimens versus the number of cycles at which the stiffnesses were taken normalized by the total life of the specimen. T-C stiffness was normalized by its value on the first cycle of a test. All the specimens tested had very similar normalized stiffness versus normalized life histories, regardless of load level or eventual number of cycles to failure. Therefore, life can be characterized in terms of stiffness change [8].

Further, the life of the specimens can be divided into three stages according to the normalized stiffness versus normalized life history:

TABLE 1—Stress versus cycles to fail specimens in fatigue.

Cyclic Stress, MPa (ksi)	Normalized to Static Compressive Strength	Life, cycles
207 (30.0)	0.714	2 410
186 (27.0)	0.643	9 350
185 (26.8)	0.638	5 760
183 (26.6)	0.633	5 470
170 (24.6)	0.586	232 260
170 (24.6)	0.586	66 540
169 (24.5)	0.583	373 260
160 (23.2)	0.552	181 380
159 (23.1)	0.550	215 500
155 (22.5)	0.536	166 210
155 (22.5)	0.536	882 660
144 (20.9)	0.498	6 475 100

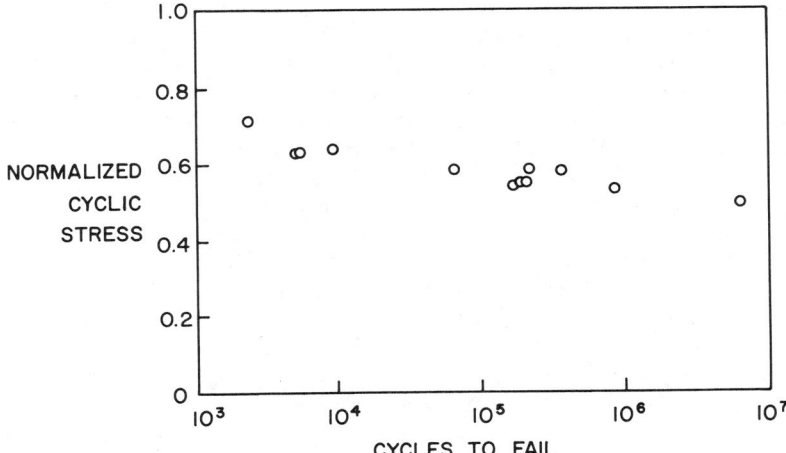

FIG. 4—Normalized cyclic stress versus log cycles to fail for AS4/PEEK specimens fatigue tested to failure.

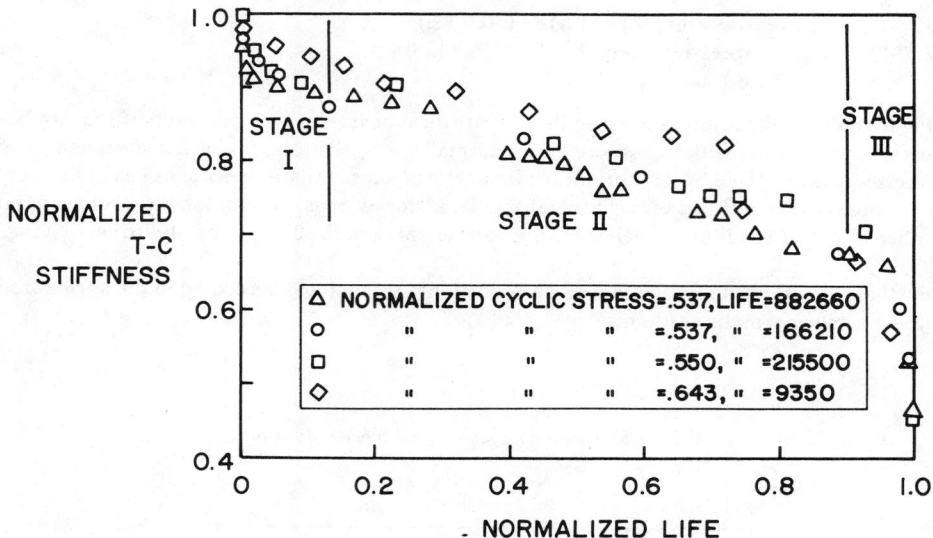

FIG. 5—*Normalized stiffness versus normalized life for several specimens fatigued to failure showing Stage I, Stage II, and Stage III of life.*

Stage I—Stage I is characterized by a rapid decrease in stiffness and lasting for about the first 10 to 15% of life.

Stage II—Stage II is where the stiffness degradation is not so rapid and lasts to about 90% of life.

Stage III—Stage III is characterized by a rapid degradation in stiffness associated with a rapid development of fatigue damage prior to failure.

In addition to calculating a T-C stiffness for each stiffness sampling point, a zero-tension (0-T) or a zero-compression (0-C) stiffness can be calculated as well, as shown in Fig. 3. This enables us to determine if stiffness degradation is occurring equally in tension and in compression or if the degradation is occurring more in one mode than in the other.

A stiffness ratio is defined and calculated by dividing the tension stiffness by the compression stiffness. A stiffness ratio of 1 indicates that the tension and compression stiffness are equal. A stiffness ratio greater than 1 indicates that the tension stiffness is greater than the compression stiffness. A stiffness ratio less than 1 indicates that the compression stiffness is greater.

For most of the specimens, the stiffness ratio at the beginning of life was slightly greater than 1; and the stiffness ratio remained greater than 1 throughout the test. However, in some cases, particularly those with high cyclic loads and short lives, the stiffness ratio was initially slightly greater than 1 but decreased to a value less than 1 during the life of the specimen; indeed, it continued to decrease as the test proceeded until failure. The decreasing stiffness ratio suggests a different damage process occurring in the specimens cycled at high cyclic levels. This was confirmed by the radiography, deply, and sectioning studies.

Figure 6 illustrates the stiffness ratio versus normalized life history for two specimens, one fatigued at 64.3% of the static compressive strength for 9350 cycles before failure and the other fatigued at 53.6% of the static compressive strength for 166 210 cycles.

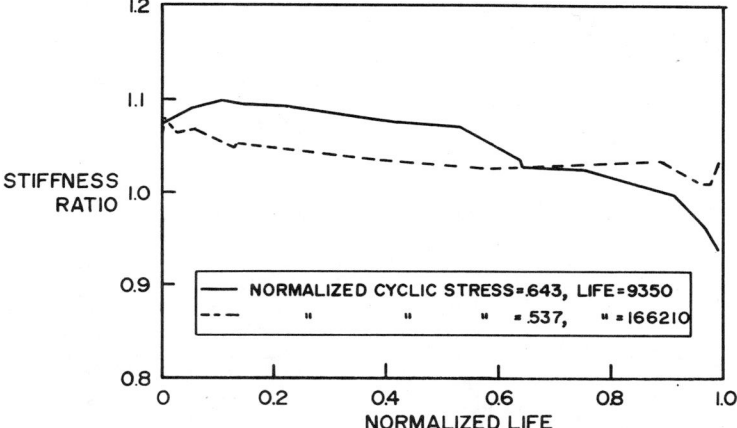

FIG. 6—*Stiffness ratio versus life for a specimen tested at a high cyclic stress and for a specimen tested at a low cyclic stress.*

Residual Strength Test Results

In order to determine the effect of the fatigue damage on the residual strength of the specimens as fatigue damage accumulated, several specimens were cycled at 64.3 and 52.4% of the static compressive strength. Stiffness was monitored in real time during the test, and testing was terminated on three specimens for both the high and low stress level after they suffered a 5% T-C stiffness loss, a 15% T-C stiffness loss, and a 30% T-C stiffness loss. This was designed to provide three specimens from each stress level in the midst of Stage I, the midst of Stage II, and at an advanced state of damage near the beginning of Stage III. One specimen at each stage of life was to be tested for residual tensile strength, one for residual compressive strength, and one specimen was to be sectioned and deplied in order to examine the nature of the damage.

Results of the residual strength tests appear in Tables 2 and 3 and in Fig. 7. (Also included in

TABLE 2—*Residual tensile strengths.*

Normalized Cyclic Stress[a]	Cycles	Normalized Stiffness[b]	Normalized Strength[c]
0.642	1 500	0.948	0.97
0.642	2 000	0.851	0.88
0.642	4 000	0.710	0.75
0.523	6 000	0.949	0.98
0.523	120 000	0.848	1.18
0.523	719 090	0.700	1.14
0.491	1 000 000	0.850	1.24
0.445	1 000 000	0.920	1.14

[a]Normalized to 290 MPa (42.0 ksi) unfatigued compression strength.
[b]Tension-compression stiffness measured on last cycle normalized to that measured ⸱ ‛he first load cycle.
[c]Normali. . to 351 MPa (50.9 ksi) unfatigued tensile strength.

TABLE 3—*Residual compressive strengths.*

Normalized Cyclic Stress[a]	Cycles	Normalized Stiffness[b]	Normalized Strength[a]
0.642	2 500	0.944	0.98
0.642	2 750	0.833	0.91
0.642	2 880	0.610	0.85
0.523	6 000	0.953	0.93
0.523	130 000	0.855	0.89
0.523	243 000	0.690	0.79
0.500	1 000 000	0.880	0.91
0.486	1 000 000	0.870	0.82
0.443	1 000 000	0.905	0.90
0.440	1 000 000	0.910	0.89

[a]Normalized to 290 MPa (42.0 ksi) unfatigued compressive strength.
[b]Tension-compression stiffness measured on last cycle normalized to that measured on the first load cycle.

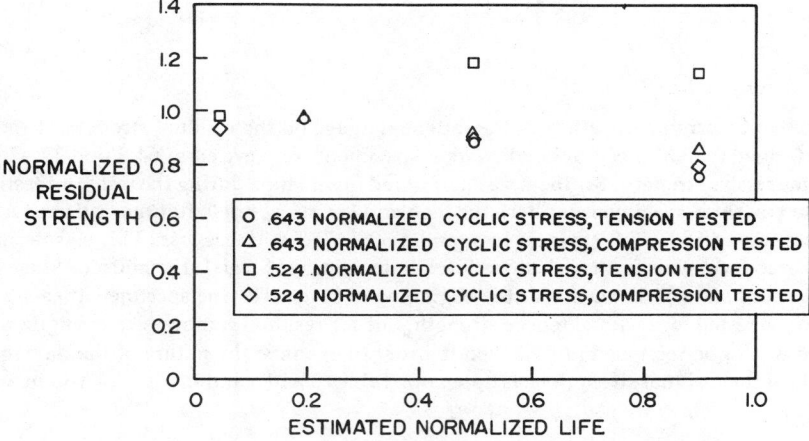

FIG. 7—*Residual strengths of specimens cyclically loaded at 64.3 and 52.4% of monotonic compressive strength plotted against estimated proportion of life.*

Tables 2 and 3 are results from tests early in the program in which some specimens were cyclically loaded to 1 000 000 cycles and tested for residual strength.) For plotting purposes, proportion of life used was estimated by comparing the stiffness versus cycles history of the specimens with those from specimens that had been fatigued to failure as shown in Ref *8*. The specimens cycled at 64.3% of the monotonic compressive strength showed either no significant change in tensile strength for a low number of cycles and a minimum amount of damage, as indicated by the 5% stiffness degradation, or a reduction in residual tensile strength at higher numbers of cycles and greater stiffness degradation. The specimens cycled at lower stresses, 52.4% of monotonic compressive strength or less, showed either no significant change in tensile strength for low numbers of cycles and small stiffness changes or actual increases in residual tensile strength. All the specimens tested for residual compressive strength showed varying amounts of strength reduction.

Results of Radiography, Deply, and Specimen Sectioning

All the fatigue specimens whose cyclic loading was halted for residual strength tests or for deplying or sectioning were X-ray radiographed in order to ascertain the damage development as cyclic loading progressed. Sample radiographs are presented in Fig. 8. According to the radiographs, damage occurs similarly in specimens cycled at both load levels until the end of Stage I. Thereafter, the specimens cycled at low cyclic stress, 52.4% of the monotonic compressive strength, show more extensive damage. That damage appears to grow both transversely and, to a greater extent, longitudinally. The radiographs, together with visual observations of the specimen under low stress cycling, suggest that most of the damage is in the form of matrix cracks and delaminations. The outer 0° plies in the vicinity of the hole can be seen to be separate from the rest of the specimen, particularly when the specimen is compressed and the outer plies buckle outward. Broken 0° fibers can be observed in the outer plies. Since matrix damage affects the 0-T and 0-C stiffnesses equally, stiffness ratio remains relatively constant for the specimens cycled at low-cyclic stresses as shown in Fig. 6.

The radiographs of the high-cyclic stress specimens in Stages II and III of life also show matrix cracks, delaminations, and broken 0° fibers at the hole. However, the extent and distribution of damage is much different in the high stress specimens than in the low stress specimens cycled to the same stages of life, but to a greater number of cycles. At midlife, matrix cracks in the 0° plies are much shorter, and the delaminated area is less than in the corresponding low cyclic stress specimens. In general, the damage region is significantly smaller, especially in the longitudinal (0°) direction. The dominant direction of damage growth during the second half of life is transverse to the 0° direction. The smaller damage region in the specimens cycled at high-stress levels contain a high density of matrix cracks in all plies, delaminations which fill in the regions between crossing matrix cracks in the adjacent plies, and broken fibers which extend beyond the boundary of the delamination zone. The broken 0° fibers contribute to stiffness degradation and particularly to 0-T stiffness degradation. The rapid 0-T stiffness degradation results in decreasing stiffness ratio for the specimens cycled at high-cyclic stresses as shown in Fig. 6.

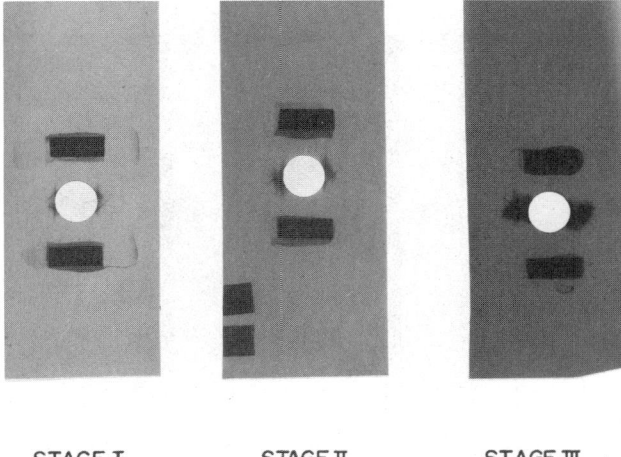

STAGE I   STAGE II   STAGE III

FIG. 8—*X-ray radiographs of specimens cyclically loaded at 64.3% of monotonic compressive strength showing damage development at various proportions of estimated life.*

Figure 10 is a photograph of a low stress specimen at midlife showing longitudinal cracks in the outer plies as well as broken fibers. The process of damage in the surface ply is influenced by the matrix cracks tangent to the hole in the surface (0°) and second (+45°) plies. The tangent matrix cracks develop early in fatigue life, as shown in Fig. 9. Fiber fractures in the surface ply initiate at the tangent matrix cracks in the surface ply near the area where they cross the tangent matrix cracks in the second ply. The crossing cracks produce local stress concentrations in the surface plies, causing a group of fibers to break locally. Fiber fracture arrests as the local concentration of stress diminishes [6,10], resulting in a second matrix crack at the end of the broken fiber group. The process of sequential fiber fractures continues as cyclic loading continues.

The path of fiber fracture increments is not necessarily perpendicular to the load axis. Most often, the path is along the tangent matrix crack in the underlying +45° ply, as shown in Fig. 10. Under completely reversed loading, fiber fractures on the macroscopic scale are common in the PEEK specimens and toughened epoxy laminates, but are not observed in brittle epoxy systems such as T300/5208 and AS4/3501-6 [6].

After being radiographed, specimens not tested for residual strength were sectioned by cutting them in half longitudinally through the center of the hole. One side of a specimen was then deplied by heating it to 700°C for 4 h in an inert environment so that the matrix was burned away using a technique patterned after the method of Freeman [11]. The individual plies then were separated and microscopically examined for broken fibers, particularly broken 0° fibers.

Broken 0° fibers were observed in the two-surface plies of specimens cycled at high and low stress levels. However, the deply results showed that specimens cycled at the high stress level also suffered fiber fractures in the first interior 0° plies (plies Number 8 and 25 in the lamination sequence). Fiber fractures were not observed in the interior 0° plies of specimens cycled at low stress levels until very late in life (Stage III). Figure 11 shows normalized tensile stiffness at the termination of a test plotted against the proportion of broken 0° fibers in all the 0° plies. In general, the tensile stiffness decreases as the number of 0° fiber fractures increases.

Figure 5 shows a rapid loss in T-C stiffness during the final stage of fatigue life. The tensile and compressive components of stiffness change during Stage III; however, the tensile component changes more rapidly during the final stage of the high stress test, as shown in Fig. 6. The rate of decrease in tensile stiffness, or tensile stiffness loss per cycle, is an indicator of damage

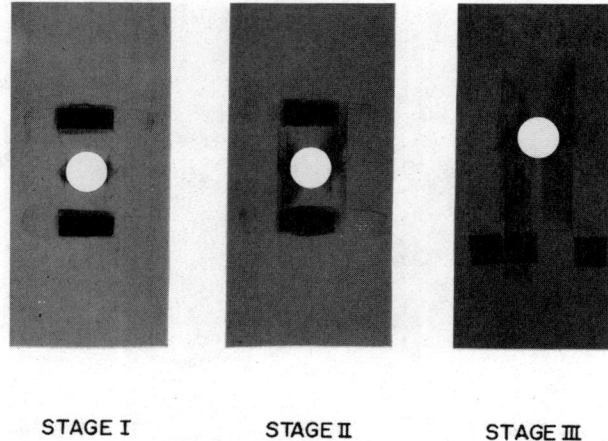

STAGE I   STAGE II   STAGE III

FIG. 9—*X-ray radiographs of specimens cyclically loaded at 52.4% of monotonic compressive strength showing damage development at various proportions of estimated life.*

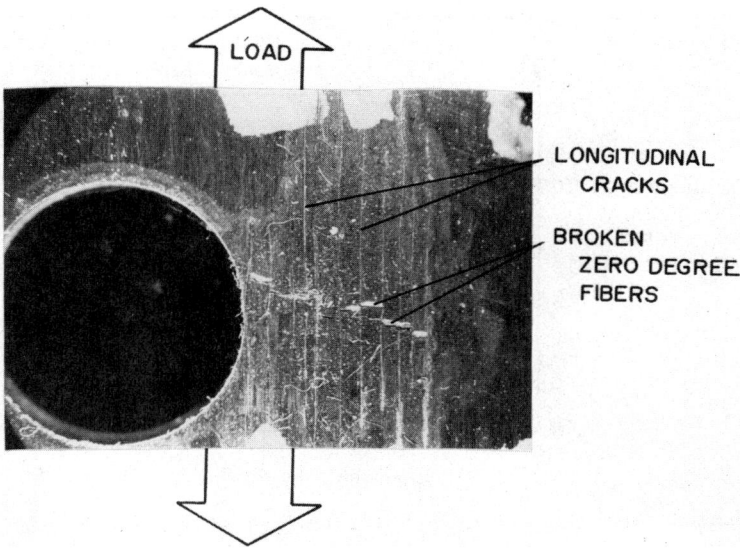

FIG. 10—*Photograph of a low load specimen at midlife showing longitudinal cracks and broken fibers in the outer ply.*

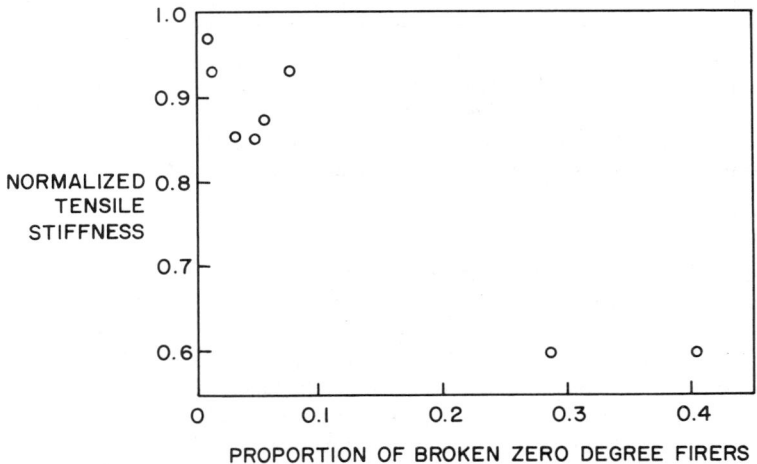

FIG. 11—*Normalized tensile stiffness versus proportion of broken zero degree fibers in specimens tested at high and low cyclic stress levels.*

accumulation rate near the end of life. It is calculated as the change in stiffness over the final sampling interval divided by the number of cycles during the final interval before the residual strength test.

Figure 12 shows the residual tensile strength, normalized with respect to the 351 MPa (50.9 ksi) initial tensile strength, versus the tensile stiffness loss per cycle prior to the residual strength test. The residual strengths of specimens tested at the low cyclic stresses and having small ten-

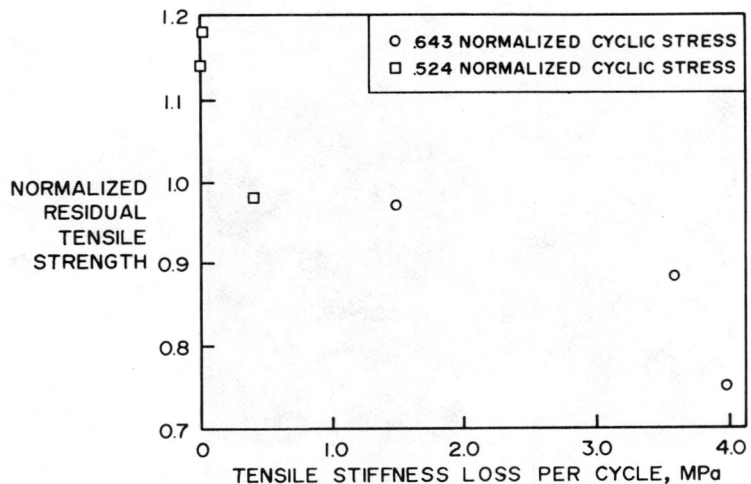

FIG. 12—*Normalized residual tensile strength versus tensile stiffness loss per cycle for specimens at various stages of life.*

sile stiffness change rates were either greater than or only slightly less than the initial tensile strength. However, the residual tensile strengths of specimens tested at high cyclic stresses were as much as 25% less than the initial strength following a very rapid decrease in tensile stiffness.

The correlations between tensile stiffness reduction and broken 0° fibers (Fig. 11) and rate of tensile stiffness loss and residual tensile strength (Fig. 12) suggest a relationship between residual tensile strength and broken 0° fibers. Although such a general relationship appears intuitive, the exact relationship between residual tensile strength and number of fiber fractures in 0° plies has not been established. Both measurements require destruction of specimens, especially if the broken fibers on interior plies are to be included in the count.

Discussion

During completely reversed cyclic loading of notched AS4/PEEK specimens, matrix cracks, delaminations, and fiber fractures were initiated at the hole. The extent of the damage zone, its direction of growth, and the modes of damage are dependent on the stress level. At low cyclic stresses, the dominant growth direction is longitudinal and specimens fail in compression. The failure mode was confirmed by the crushed nature of the failed specimens. At high cyclic stress levels, the dominant damage growth direction is transverse and specimens fail in tension. The failure mode was suggested by the change in tensile stiffness and confirmed by separation of the failed specimens. The failure mode transition occurs between a normalized stress of 0.52 and a normalized stress of 0.64 (10^4 to 10^5 cycles).

At low cyclic stress levels, damage (in the form of matrix cracks and delaminations) increases the effective radius of curvature of the hole, thereby reducing the local stresses at the hole and redistributing the load in the ligaments of material on each side of the hole [6,10]. The result is an increase in tensile strength of the notched laminate. Generally, under low cyclic stress conditions, PEEK composite specimens fail in compression due to delamination and the attendant loss of stiffness. The interior plies of material cannot sustain the compressive loads. Although broken 0° fibers were observed in specimens cycled at the lower stress levels, the combination of reduced stress concentration factor and low applied cyclic stresses is not enough to cause tensile failure.

At high cyclic stresses, damage develops transversely from the hole and reduces the load-carrying capability of the ligaments. During the relatively short fatigue lives, the cycle-dependent development of delamination in the tough, thermoplastic matrix is not sufficient to reduce the stress concentration at the hole and redistribute the load path away from the hole. High local stresses at the hole, due to the combination of a high stress concentration and the magnitude of the applied cyclic stress, produce fiber fractures which progress sequentially across the ligaments, thus reducing their tensile strength and causing a tensile failure mode.

The suppression of delamination due to the high interlaminar fracture toughness of the PEEK thermoplastic matrix does not permit the beneficial consequences of damage (tensile strength increase) to occur in the notched specimens subjected to high cyclic stresses. Consequently, the AS4/PEEK specimens fail in tension at fatigue lives less than those of epoxy matrix specimens which failed in compression as reported previously [5].

The optimum design of high-performance, damage-tolerant composite structures to sustain long-term cyclic loading cannot be based on the use of tough matrix composite materials alone. Long-term damage tolerance must be engineered through optimization of fiber and matrix properties, selection of laminate arrangement, choice of geometry, and careful attention to damage modes and failure modes.

Acknowledgments

This work was performed under NASA Grant NAG-1-343, monitored by T. K. O'Brien of the Structures Laboratory, U.S. Army Research and Technology Laboratories, NASA Langley Research Center. The authors sincerely appreciate the support of this program by these organizations.

References

[1] O'Brien, T. K., "Fatigue Delamination Behavior of PEEK Thermoplastic Composite Laminates," *Proceedings of the First Technical Conference of the American Society for Composites*, Dayton, OH, 1986.

[2] Newaz, G. M., "Advanced Thermoplastic Composite," *ASTM Standardization News*, Vol. 15, No. 10, 1987.

[3] Frisch, E. E., "Composites in Medical Devices," *ASTM Standardization News*, Vol. 15, No. 8, 1987.

[4] O'Brien, T. K. and Murri, G. B., "Interlaminar Shear Fracture Toughness and Fatigue Thresholds for Composite Materials," *Composite Materials: Fatigue and Fracture, Second Volume, ASTM STP 1012*, ASTM Philadelphia, 1989.

[5] Simonds, R. A., Bakis, C. E., and Stinchcomb, W. W., "Effects of Matrix Toughness on Fatigue Response of Graphite Fiber Composite Laminates," *Composite Materials: Fatigue and Fracture, Second Volume, ASTM STP 1012*, ASTM, Philadelphia, 1989.

[6] Reifsnider, K. L., Bakis, C. E., Yih, H. R., and Stinchcomb, W. W., "The Mechanics of Microdamage in Notched Composite Laminates," presented at the ASME Winter Annual Meeting, Boston, MA, 1987.

[7] O'Brien, T. K., "Analysis of Local Delaminations and Their Influence on Composite Laminate Behavior," *Delamination and Debonding of Materials, ASTM STP 876*, W. S. Johnson, Ed., American Society for Testing and Materials, Philadelphia, 1985, pp. 282–297.

[8] Bakis, C. E. and Stinchcomb, W. W., "Response of Thick, Notched Laminates Subjected to Tension-Compression Cyclic Loads," *Composite Materials: Fatigue and Fracture, ASTM STP 907*, H. T. Hahn, Ed., American Society for Testing and Materials, Philadelphia, 1986, pp. 314–334.

[9] Bakis, C. E., Simonds, R. A., and Stinchcomb, W. W., "A Test Method to Measure the Response of Composite Materials Under Reversed Cyclic Loads," *Test Methods and Design Allowables for Fibrous Composites: Second Volume, STP 1003*, ASTM, Philadelphia, 1989.

[10] Kress, G. R. and Stinchcomb, W. W., "Fatigue Response of Notched Graphite/Epoxy Laminates," *Recent Advances in Composites in the United States and Japan, ASTM STP 864*, J. R. Vinson and M. Taya, Eds., American Society for Testing and Materials, Philadelphia, 1985, pp. 173–196.

[11] Freeman, S. M. in *Composite Materials: Testing and Design (Sixth Conference), ASTM STP 787*, I. M. Daniel, Ed., American Society for Testing and Materials, Philadelphia, 1982, pp. 50–62.

Rodney R. Hoffman, [1] *Ronald P. Tye,* [2] *and Joseph G. Chervenak* [2]

Investigation into Thermal Conductivity of Composite Materials for Electronics Packaging

REFERENCE: Hoffman, R. R., Tye, R. P., and Chervenak, J. G., "**Investigation into Thermal Conductivity of Composite Materials for Electronics Packaging,**" *Advances in Thermoplastic Matrix Composite Materials, ASTM STP 1044,* G. M. Newaz, Ed., American Society for Testing and Materials, Philadelphia, 1989, pp. 146–153.

ABSTRACT: As part of a two-year investigation into the use of composite materials for military electronic equipment enclosures, the thermal conductivity of composite materials was determined by testing and predicted by an analytical technique. The objectives included identifying methods of improving thermal conductivity of composites and utilization of a prediction method. The comparative test and the Lewis-Nielson semitheoretical prediction methods were used and are described. The materials evaluated were a thermoplastic polymer with several discontinuous filler types. The analytical predictions were in good agreement with the test results except at high filler levels, but do not take into account anisotropic material properties or temperature effects. Test and prediction results are presented.

KEY WORDS: composite materials, thermal properties, test results

Magnavox Electronic Systems Co. is a major supplier of electronic equipment to the Department of Defense and other agencies. For most airborne and ground-based equipment, reduction of weight is an important goal. Reinforced thermoplastic composite materials can reduce equipment weight when used to replace metals as the enclosure material. They can often reduce the product cost also, and therefore their use is being actively investigated by Magnavox.

Aluminum is the material most often used for the electronic equipment enclosure. Switching to a thermoplastic material introduces a number of problems, one of which is low thermal conductivity when compared to aluminum. For sealed units such as radios and hand-held computers, the temperature rise across the housing is a part of the total temperature rise from the ambient to each electronic component's junction temperature. Since the reliability of electronic equipment is reduced as the component junction temperatures rise, the thermal conductivity of the housing material is an important design parameter.

An investigation was undertaken at Magnavox to determine which, if any, thermoplastic resin and discontinuous filler combinations are suitable as electronic equipment housings. While other material properties were considered during this investigation, only the work relating to thermal conductivity is presented. The objectives of the investigation were to: (1) identify those filler materials and/or types which effectively increase the thermal conductivity, and (2) select an analytical method to predict the thermal conductivity of any filler and matrix combination.

The approach taken to accomplish these objectives consisted of obtaining samples of a thermoplastic polymer (Nylon 6/6) with carbon fiber, aluminum flake, and tungsten powder fillers.

[1] Magnavox G&I Electronics Co., Fort Wayne, IN 46808.
[2] Dynatech R/D Co., Cambridge, MA 02139.

146

These materials were then subjected to thermal conductivity testing using the comparative method. Predictions were made of the material thermal conductivity using the Lewis and Nielson semitheoretical model. The prediction results and the test results were then compared to determine the accuracy of the prediction method.

Analytical Method

A number of theories and methods have been proposed to predict the thermal conductivity of composite materials. Each model works very well for its own particular set of idealized situations, but no one model has been proven to be effective for all combinations of continuous and discontinuous phases. However, for a thermoplastic matrix with a solid particulate filler, the Lewis and Nielson semitheoretical method has been found to best fit the experimental data [1].

The Lewis and Nielson semitheoretical method includes the thermal conductivity of both the continuous phase (thermoplastic resin) and the discontinuous phase (filler). It also takes into account the shape of the filler particles, the amount of filler, and the theoretical maximum amount of filler which may be packed into the matrix. It assumes the two-phase system is identical in all directions. The Lewis and Nielson semitheoretical model is written as

$$k = k_c((1 + AB\phi)/(1 - B\psi\phi)) \qquad [2,3] \qquad (1)$$

where

k = thermal conductivity of two phase system,
k_c = thermal conductivity of the continuous phase (matrix),
A = a constant relating the shape and orientation of the discontinuous phase (filler),
B = a constant relating the relative thermal conductivities of the two phases,
ϕ = volume fraction of the discontinuous phase, and
ψ = a constant relating the maximum packing fraction of the discontinuous phase.

The rationale for selecting the numerical values of A, B, and ψ are described in detail in Refs 2 and 3.

Test Method

The proposed ASTM test method "Guarded, Comparative, Longitudinal Heat Flow Technique" was used to measure the thermal conductivity [4]. This test method employs a "stack" consisting of: upper heater, top reference, sample, bottom reference, lower heater, and heat sink as shown in Fig. 1. Thermal insulation and a cylindrical guard with matching temperature gradient are used to maximize linear heat flow through the stack and minimize heat flow in the radial direction.

The heat flow circuit is then the analogue of an electric circuit with resistors in series as shown on the right of Fig. 1. Electric current (I) is analogous to heat flow, and voltage drop (ΔV) is analogous to temperature drop. The top and bottom references of known thermal resistance are used to measure the heat flow into and out of the sample. When the heat flow reaches steady state and the heat into the sample agrees with the heat out of the sample, the thermal conductivity of the sample can be found from

$$k = \frac{Q}{\left(\dfrac{\Delta T}{\Delta x}\right)_{sample}} \qquad (2)$$

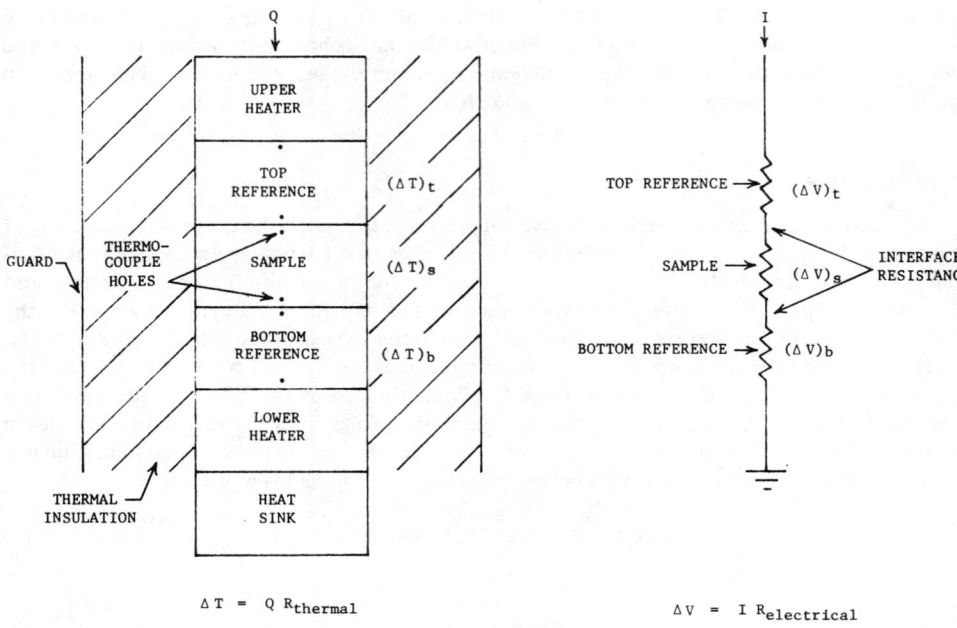

$$\Delta T = Q\,R_{thermal}$$

$$\Delta V = I\,R_{electrical}$$

FIG. 1—*The comparative method of measuring thermal resistance and its equivalent electric circuit.*

where

$$Q = \text{heat flow per cross-sectional area, and}$$
$$(\Delta T/\Delta x)_{sample} = \text{measured thermal gradient in the sample.}$$

For the composite materials measured here, 0.25 mm grooves were milled into the surface and butt-welded type-K thermocouples were cemented in place. Pyrex 7740 ($k = 1.12$ W/m/ °K at 40°C) was used as the reference material. Dow Corning 340 heat sink compound was used at the reference-to-sample interfaces in order to reduce the contact thermal resistance. Before the data was taken, a qualification run was performed on Pyroceram 9606 ($k = 3.92$ W/m/°K at 40°C) and agreement with the known value of the thermal conductivity was within 2%. Tests on multiple specimens showed the standard deviation due to reproducibility of the measurement and sample variation to be on the order of 5 to 8%.

Materials

Three types of materials were obtained for thermal conductivity testing. Nylon 6/6 was used as the thermoplastic matrix for all three fillers, and the samples were injection molded by LNP Corp., Malvern, Pennsylvania. The fillers used were chopped polyacrylonitrile (PAN) carbon fiber, aluminum flake, and tungsten powder.

The PAN carbon fiber was added to the matrix in by-weight filler levels of 0%, 15%, 20%, 30%, 40%, 45%, and 60%. The average diameter and length of the fiber was 8.0 m and 1.60 mm, respectively. The axial thermal conductivity was 13.8 W/m/°K and the density was 1.80 g/cm^3.

K102H aluminum flake was obtained from Transmet Corp., Columbus, Ohio. The average size was 1.2 mm long, 1.0 mm wide, and 0.03 mm thick. The density was 2.70 g/cm³, and the thermal conductivity was 209.4 W/m/°K. The flakes were added to Nylon 6/6 in by-weight filler levels of 0%, 15%, 30%, and 40%.

Tungsten powder in by-weight filler levels of 0%, 45%, 60%, and 75% was added to Nylon 6/6. The density of the tungsten was 19.24 g/cm³, while the thermal conductivity was 167.2 W/m/°K. The powder platelets had an average length, width, and thickness of 1.25, 1.00, and 0.125 mm, respectively.

Results

Figures 2, 3, and 4 summarize the results of the testing program and the analytical predictions for Nylon 6/6 with tungsten powder, aluminum flake, and carbon fiber filler, respectively. A second-order polynomial curve was used to illustrate the prediction data. The results presented in these figures apply to the molded sheet material plane and at a nominal temperature of 40°C. The numerical values of A, B, and ψ used to calculate the predicted thermal conductivity are shown in Fig. 5.

Further testing was performed on the Nylon 6/6 carbon fiber composite material. Figure 6 shows how the thermal conductivity varies with temperature and percentage filler level for both the material plane and transverse directions. These two directions are illustrated in Figure 7.

Conclusion

For all three types of fillers, the thermal conductivity increased as the filler level was increased. This was expected since the thermal conductivity of all three fillers is much greater

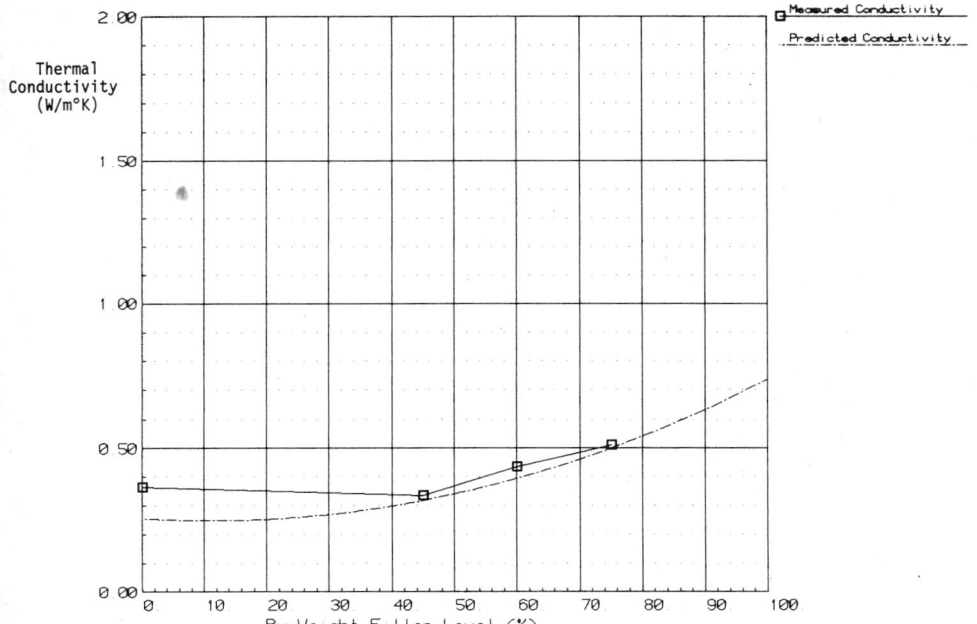

FIG. 2—*Nylon 6/6 with tungsten powder filler.*

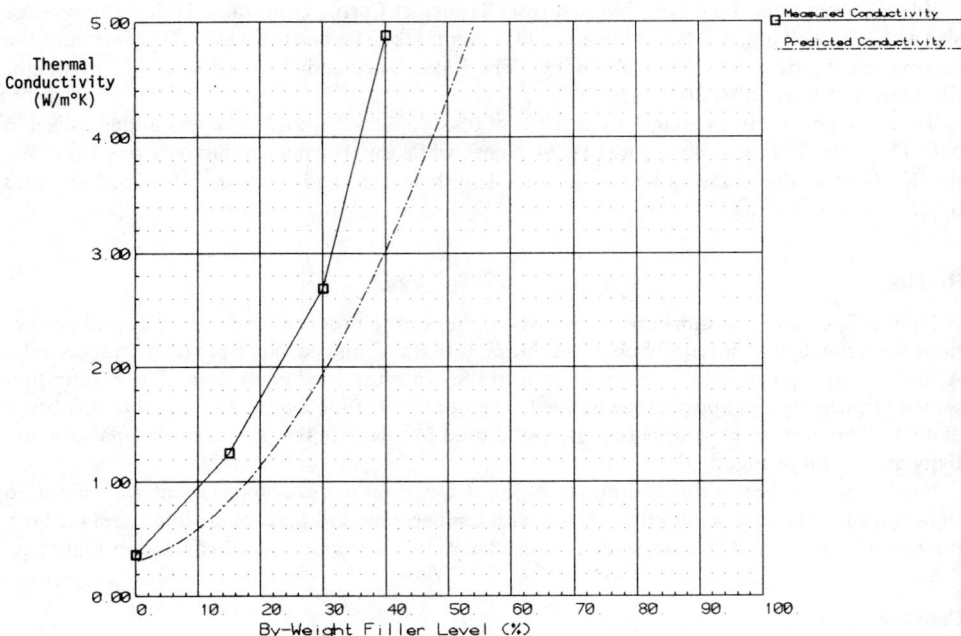

FIG. 3—*Nylon 6/6 with aluminum flake filler.*

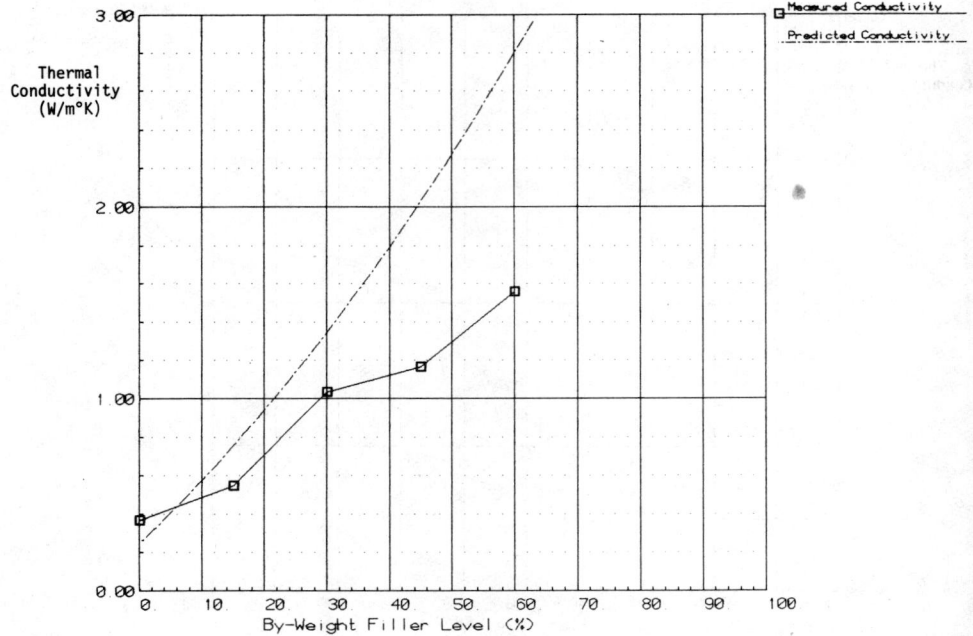

FIG. 4—*Nylon 6/6 with carbon fiber filler.*

FILLER MATERIAL	A	B	ψ *
Tungsten Powder	4.93	0.991	1.044
Aluminun Flake	36.0	0.959	1.270
Carbon Fiber	396.0	0.123	1.019

* Values at 30% by-weight filler level are shown.
ψ = 1 + nφ, where n is a function of the material
maximum packing fraction.

FIG. 5—*Numerical values of prediction parameters.*

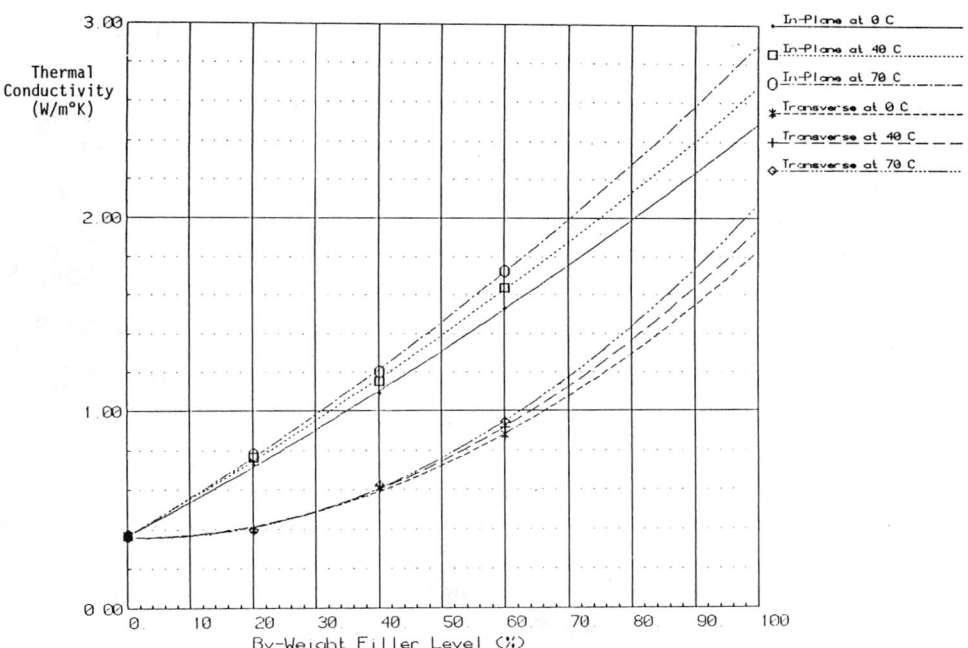

FIG. 6—*Nylon 6/6 with carbon fiber filler.*

than that of the matrix. However, the rate of increase varied widely, from nearly 0% for the tungsten powder to 1229% for aluminum flakes at the 40% filler level. Although part of this variance is due to the difference in thermal conductivity between tungsten and aluminum, it is also due to the different filler shapes.

It has been shown that the thermal and electrical conductivity of a two-phase composite material increases dramatically when the conductive filler particles are able to form an interconnecting network [5]. The filler level at which this occurs decreases as the aspect ratio of the filler particle increases. Therefore, a high aspect ratio filler particle, such as a flake or fiber, will

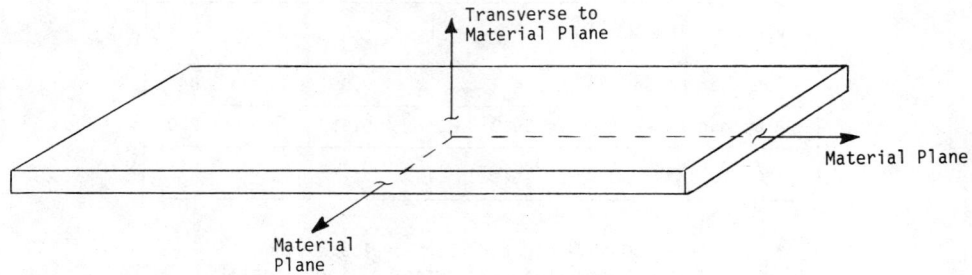

FIG. 7—*Measurement directions.*

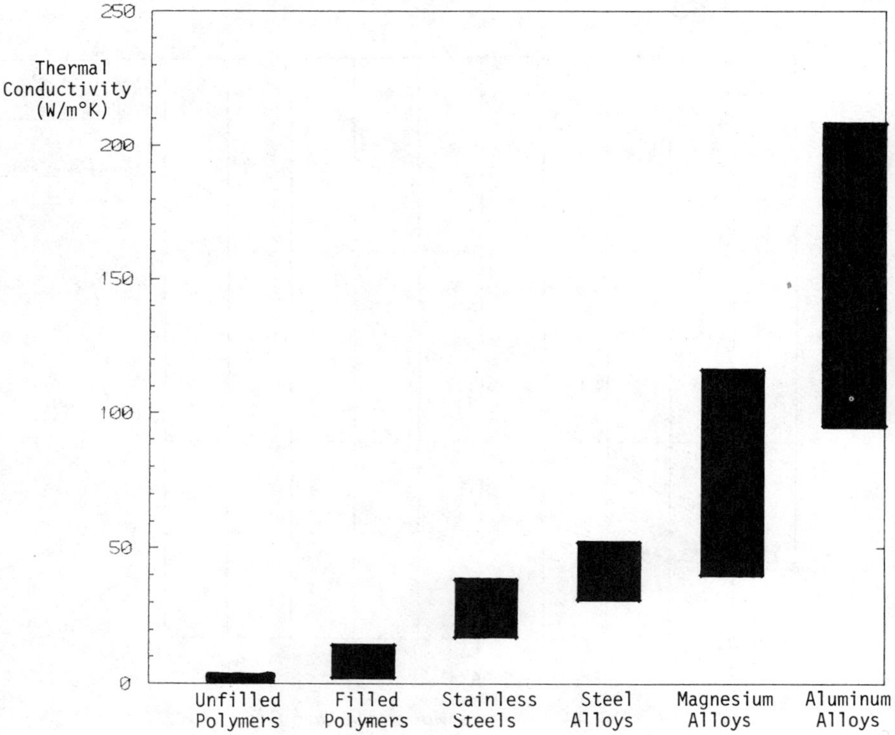

FIG. 8—*Representative thermal conductivities.*

increase the composite thermal conductivity much more effectively than a low aspect ratio filler.

The testing performed on the carbon fiber composite shows that the variation of thermal conductivity with temperature is small. However, the direction in which the conductivity was measured is important. The conductivity in the material plane is about twice the conductivity perpendicular to the material plane. This result is caused by the injection molding process, in which a resin-rich layer is formed next to the mold. Since heat must be conducted through this layer before reaching the conductive fiber network, the conductivity of the composite is sharply reduced in that direction.

The prediction method did not perform as well as expected. Predicted conductivities were significantly higher than measured values for PAN carbon fiber filler levels of 30% and higher. The conductivity of aluminum flake was consistently underestimated over the entire filler range. These discrepancies are probably due to the actual filler particle orientation being somewhat ordered, while the assumption of randomly oriented particles was used in the prediction. In addition, the particle shape can be changed during handling processes such as mixing and injection into the mold. Until more expertise is gained, this prediction method should only be used to generate rough estimates of the actual thermal conductivity.

It has also been concluded that thermoplastic matrix composites, while much more conductive than the matrix material, are still much less conductive than other materials used in electronics packaging. Figure 8 shows typical thermal conductivities of several common materials. However, despite these low conductivities, composite materials can perform as satisfactorily as electronic equipment cases, depending on the application.

For example, a typical hand-held radio may have a surface area of 0.065 m^2, a case thickness of 2.54 mm, and a total power dissipation of 50 W. For a die-cast aluminum case, the temperature rise across the case is 0.02°C. Switching to a carbon fiber composite case will increase this temperature rise to 1.0°C. However, a 1°C temperature rise is acceptable when compared to a total rise of 50°C between the ambient air and a component junction temperature. Note that an all plastic case would have a temperature rise across the case of 5°C.

Thus, thermoplastic matrix composite materials, when filled properly, can be used for electronic equipment cases. Many other factors, such as impact resistance, strength, and electromagnetic shielding, need to be considered. However, from a thermal conductivity standpoint, this program has shown that thermoplastic composite materials have the potential of successful implementation into military electronic enclosures.

References

[1] Progelhof, R. C., Throne, J. L., and Ruetsch, R. R., *Polymer Engineering and Science,* Vol. 16, No. 9, September 1976, p. 620.
[2] Nielson, L. E., *Industrial Engineering Chemical Fundamentals,* Vol. 13, No. 1, 1974, p. 18.
[3] Nielson, L. E., *Journal of Applied Polymer Science,* Vol. 17, 1973, p. 3819.
[4] Mirkovich, V. V., *Journal of the American Ceramic Society,* Vol. 48, 1965, p. 387.
[5] Bigg, D. M., *Polymer Engineering and Science,* Vol. 19, No. 16, December 1979, p. 1189.

Frederick A. Myers[1]

Stress-State Effects on the Viscoelastic Response of Polyphenylene Sulfide (PPS) Based Thermoplastic Composites

---

REFERENCE: Myers, F. A., "**Stress-State Effects on the Viscoelastic Response of Polyphenylene Sulfide (PPS) Based Thermoplastic Composites,**" *Advances in Thermoplastic Matrix Composite Materials, ASTM STP 1044*, G. M. Newaz, Ed., American Society for Testing and Materials, Philadelphia, 1989, pp. 154–182.

ABSTRACT: The viscoelastic properties of polyphenylene sulfide resin based carbon fiber-reinforced composites have been examined using torsion, centrosymmetric disk (CSD), and three-point and four-point flexure testing modes. Laminates fabricated with carbon fabric and a quasi-isotropic layup of continuous carbon fiber have been considered. Dynamic mechanical tests have shown that these materials exhibit a glass transition interval (T_g) from 50 to 150°C and a melting transition near 280°C. The storage component of the modulus or stiffness decreases from room temperature to 175°C by an amount which varies with the imposed stress or strain state. Fiber-dominated modes (three-point and four-point flexure) exhibit a smaller percentage loss in modulus or stiffness than resin-dominated modes (torsion and CSD). Arrhenius plots obtained from frequency sweep data in the glass transition region indicate the fabric-reinforced material and quasi-isotropic layup have activation energies of 130 and 105 kcal, respectively, independent of testing mode. Superposition of frequency/temperature or time/temperature data was achieved in dynamic mechanical and transient (that is, creep and relaxation) tests, respectively. Projection of the ten-year modulus, stiffness, or compliance from transient tests suggests that the response is stress-state dependent with greater change observed for resin-dominated testing modes.

KEY WORDS: carbon fiber composites, thermoplastic composites, composites, viscoelasticity, creep, relaxation, dynamic mechanical testing, time-dependent properties, time-temperature superposition, frequency dependence, glass transition temperature

Due to their processability, damage tolerance, and chemical and moisture resistance, thermoplastic resin based fiber-reinforced advanced composites have received considerable attention during the last few years [1–5]. In many applications, these materials are currently being evaluated as replacements for composites containing thermosetting matrices such as epoxies and polyimides.

However, thermoplastic-based composites present unique challenges to engineers intent on utilizing these materials for long-term structural applications. For example, compared to amorphous thermosetting resins which are utilized below their glass transition zones, thermoplastics can display varying degrees of crystallinity and relatively low glass transitions that need to be considered when attempting to predict response over long periods of time. In particular, there is a question as to the extent the time-dependent processes associated with the glass transition limit utilization as engineering materials.

One of the most characterized thermoplastic materials has been polyphenylene sulfide (PPS). Thermal, mechanical, chemical, and rheological properties have been reported for composites fabricated using PPS resin [6–11]. In this report, the viscoelastic properties of carbon fiber-

---

[1]Senior staff engineer, Armco Research Center, Middletown, OH 45043.

reinforced PPS laminates will be discussed. Effects of thermal transitions and imposed stress or strain state will be examined. The intent is to indicate that the behavior of thermoplastic-based composites is more complex than conventional thermoset-based composites and that the individual characteristics of the base resins need to be included in design decisions.

Procedure

The materials examined were removed from nominal 1.7-mm-thick plain weave carbon fabric-reinforced Ryton laminates supplied by Phillips Petroleum Co. (referred to as Ryton laminate in text) and from 24-ply $[\pm 45, 0, 90, \pm 45]_{2S}$ approximately 4-mm-thick continuous carbon fiber-reinforced polyphenylene sulfide based laminates (referred to as PPS laminate in text). The fiber volume content in the fabric laminate was approximately 55% and the continuous fiber laminate 60%. Micrographs showing the cross sections of both materials are presented in Figs. 1 and 2.

The thermal response of the composites was examined with a Perkin Elmer DSC-2 differential scanning calorimeter (DSC) in an air atmosphere at a heating rate of 10°C/min. Viscoelastic analyses were conducted using a 7700 Rheometrics Dynamic Spectrometer (RDS) 7700 in the forced torsion mode with rectangular samples approximately 35 by 12 mm by the specimen thickness. In addition, flexure tests were performed using a DynaStat Viscoelastic Analyzer with three-point (41.3 by 12.5 mm), four-point (48 by 12.5 mm), and centrosymmetric disk (25 mm diameter) geometries. Both viscoelastic instruments have temperature chambers which permit controlled heating and cooling.

The Ryton fabric-reinforced samples were placed in the RDS or DynaStat with the warp fibers oriented in the length direction with the fill fibers being oriented at 90° in the width direction. Similarly, the quasi-isotropic PPS continuous fiber laminate was tested with the 0° plies oriented in the length direction and the 90° plies having the width orientation.

Data for the dynamic mechanical tests (that is, constant applied sinusoidal deformation or load) on the RDS are presented in terms of G' and G'', the storage (elastic) and loss (viscous) components, respectively, of the complex shear modulus, G^* (that is, shear stress/shear strain), whose magnitude is related to G' and G'' by

$$G^* = [(G')^2 + (G'')^2]^{1/2}$$

Similarly, three-point and four-point flexure DynaStat data are given by the storage, M', and loss, M'', components of the complex flexural modulus, M^*.

Transient relaxation (that is, constant deformation) data from the RDS are presented as the shear relaxation modulus $G(t)$, which is the ratio of the shear stress to shear strain. DynaStat results are expressed in terms of $M(t)$ for relaxation or $D(t)$ for creep (that is, constant load), which represent the flexural modulus (stress/strain) or compliance (strain/stress), respectively.

The results for the torsional and flexure tests are presented as moduli or compliances calculated for the particular geometry assuming isotropic material properties. In the case of the centrosymmetric disk (CSD) tests, the data are given in terms proportional to the stiffness of the material. Values for the storage component and loss component of the complex stiffness are presented as E' and E'', respectively, whereas relaxation data are expressed as $E(t)$.

Typical operating conditions for the RDS during dynamic tests consisted of imposing a nominal shear strain of 0.2% over a frequency range of 1.0 to 100 rad/s. DynaStat dynamic tests were conducted in the load control mode and covered a similar frequency domain. During these tests, the temperature for both instruments was increased in a stepwise fashion. Data were obtained after an induction period at each temperature to allow for equilibration.

Transient relaxation tests were performed by imposing a constant deformation on the sample

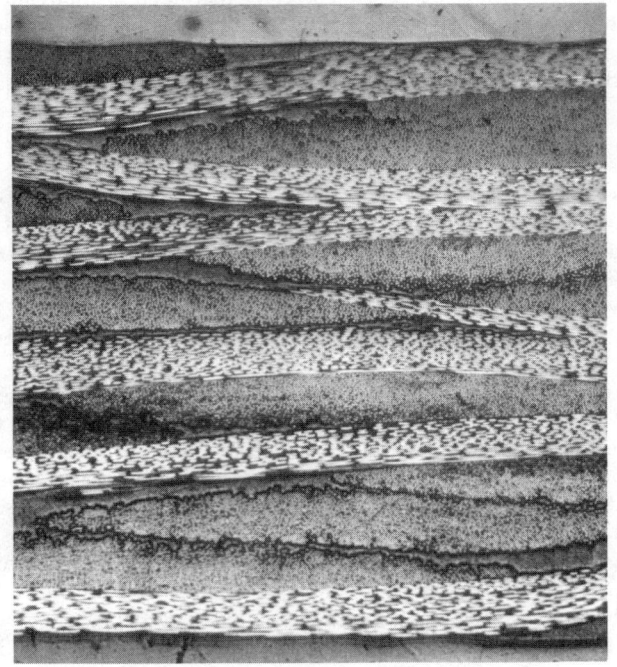

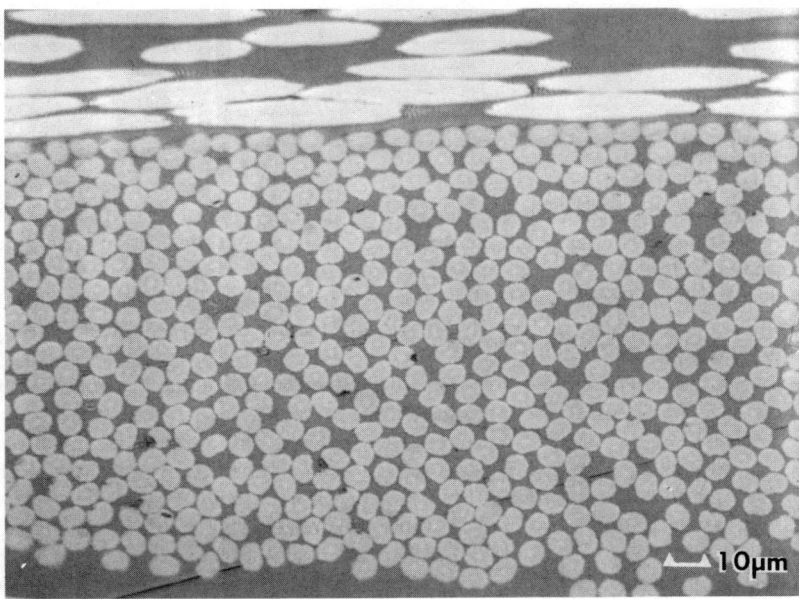

FIG. 1—*Light micrographs of polished cross sections of Ryton carbon fabric reinforced laminate indicating good fiber wetout.*

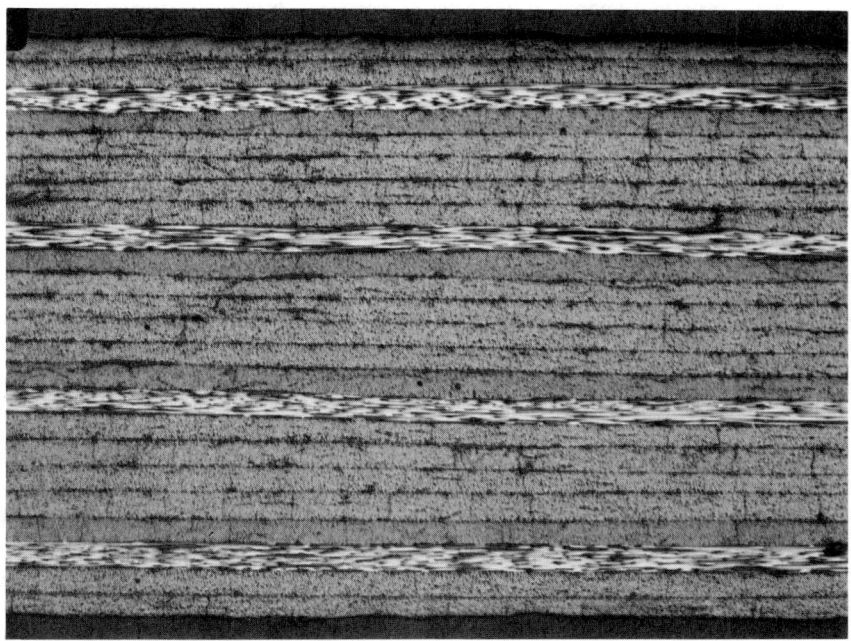

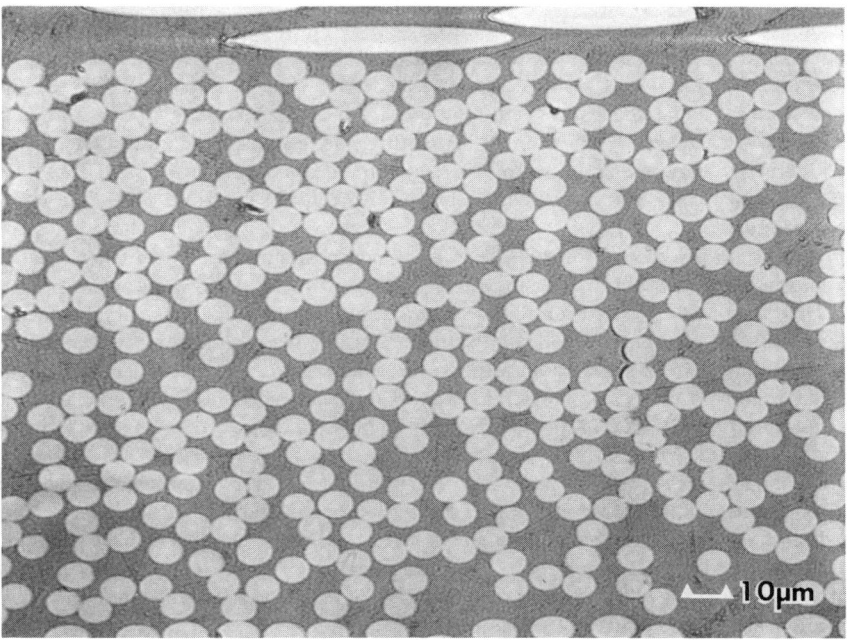

FIG. 2—*Light micrographs of polished cross sections of PPS continuous carbon fiber reinforced laminate indicating good fiber wetout.*

over a range of isothermal temperatures, whereas creep involved imposition of constant loads. Test duration in most cases was 1000 s at each temperature.

Results and Discussion

Dynamic Mechanical Tests

An initial temperature sweep of the as-received Ryton carbon fabric reinforced laminates from room temperature to 300°C appears as shown in Fig. 3. There is a precipitous decrease in the G' storage component of stiffness and sharp maximum in the loss component, G'', near 80°C. With continued heating, G' recovers and G'' passes through a second, smaller maximum near 120°C and then steadily decreases until a final transition is observed near 280°C.

After cooling to room temperature by 5°C/3-min steps, a second scan of the same material appears as shown in Fig. 4. The room temperature moduli are slightly greater than the original values. In addition, the response during heating exhibits a more conventional transition near 90°C and the strong G'' peak seen in Fig. 3 at 80°C is not present. The sharp decrease in stiffness at high temperatures is again observed near 280°C.

This behavior is reflected in the DSC scans presented in Fig. 5. During the initial heating, a glass transition is observed at 86°C followed by a crystallization exotherm near 120°C. The high temperature melting peak endotherm occurs at 277°C. During reheat, the glass transition region near 88°C is less well defined and the crystallization exotherm is absent. The melting peak is observed at approximately the same temperature (276°C) that was recorded during the initial heating.

These results indicate that the Ryton fabric laminate as received had a lower level of crystallinity than was present during reheating, presumably a consequence of the processing history imposed by Phillips. It is known that PPS crystallizes readily at temperatures above T_g [12]. This is most likely what occurred during initial heating and was responsible for the recovery in G' and maximum in G'' near 120°C seen in Fig. 3. Once crystallized, the response appears relatively stable below 175°C as indicated by the results seen in Fig. 6. In this case, the as-received sample was heated from room temperature to 175°C, cooled (5°C/3-min steps), reheated, and finally recooled. After the initial heating, the response on the subsequent cycles was nearly identical. Therefore, in order to assure adequate crystal formation, all subsequent testing on Ryton fabric laminates was conducted after an initial heat-and-cool cycle from room temperature to 175°C.

Temperature sweep results for the Ryton fabric composite during reheat (that is, after crystallization) at 1.0 rad/s obtained using three-point flexure on the DynaStat are presented in Fig. 7. Note that while a transition region is apparent over the same interval as determined in the RDS torsional results for a reheated sample (Fig. 4), the magnitude of the change in the storage component, M', from room temperature to 175°C is much less ($\sim 20\%$ decrease in flexural storage M' versus a 75% decrease in torsional storage G'). This response, which was similar for both three-point and four-point loading, is associated with the flexure testing being a more fiber-dominant mode of deformation.

An intermediate case is shown in Fig. 8 where the sample was tested as a centrosymmetric disk (CSD). Note that the glass transition region is observed over a slightly higher temperature range than the other tests. This is probably associated with lower rates of heat transfer in the CSD fixture. The decrease in storage E' stiffness component from room temperature to 175°C is approximately 50%. This indicates that the CSD geometry is a less fiber-dominated form of deformation than three-point or four-point flexure.

RDS temperature scans of the as-received (initial) and reheated continuous fiber PPS quasi-isotropic laminate are presented in Fig. 9. This material has glass and melting transitions similar to those observed in Fig. 4 for the fabric-reinforced Ryton samples. However, the G'' loss

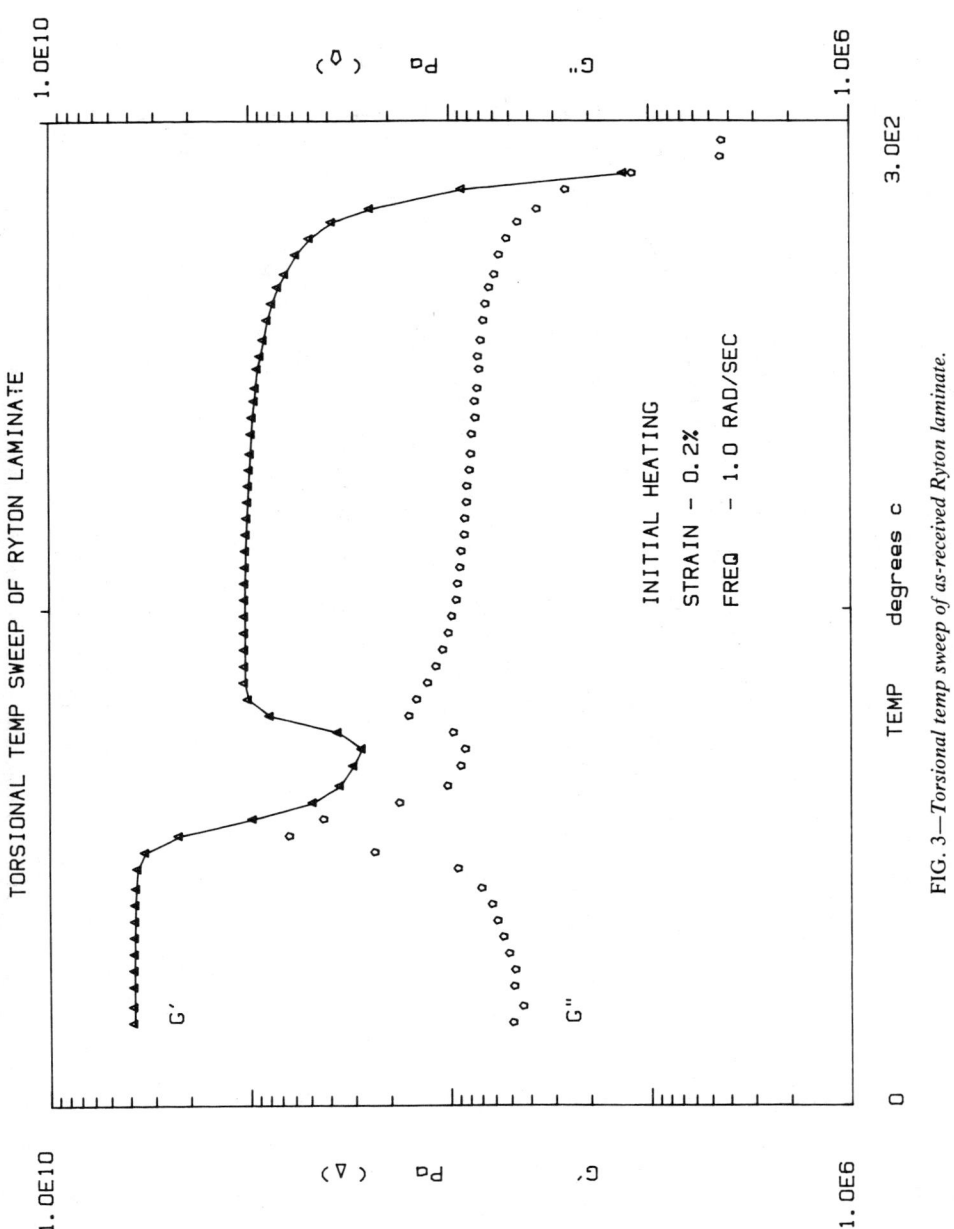

FIG. 3—*Torsional temp sweep of as-received Ryton laminate.*

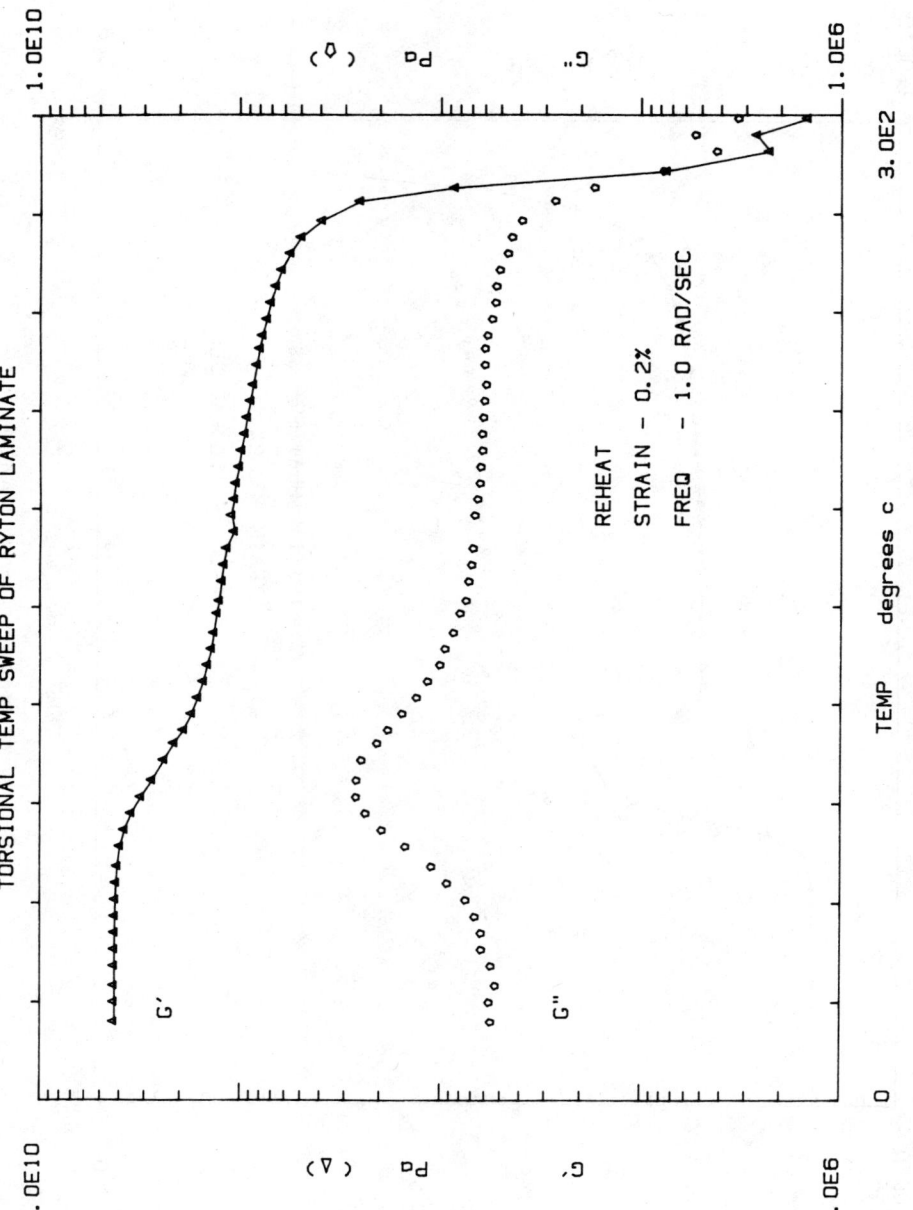

FIG. 4—*Torsional temp sweep of Ryton laminate during reheat.*

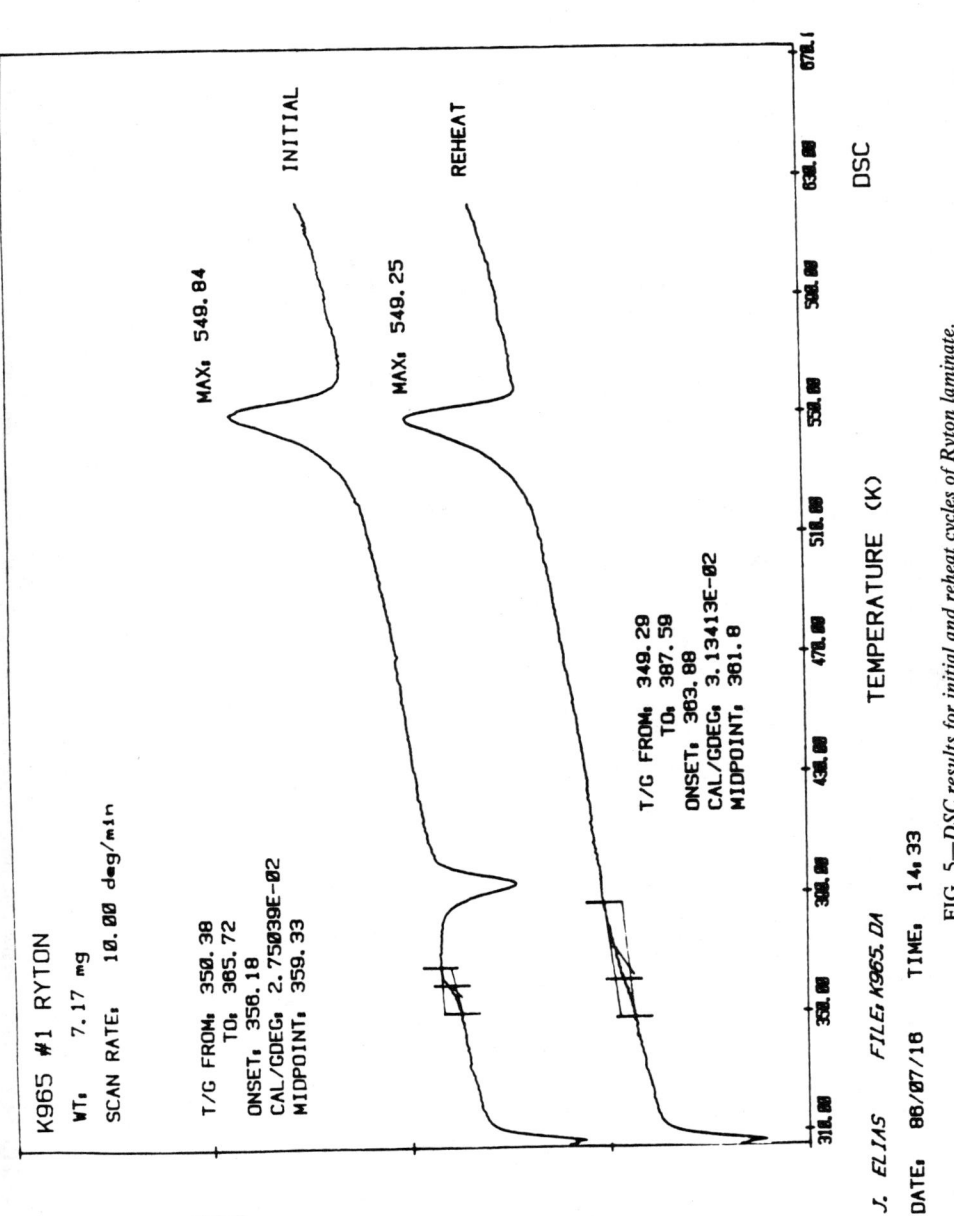

FIG. 5—*DSC results for initial and reheat cycles of Ryton laminate.*

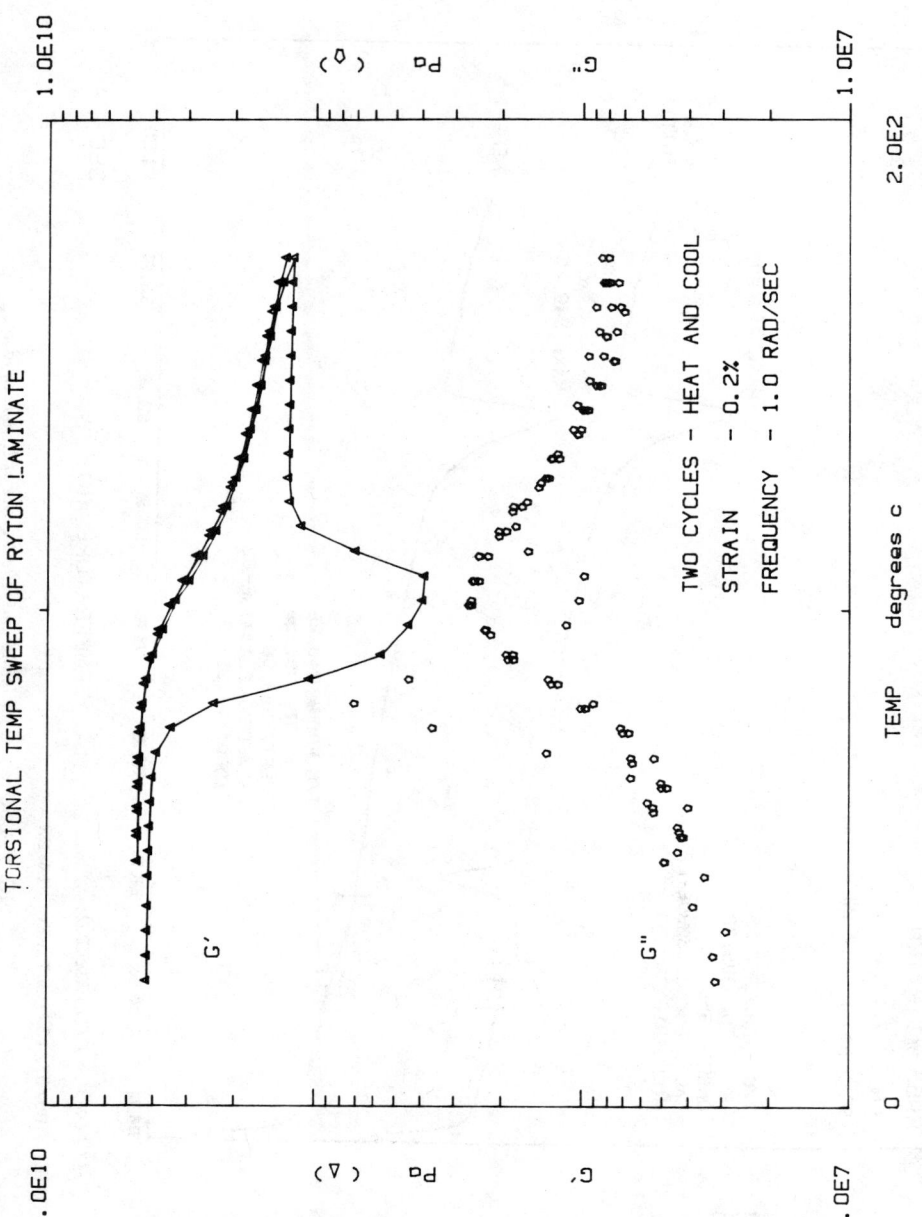

FIG. 6—*Torsional temp sweep of Ryton laminate during two-cycle heat-and-cool sequence in temperature interval 25 to 175°C.*

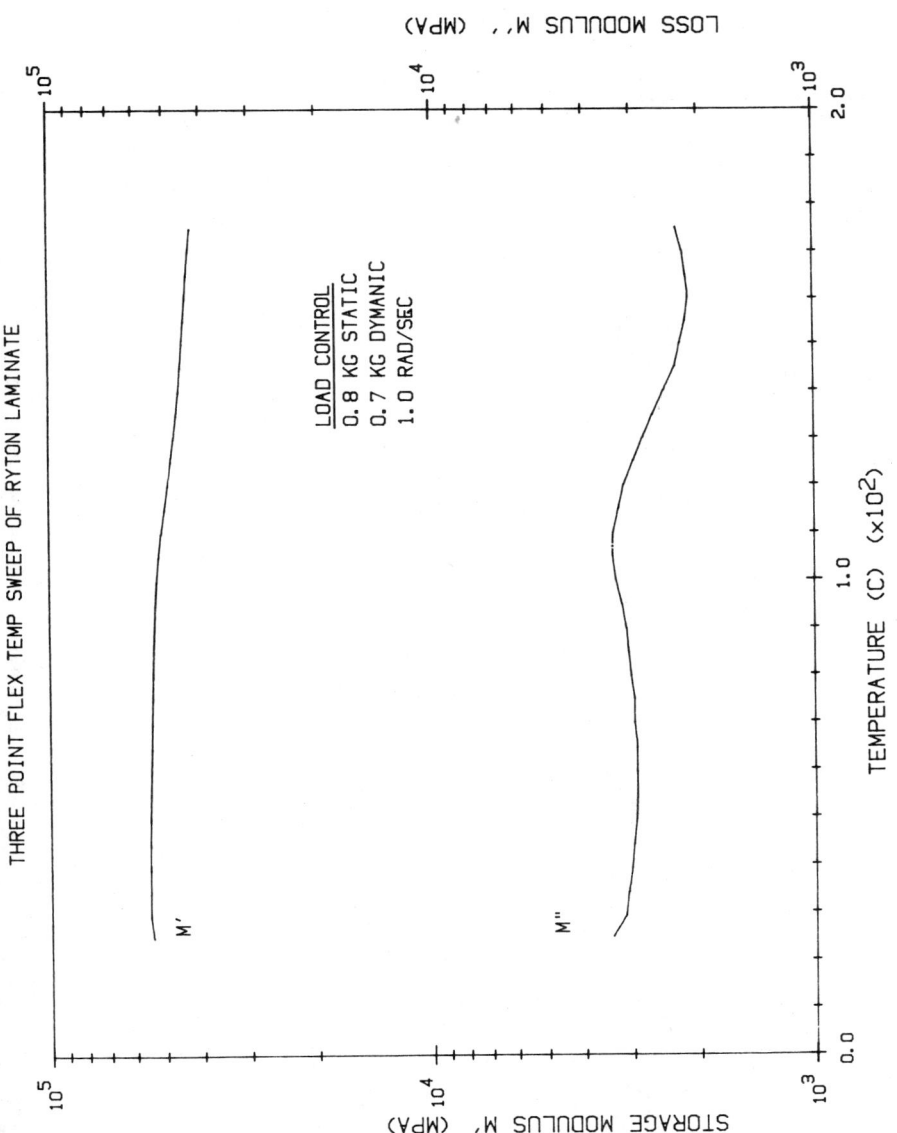

FIG. 7—*Three-point flexure temp sweep of Ryton laminate during reheat.*

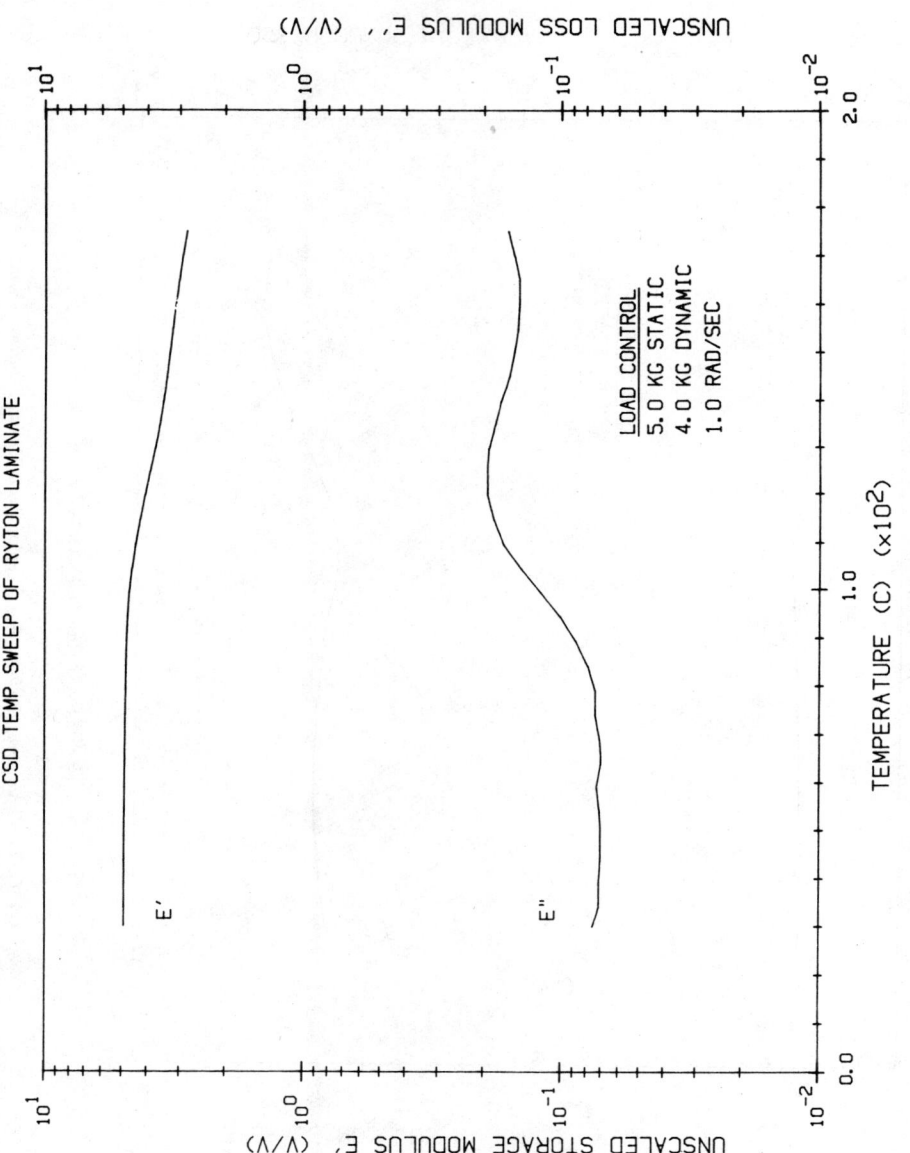

FIG. 8—*CSD temp sweep of Ryton laminate during reheat.*

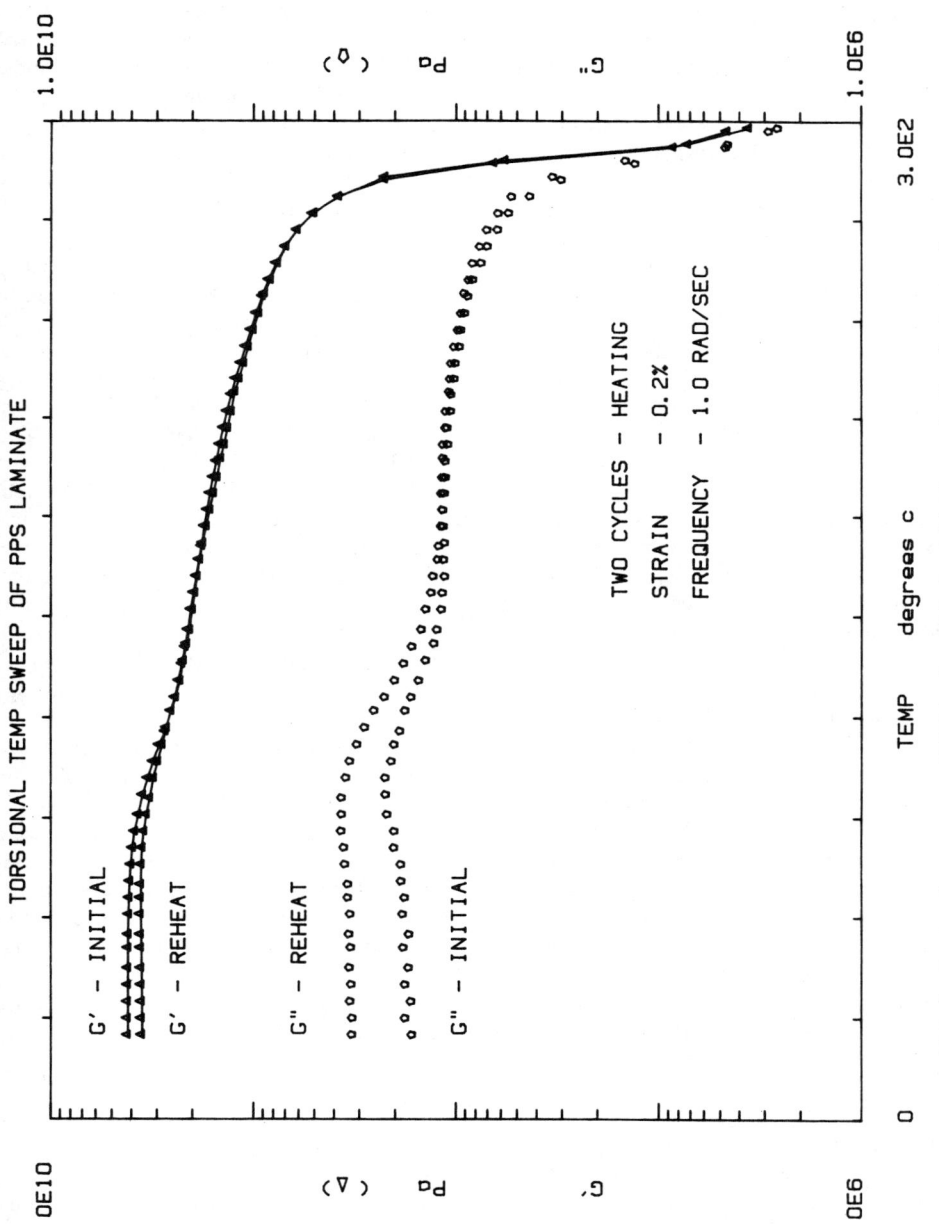

FIG. 9—*Torsional temp sweep of PPS laminated during initial (as-received) and reheat cycles.*

component peak near 105°C is at a slightly higher temperature and is less distinct. The response indicates that the as-received material is highly crystalline (no sharp drop in G' or increase in G'' near 80°C during the initial heating).

The DSC analyses of this sample is presented in Fig. 10. A small exotherm is observed near 115°C in contrast to the more well-defined peak near 120°C in Fig. 5. This result agrees with the RDS data (Fig. 9) and suggests that the crystallinity is higher in the as-received continuous fiber PPS sample than in the as-received Ryton fabric-reinforced composite.

Three-point flexural test results from room temperature to 175°C for this material are presented in Fig. 11. The response is similar to the RDS data in that the glass transition area is centered near 110°C. Percentage loss in the magnitude of M' from room temperature to 175°C is 30% for the three-point flexural data compared to approximately 55% for the torsional G' results seen in Fig. 9.

A similar scan of the continuous fiber PPS laminate on the DynaStat using four-point loading is presented in Fig. 12. Compared to the three-point loading results (Fig. 11), a smaller decrease in storage modulus, M', is observed (20 versus 30%). This is to be expected in that the sample is subjected to pure bending between the center loading noses in the four-point test. Therefore, shear is suppressed in this region. Consequently, the response in four-point loading is expected to be more fiber dominated, and transitions associated with the resin system should be less noticeable relative to the more resin-dominated test modes (that is, torsion and CSD). This result is consistent with that observed for the fabric-reinforced Ryton laminate.

In addition to constant frequency tests, both the Ryton and PPS laminates have been subjected to frequency sweeps at temperatures in the glass transition region (50 to 150°C). A typical result for the G' and G'' components of Ryton fabric-reinforced composite tested in the torsional mode are shown in Fig. 13. As expected, higher frequencies shift the curves in the direction of higher temperatures. This data can be cross plotted, such as shown in Fig. 14, for the G'' component to present the results as a function of frequency at constant temperature. Individual curves can then be shifted along the frequency axis to form a master curve covering 17 decades, such as presented in Fig. 15. A similarly obtained master curve for the G' storage component is presented in Fig. 16.

Frequency sweep data can also be utilized to determine activation energies for the transitions. Taking the maxima in the G'', E'', and M'' curves as the reference points, the activation energies for the Ryton and PPS samples were determined for the various geometries from the slopes of the Arrhenius plots shown in Figs. 17 and 18. The values for the Ryton laminate were approximately 130 kcal, whereas those for the PPS material are near 105 kcal. These are considered accurate to ±10 kcal. Note that there is agreement between the various test conditions, suggesting that the same mechanisms are responsible for the observed transitions. The shift in data to higher temperatures for the CSD geometry is believed to be due to the lower thermal transport of the test fixture. This results in a greater temperature differential between the sensing probe and sample than is observed for the other test conditions and possibly temperature gradients within the sample itself.

Transient Tests

Relaxation and creep tests were performed on the fabric-reinforced Ryton and continuous fiber-reinforced PPS laminates using the torsion, flexure, and CSD geometries. Typical results are presented in Fig. 19. The isothermal curves were obtained by subjecting samples to a constant torsional deformation of 0.2% and monitoring the decrease in torque over a period of 1000 s. The temperature range (50 to 150°C) coincides with the glass transition zone of the material as determined by the dynamic mechanical tests discussed earlier.

Once these data are obtained, the individual curves can be shifted relative to a reference curve to form a master curve such as shown in Fig. 20. These results indicate that with the imposed

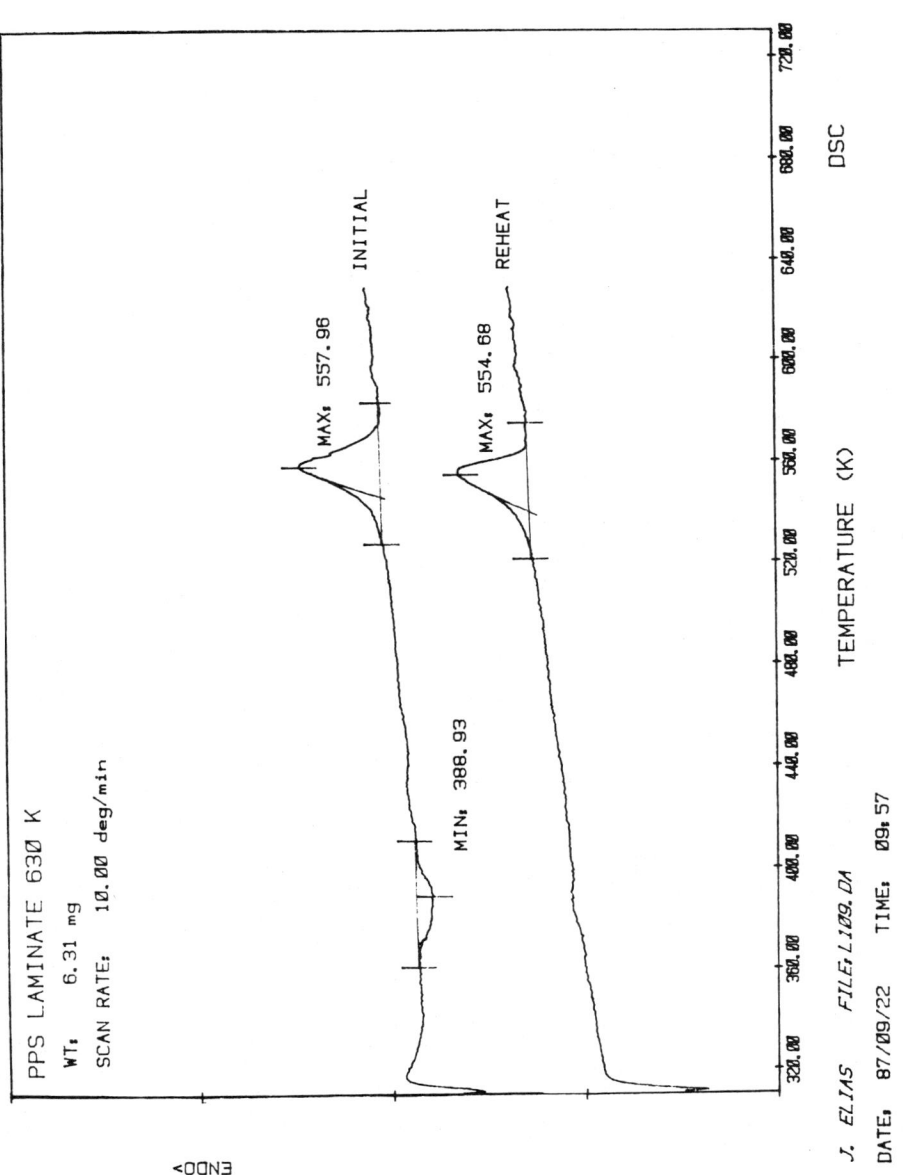

FIG. 10—*DSC results for initial and reheat cycles for PPS laminate.*

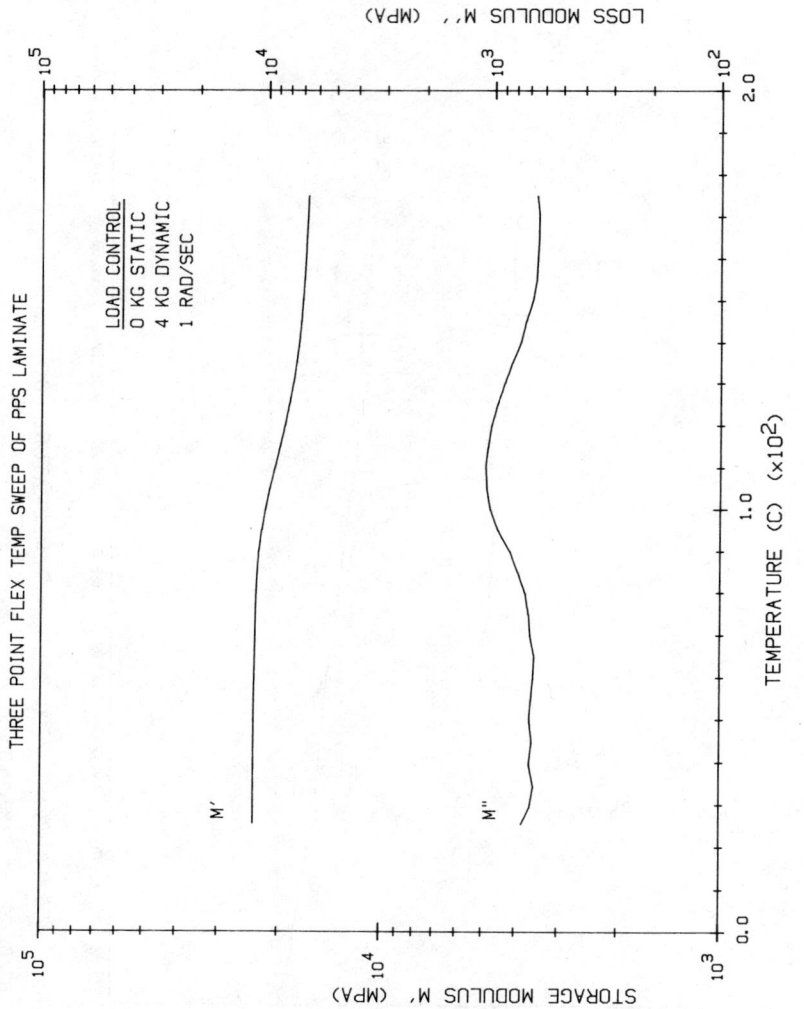

FIG. 11—*Three-point flex temp sweep of PPS laminate.*

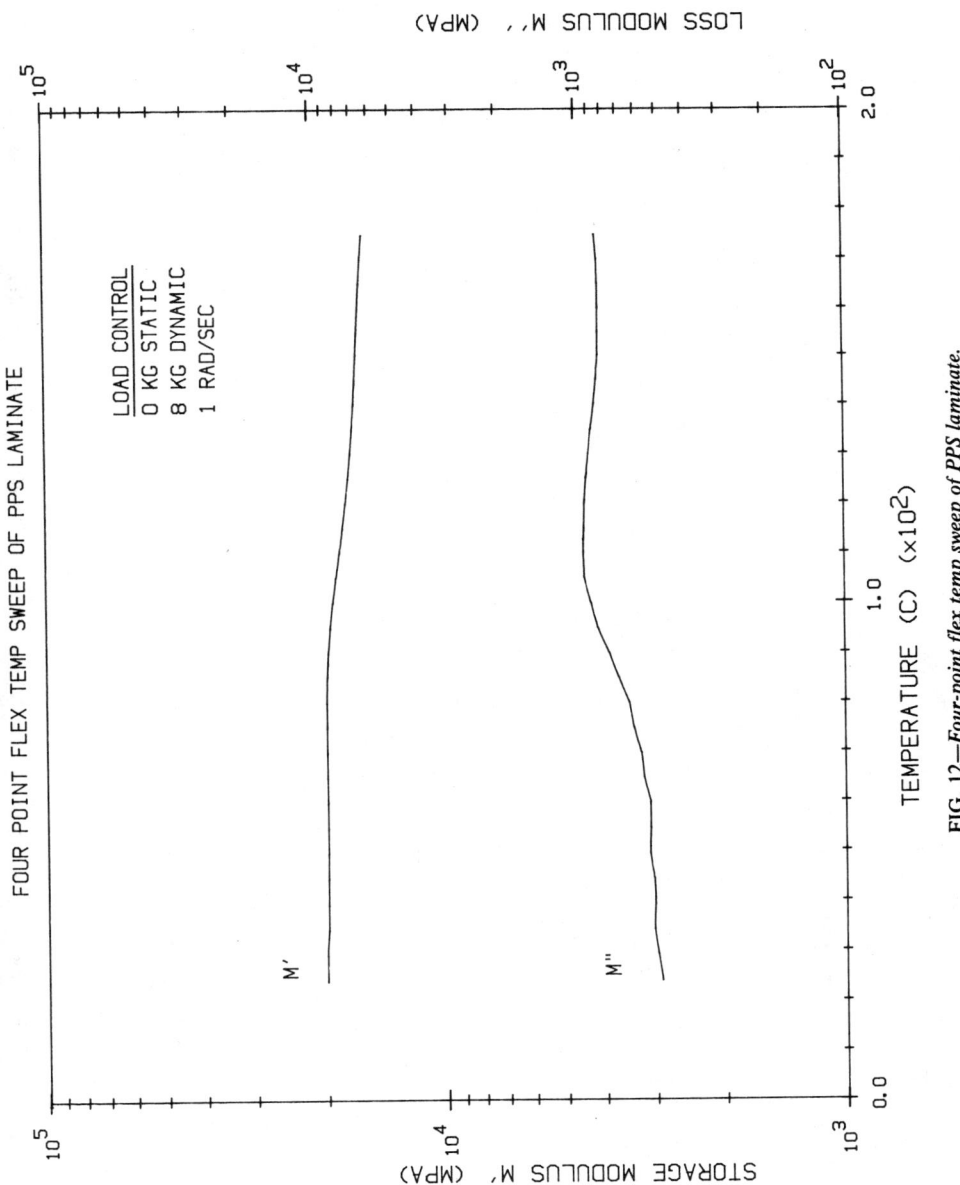

FIG. 12—*Four-point flex temp sweep of PPS laminate.*

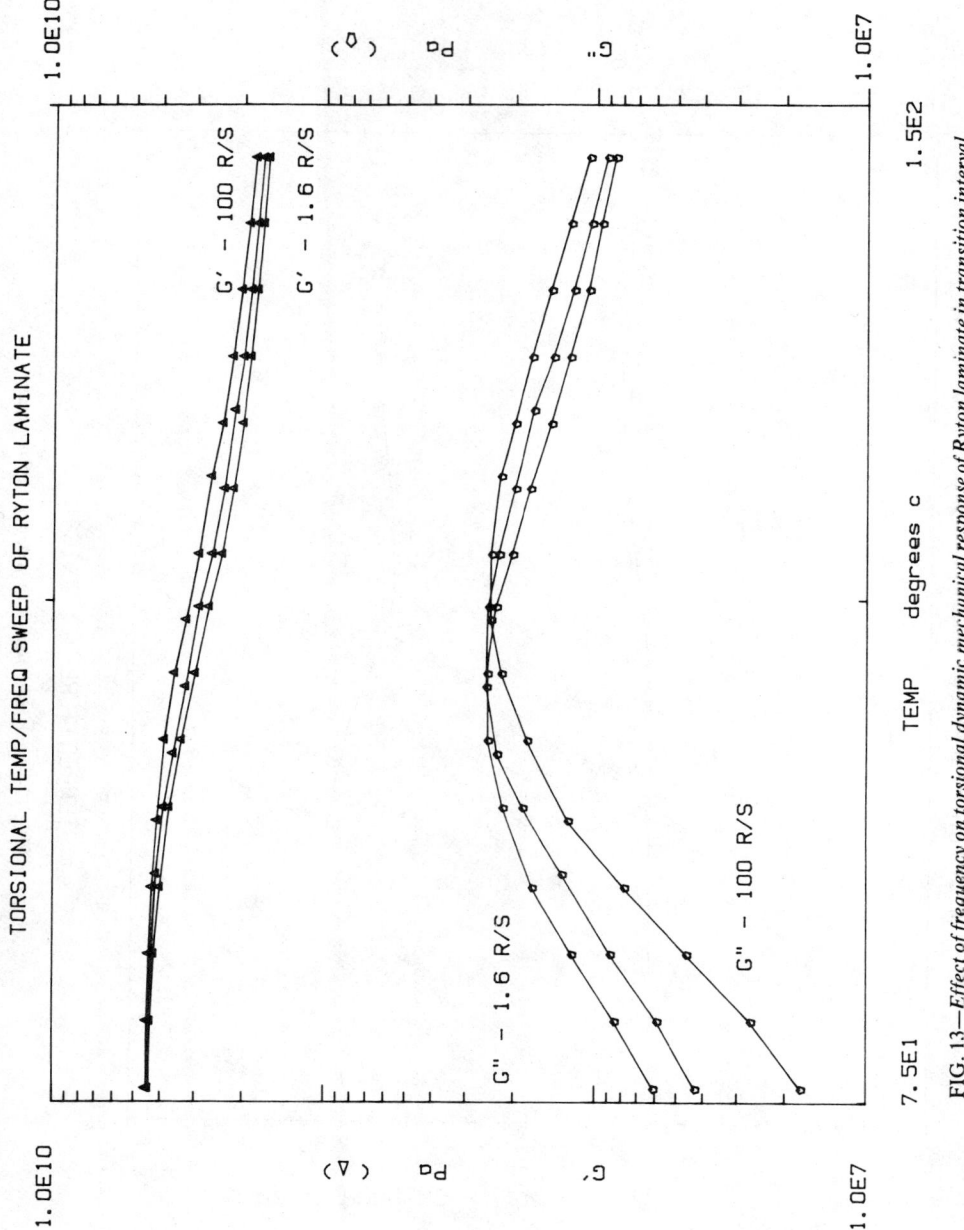

FIG. 13—*Effect of frequency on torsional dynamic mechanical response of Ryton laminate in transition interval.*

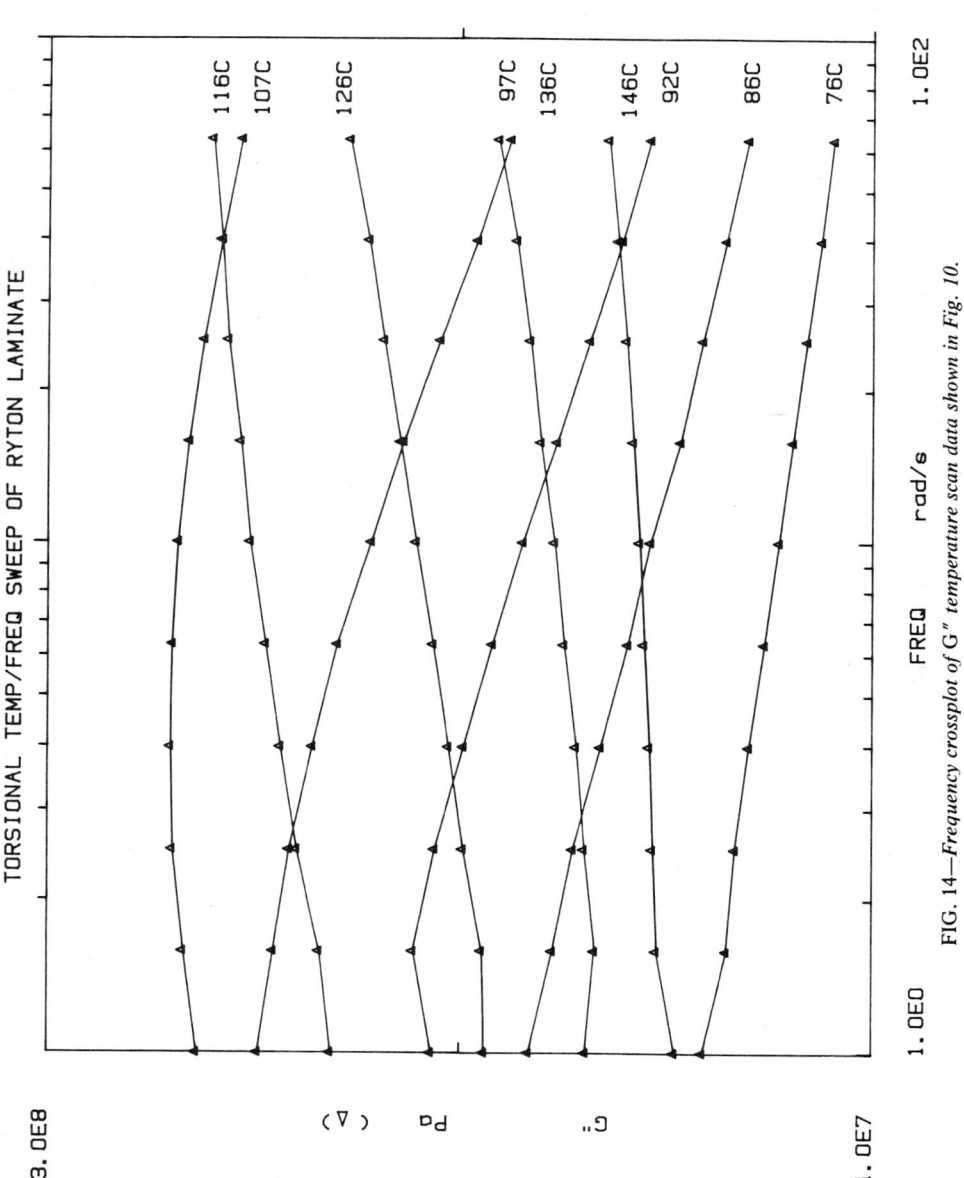

FIG. 14—*Frequency crossplot of G" temperature scan data shown in Fig. 10.*

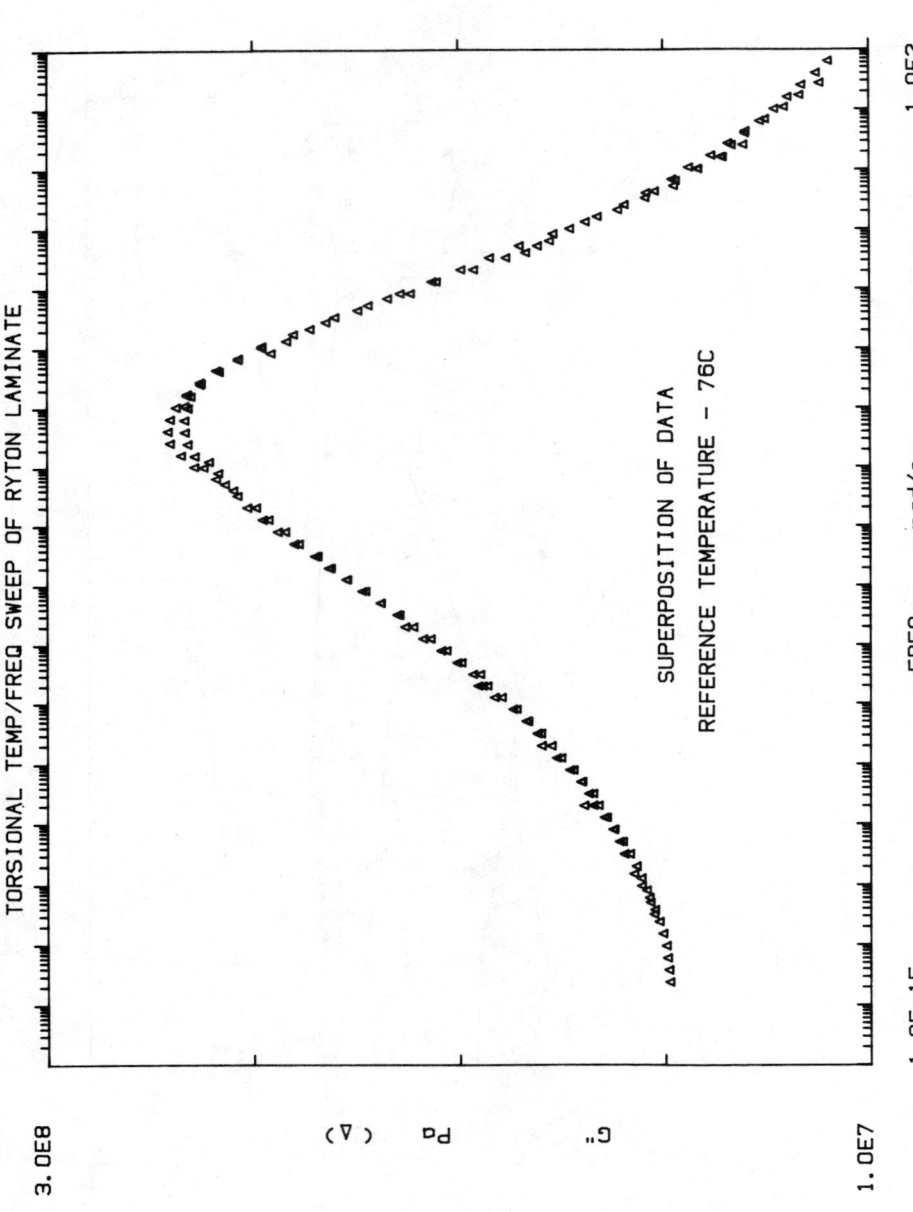

FIG. 15—Master curve obtained by shifting G" curves in Fig. 11 along frequency axis relative to reference curve at 76°C.

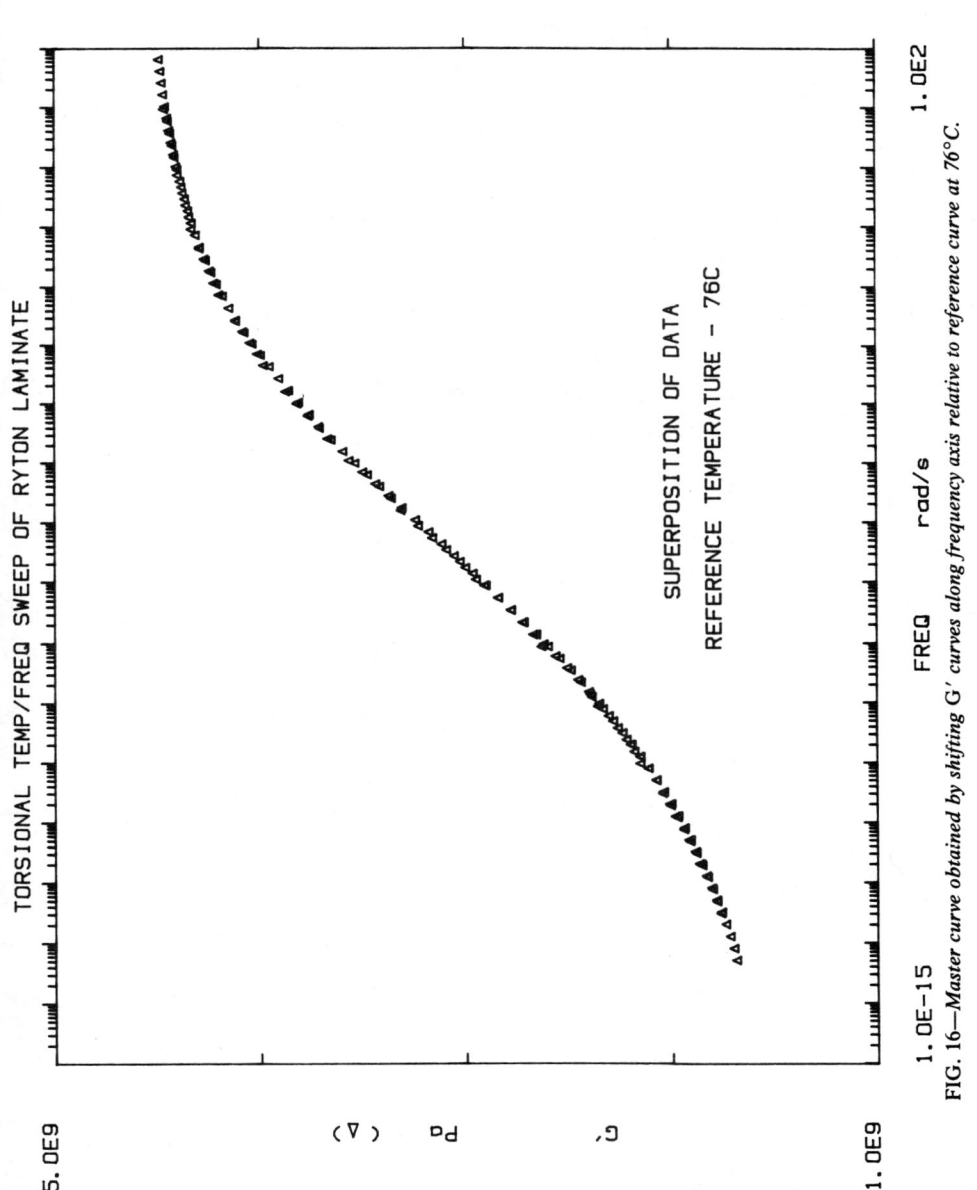

FIG. 16—*Master curve obtained by shifting G′ curves along frequency axis relative to reference curve at 76°C.*

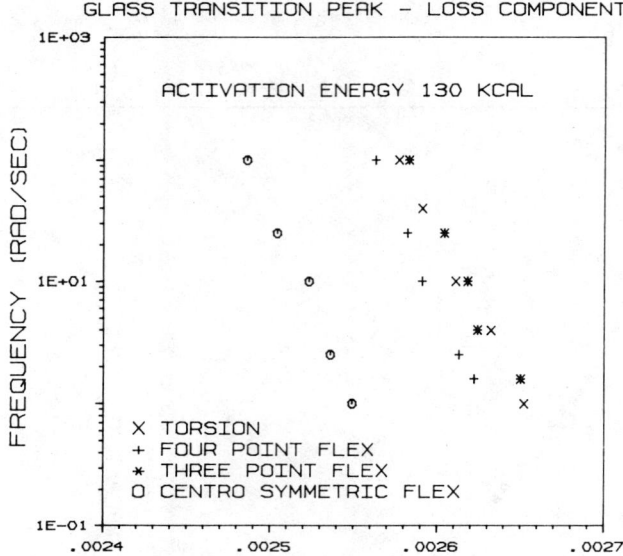

FIG. 17—*Arrhenius plot of peak in loss component curve for Ryton laminate.*

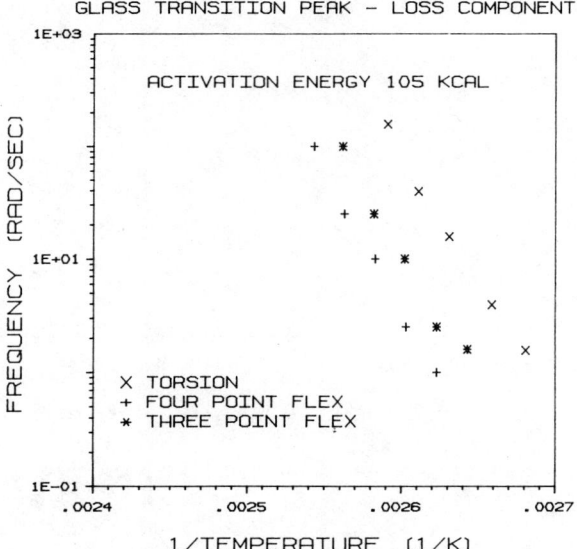

FIG. 18—*Arrhenius plot of peak in loss component curve for PPS laminate.*

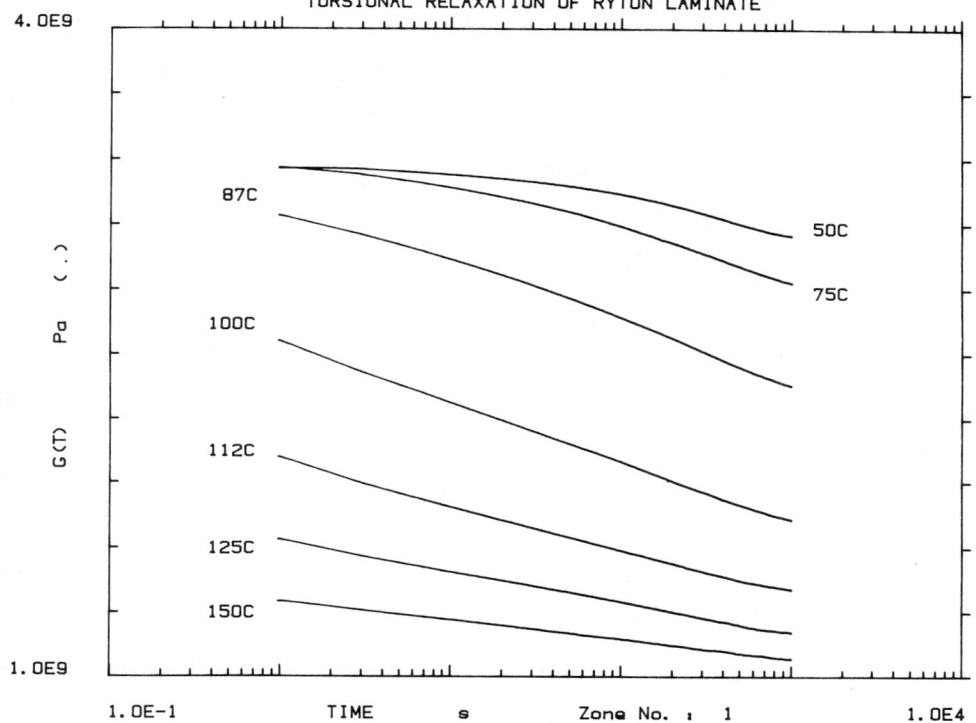

FIG. 19—*Torsional relaxation results for Ryton laminates.*

constant 0.2% strain, over a period of 10 years (3×10^8 s) the shear modulus (or stiffness) for this material should relax approximately 50% from its initial, short-term value at the reference temperature of 75°C.

When the fabric-reinforced Ryton laminate was tested in three- and four-point flexure using a constant deformation of 0.037 and 0.05 mm, respectively, the relaxation master curves shown in Fig. 21 were obtained. The results indicate that the degree of relaxation is less than that observed in the torsion case. With a reference temperature of 75°C, the amount of relaxation over the ten-year interval is approximately 25% for three-point flexure and 20% for four-point flexure compared to 50% for the torsion case.

Similar data obtained in the centrosymmetric mode for the Ryton laminate are shown in Fig. 22. Test conditions consisted of applying a 0.1-mm deformation to the circular disk. The stiffness decreased slightly less than 40% over the 3×10^8 s interval. Better temperature agreement is expected between the CSD geometry and other testing modes during transient loading since the times involved for equilibration between temperature steps is considerably longer (hours versus minutes).

The ranking of the torsion, three-point and four-point flexure, and CSD data with respect to the magnitude of the loss in modulus or stiffness as determined by the superposition process is similar to changes observed in the storage components of the stiffnesses during dynamic mechanical testing.

Creep tests were also performed on the Ryton fabric-reinforced laminates in the three-point

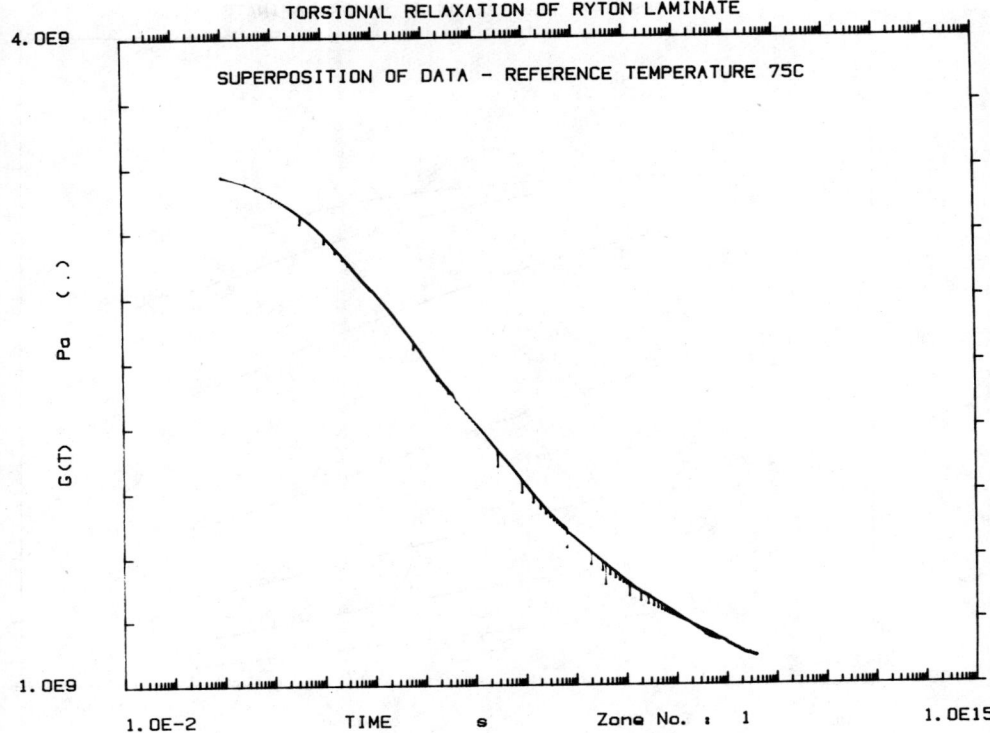

FIG. 20—*Master curve obtained by shifting isothermal relaxation curves in Fig. 19 along time axis relative to reference curve at 75°C.*

and four-point flexure and CSD modes. Flexural results are presented in Fig. 23. As observed in the case of stress relaxation, the amount of long-term creep as determined by time/temperature superposition was greater in three-point flexure than four-point flexure. CSD test results during creep were greater than those measured for three-point flexure.

Relaxation and creep data were also obtained for the PPS quasi-isotropic laminates in the four-point flexural mode. Results are shown in Fig. 24. As can be seen, the time/temperature superposition indicates that the magnitude of compliance increase and modulus decrease is nearly the same with slightly greater long-term change observed in the creep tests.

Conclusions

These viscoelastic analyses indicate that the Ryton and PPS thermoplastic composites discussed previously exhibit melting points near 280°C and glass transition temperatures near 100°C. Even though the materials can exhibit a high degree of crystallinity, the presence of the glassy phase and onset of segmental mobility near T_g can result in significant time-dependent phenomena in this temperature region.

Tests performed in torsion, three-point and four-point flexure, as well as centrosymmetric flexure have shown that the amount of modulus decrease in the T_g interval (50 to 150°C) or degree of change in stiffness or compliance as predicted by time/temperature superposition techniques is stress- or strain-state dependent. As would be expected, fiber-dominated modes

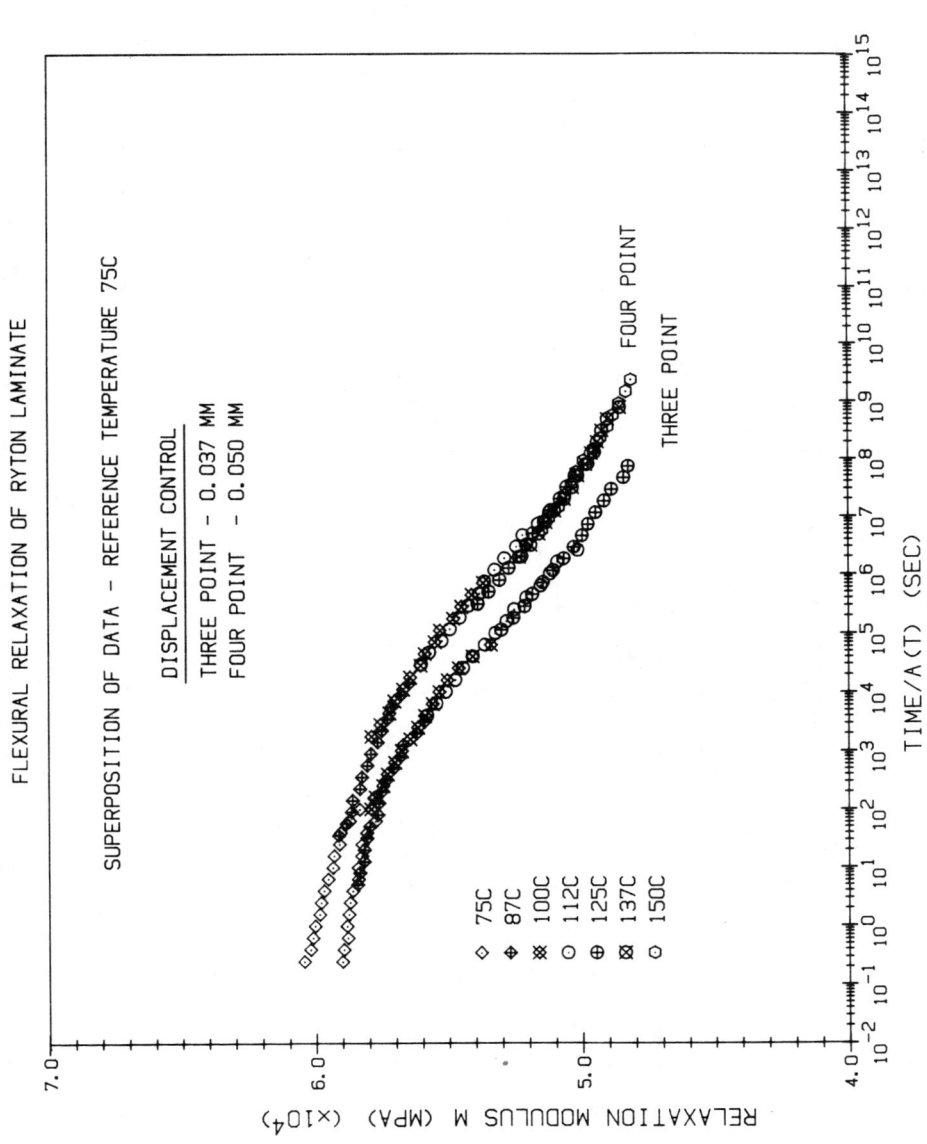

FIG. 21—*Relaxation master curves for Ryton laminate in three-point and four-point flexure.*

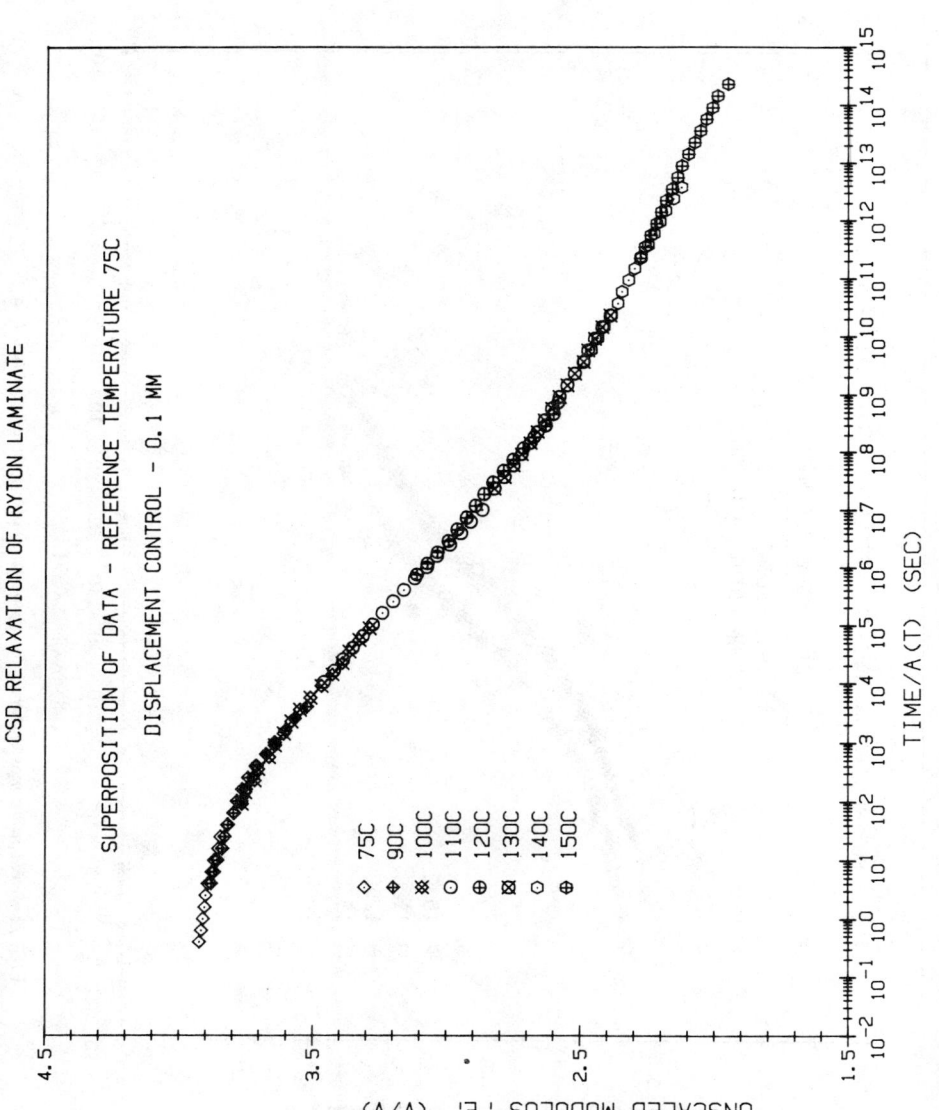

FIG. 22—*Relaxation master curve for Ryton laminate obtained during CSD flexure.*

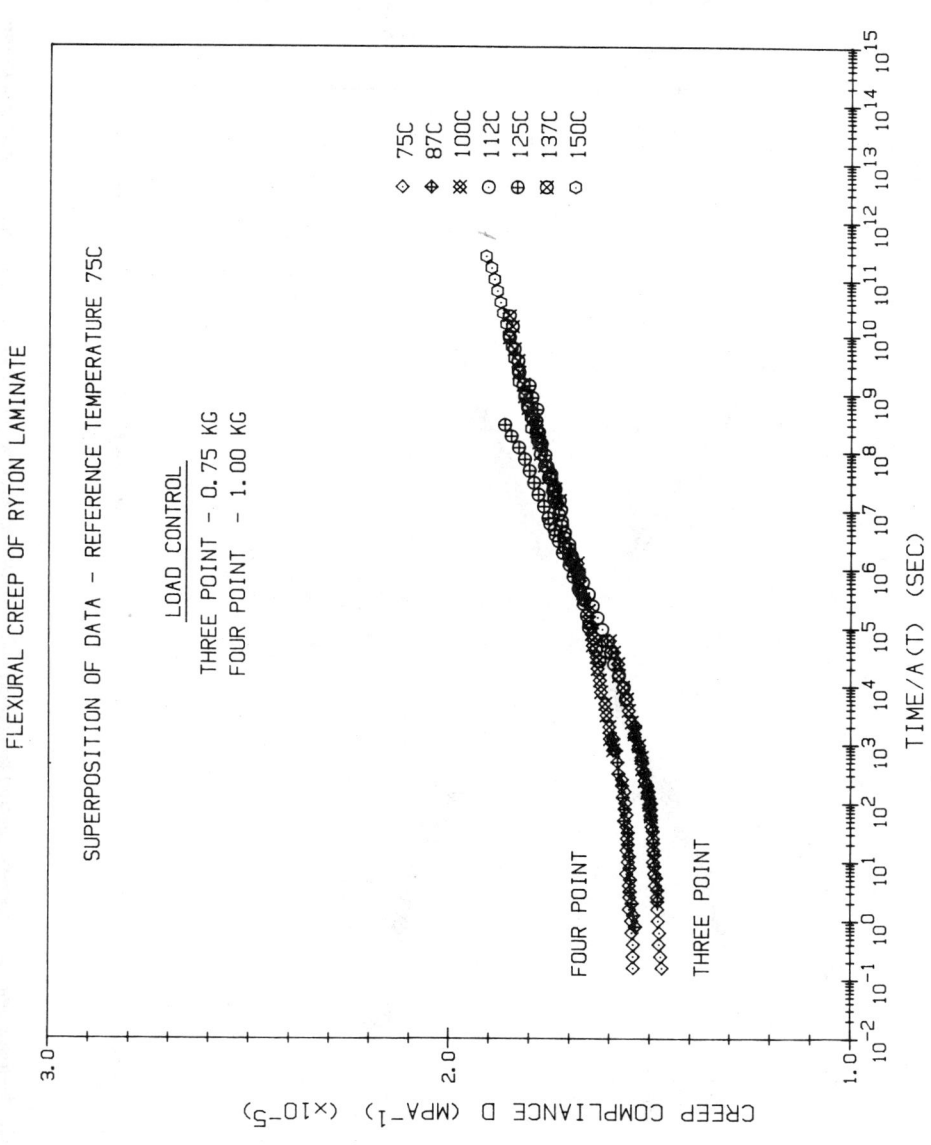

FIG. 23—*Creep master curves for Ryton laminate in three-point and four-point flexure.*

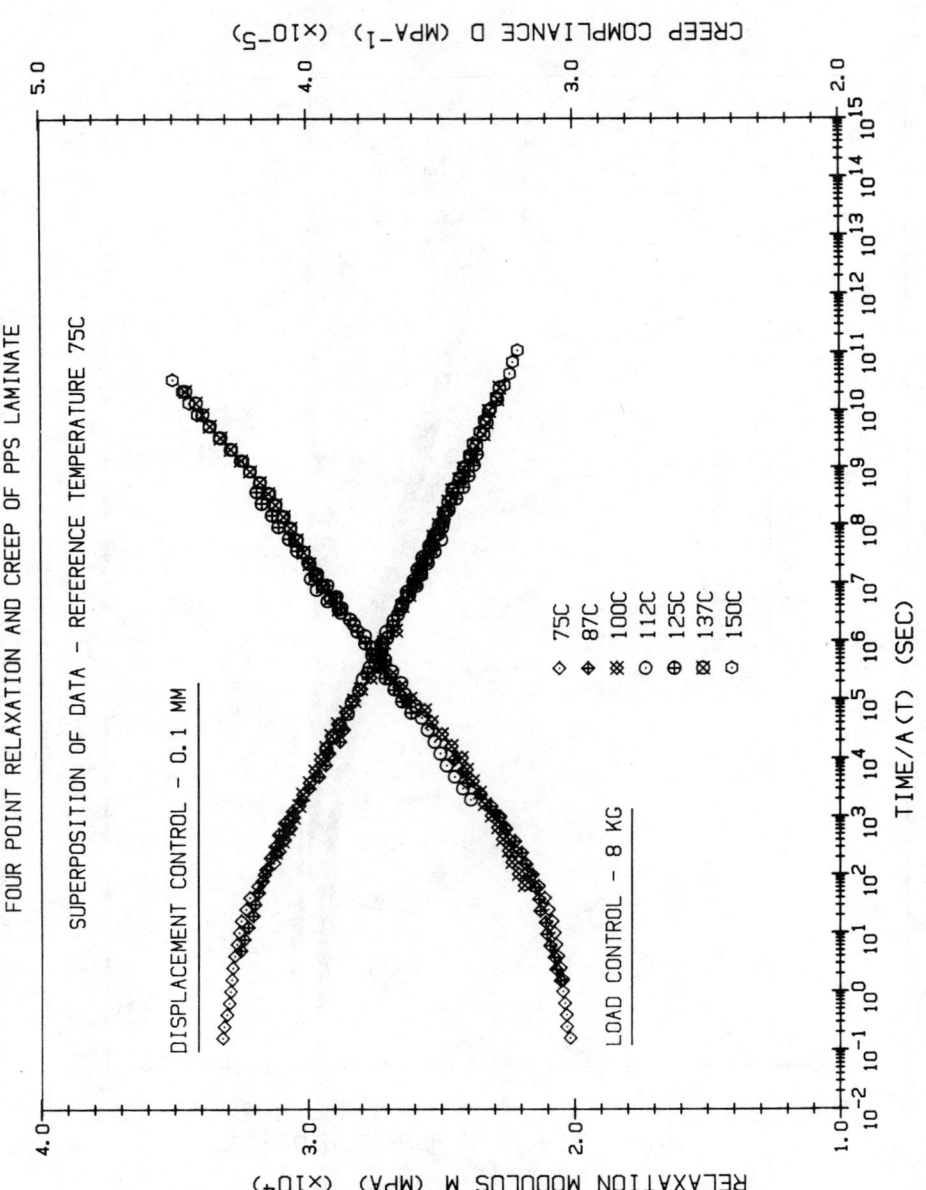

FIG. 24—*Relaxation (displacement control) and creep (load control) master curves for PPS laminate obtained by four-point bending.*

TABLE 1A—*Dynamic mechanical tests: Change in storage modulus and stiffness, 25 to 175°C-1.0 rad/s.*

	Ryton	PPS
Torsion	75%	50%
CSD	50%	...
Three-point flex	20%	30%
Four-point flex	20%	20%

TABLE 1B—*Transient tests: Change in superposition modulus, compliance, stiffness, 1 s to 10 year-75°C reference.*

	Ryton		PPS	
	Relax	Creep	Relax	Creep
Torsion	50%	...	...	...
CDS	40%	45%	...	...
Three-point flex	25%	30%	...	...
Four-point flex	20%	15%	25%	40%

NOTE: Values for dynamic mechanical and transient tests can vary by ±5%.

(three-point and four-point bending) exhibit lower degrees of change than observed for resin-dominated shear tests. A summary of the results is presented in Table 1.

Other tests have been conducted at lower deformations and loads. The results were similar to those shown in Table 1 with respect to the degree of change in stiffness or compliance. Therefore, this suggests that the response is representative of that in the linear viscoelastic region of behavior. Consequently, the amount of change in moduli, compliances, or stiffnesses are the lowest that would be expected. Higher loads and/or deformations in the nonlinear region would result in greater change in the respective properties over the same interval.

In summary, when attempting to predict the response of Ryton or similar thermoplastic-based composites, attention needs to be given both to the material of interest and transition temperatures relative to the design parameters as well as the particular state of stress or strain to which they are subjected. Significant property change can occur due to the glass transition phenomena associated with the amorphous phase in these high-melting-temperature polymers. Similar results can be expected in other thermoplastic-based composites.

Tests are currently in progress to evaluate the response of crystalline polyetheretherketone (PEEK) and amorphous polyimide (KII) based materials. These results will be compared to 176°C (350°F) state-of-the-art epoxies and high-performance imides such as V-378A and PMR-15.

References

[1] Browning, C. E., "The Role of the Polymeric Matrix in the Processing and Structural Properties of Composite Materials," Seferis and Nicholais, Eds., Plenum, 1981, pp. 231–243.
[2] Willats, D. J., *SAMPE Journal,* Vol. 20, No. 5, 1984, pp. 6–10.
[3] Cogswell, F. N. and Leach, D. C., *Plastics and Rubber Processing and Applications,* Vol. 4, No. 11, 1984, pp. 271–276.

[4] Sheppard, C. H. and House, E. E., "Development of Improved Graphite Reinforced Thermoplastic Composites," Report D180-25470-1, Naval Air Systems Command, Washington, DC, December 1981.

[5] Whitaker, R. B., Nease, A. B., and Yelton, R. O., "Characterization of Polyetherether Ketone and Other Engineering Thermoplastics," Report DE84-010668, Department of Energy, Washington, DC, April 1984.

[6] O'Connor, J. E., Ma, C. C., and Lou, A. Y., 39th SPI, January 1984, Society of the Plastics Industry, New York, NY, 11-E, pp. 1–5.

[7] Martin, C. C., O'Connor, J. E., and Lou, A. Y., 29th SAMPE Symposium, Society for the Advancement of Materials, Covina, CA, April 1984, pp. 753–764.

[8] Murtha, T. P., Ma, C. C., South, A., Walker, J. H., and Brady, D. G., 29th SAMPE Symposium, April 1984, Society for the Advancement of Materials, Covina, CA, pp. 870–879.

[9] Brady, D. G., Murtha, T. P., Walker, J. H., South, A., and Ma, C. C., SPE ANTECH, Society of Plastics Engineers, Brookfield Center, CT, Vol. 30, 1984, pp. 690–693.

[10] Ma, C. C., 30th SAMPE Symposium, March 1985, Society for the Advancement of Materials, Covina, CA, pp. 543–553.

[11] Ma, C. C., Hsia, H., Liu, W., and Hu, J., *Polymer Composites*, Vol. 8, No. 4, August 1984, pp. 256–264.

[12] Brady, D. G., *Journal of Applied Polymer Science*, Vol. 20, 1976, pp. 2541–2551.

H. Thomas Hahn,[1] *Kenneth L. Jerina,*[2] *and Pierre Burrett*[3]

Fiber Orientation and Fracture Morphology in Short Fiber-Reinforced Thermoplastics

REFERENCE: Hahn, H. T., Jerina, K. L., and Burrett, P., **"Fiber Orientation and Fracture Morphology in Short Fiber-Reinforced Thermoplastics,"** *Advances in Thermoplastic Matrix Composite Materials, ASTM STP 1044,* G. M. Newaz, Ed., American Society for Testing and Materials, Philadelphia, 1989, pp. 183–198.

ABSTRACT: Spatial fiber distribution and fracture morphology were studied for three injection-molded thermoplastic composites: PPS, PAI, and PEEK reinforced with carbon fibers. Specimens were dog bone-shaped tensile coupons and flex bars. Both optical and scanning electron microscopes were used. In general, most fibers were found to be aligned in the mold-fill direction except in the core. The fibers in the core tended to be normal to the mold-fill direction. Yet, the exact distribution pattern varied from material to material and depended upon the gate design. Examinations of fracture surfaces indicated that the best interfacial bonding was in the PEEK composite followed by the PAI composite. In the core, fracture surfaces generally followed the fiber orientations. The PPS composite showed the most improvement over the matrix properties in both static and fatigue strengths despite poor interfacial bonding. The least improvement was observed of the PAI composite in which matrix cracks were seen to cut through fibers. Nevertheless, the strongest matrix, PAI, yielded the strongest composite, and the weakest matrix, PPS, yielded the weakest composite.

KEY WORDS: injection-molded composites, injection-molded thermoplastics, fiber-reinforced thermoplastics, fiber orientation distribution, fracture morphology, short fiber composites

During injection molding of fiber-reinforced polymers, fibers tend to be oriented preferentially. This orientation results in anisotropy in the molded products. The factors influencing fiber orientation are known to include the gate design, the viscosity of the melt, the mold lubrication, the fill rate, and the shape of the mold cavity [1].

Mechanical properties of short fiber-reinforced thermoplastics usually are determined using specimens fabricated by injection molding. The fiber distributions inside these specimens are, in general, different from those in actual injection-molded products. Thus, the use of the measured properties in design must take into account the different fiber distributions.

As the polymer is injected into the mold cavity through a gate located at the center of an edge of the mold for a specimen, it diverges from the core toward the mold surfaces. Since the fibers orient themselves parallel to the flow lines, they remain normal to the mold-fill direction in the core. Furthermore, the polymer solidifies upon contact with the rather cold surfaces of the mold, leading to a skin region where the fibers are aligned parallel to the wall due to the high

[1]Professor, Department of Engineering Science and Mechanics, The Pennsylvania State University, University Park, PA 16802.

[2]Associate professor, Department of Mechanical Engineering, Washington University, St. Louis, MO 63130.

[3]Undergraduate assistant, Department of Mechanical Engineering, Washington University, St. Louis, MO 63130.

shear gradient in this region. Thus, two distinct regions are observed in the specimen: the core containing fibers mainly normal to the mold-fill direction, and the skin where the fibers are parallel to the fill direction. The relative thicknesses of the core and skin regions depend upon the shear gradient at the wall. The more intense the gradient, the thinner the skin.

A core region in a glass fiber-reinforced PET was observed by Fakirov and Fakirova [2]. They also noticed significantly fewer fibers in the core than in the skin region. Similar results were obtained by Bright et al. [3] in polypropylene containing 20% by weight short glass fibers.

The fiber orientation distribution was computed by Gillespie et al. [4] using fluid mechanics equations in conjunction with experimentally observed boundary conditions. Their results indicated the presence of core and skin regions. The core region further showed an alternating pattern of horizontal and vertical fibers. The authors claimed that the horizontal orientation was due to the velocity gradient across the width while the vertical orientation was due to the velocity gradient across the thickness.

In the present work three different high-performance thermoplastic resins were chosen to study a correlation between fiber orientation distribution and fracture morphology. These resins were polyphenylene sulfide (PPS), polyetherether ketone (PEEK), and polyamide-imide (PAI). PPS and PEEK are semicrystalline, whereas PAI is amorphous. All resins were reinforced with 30% by weight short carbon fibers. The PPS composite is available as Ryton AC 50/30 and the PAI composite is available as Torlon 7130.

Experimental Procedure

For each of the three composites, two types of specimens were used: ASTM Type I tensile coupon of dog bone shape (12 by 3 by 70 mm in gage section and 19 mm wide in grip region), and the ASTM flex bar (12 by 3 by 127 mm). The specimens were injection molded through one gate located at the center of one end of the specimen. There were two exceptions: the PAI flex bar which had its gate located in a corner, and the PAI tensile coupon in which one end was used in its entirety as a gate. The injection-molded specimens were supplied by the following

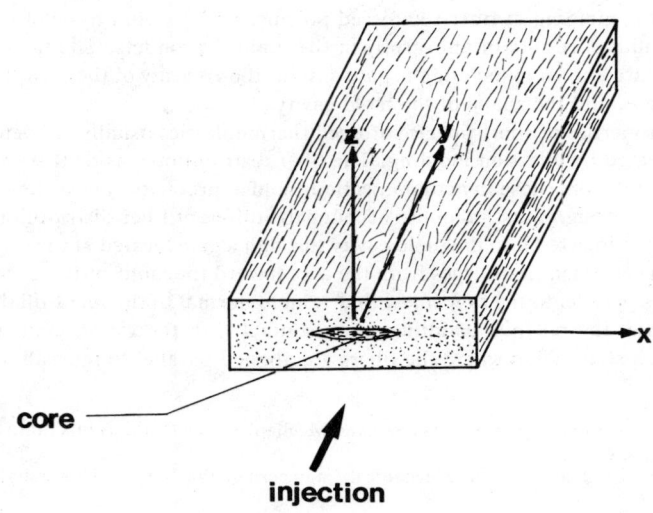

FIG. 1—*A schematic of fiber orientation distribution.*

manufacturers: Phillips Petroleum (PPS), Amoco (PAI), and LNP Corporation (PEEK). Since the exact molding conditions were not known, no attempt was made to correlate the observed fiber distributions to the molding conditions.

The PPS/C (carbon) specimens had smooth surfaces whereas the PAI/C and PEEK/C specimens exhibited rough surfaces. A smooth surface is a result of using a hot mold, and hence the molded product may have a high level of crystallinity. A rough surface is characteristic of parts molded at a low mold temperature.

Three cross sections were examined on an optical microscope at magnifications up to ×100: the midplane (*x-y* plane), the longitudinal cross section (*y-z* plane), and the transverse cross section (*x-y* plane) (Fig. 1). Since the core region was quite thin, only 10 to 20% of the total thickness, specimens were gradually polished to reveal the midplane.

Fracture surfaces were examined on specimens tested under the following loading conditions: static tension, tension-tension fatigue ($R = 0.1$), and creep. Static and fatigue tests were performed on all three composites, but creep tests were performed on PPS and PEEK only. All tests were conducted on an MTS machine. Three different crosshead speeds (2.5, 0.25, 0.025 mm/s) were used for the static tension tests. Fatigue tests were performed at a frequency of 15 Hz for PAI, 30 Hz for PPS, and 2 to 30 Hz for PEEK. Creep tests were carried out at temperatures ranging from 35 to 95°C, while the applied stress was varied from 20 to 60% of the ultimate tensile strength (UTS).

Results

Mechanical Properties

The effect of loading rate on modulus and strength was marginal for all three composites at room temperature. Therefore, all data were combined to determine the average moduli and strengths. The results are summarized in Table 1. The data indicate that carbon fibers are most effective in stiffening and strengthening PPS. PEEK/C shares the same degree of strengthening with PPS/C, whereas its stiffening is comparable to that of PAI/C.

Some results from fatigue and creep tests are shown in Figs. 2 and 3. The data for PPS/C includes those reported in Ref 5. Based on the normalized fatigue stress, the PPS/C composite exhibits the least fatigue sensitivity, that is, the smallest slope of the *S-N* curve, although it has the lowest strength. On the other hand, PEEK/C exhibits the steepest slope and its normalized fatigue strength at 10^6 cycles falls between those of PPS/C and PAI/C. At 80°C, both PPS/C and PEEK/C exhibit creep as the applied stress exceeds 20% UTS.

TABLE 1—*Moduli and ultimate tensile strengths (UTS).*

	Neat Polymer[b]		Composite[a]		Ratio	
	Modulus, GPA	Strength, MPa	Modulus, GPa	Strength, MPa	Composite/Polymer	
					Modulus	Strength
PPS	3.3	66	23 (4.4)[c]	143 (6.6)	7.0	2.2
PEEK	4.2	103	21 (3.8)	219 (8.5)	5.0	2.1
PAI	5.0	186	26 (5.1)	235 (5.2)	5.2	1.3

[a]Nine specimens were tested for each composite.
[b]The modulus and strength values for neat polymers were obtained from the open literature.
[c]Numbers inside parentheses indicate coefficients of variation in percent.

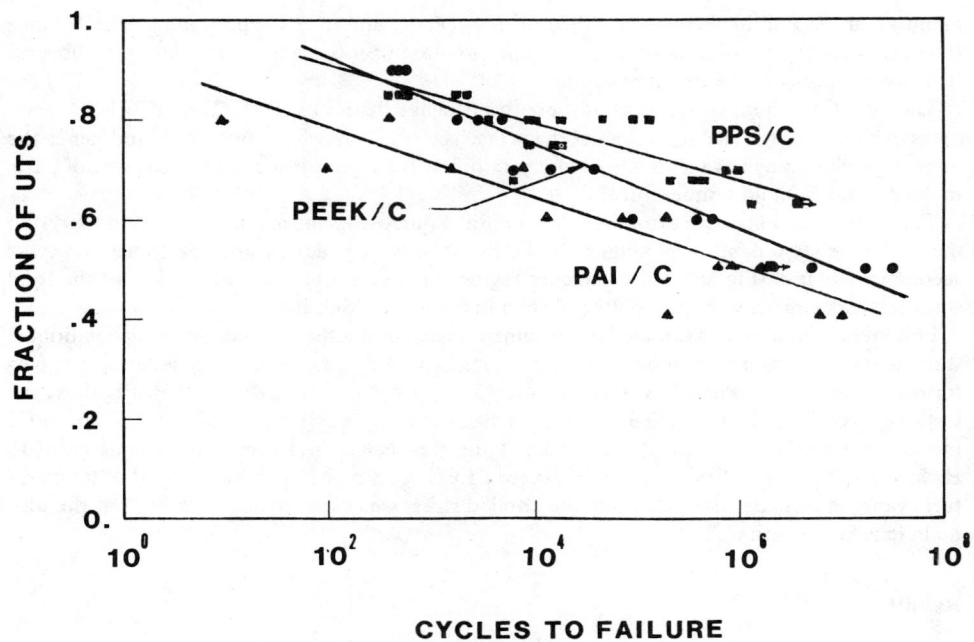

FIG. 2—*Normalized stress-life relations in tension-tension fatigue.*

Fiber Orientation Distribution

For Type I specimens, the existence of a core region, as shown schematically in Fig. 1, was confirmed for PEEK/C and PPS/C but not for PAI/C. The fracture surface of a PEEK/C specimen in Fig. 4 shows a core region which appears white in comparison with the surrounding region.

Aside from the difference in the type of matrix used, gates of different shapes were used for the three composites. The gates for both PPS/C and PEEK/C were rectangular but smaller than the specimen cross section. The gate for PPS/C was 6 mm wide and spanned only halfway through the specimen thickness from one surface, while the gate for PEEK/C was 5 mm wide and covered the entire specimen thickness. However, for PAI/C, one entire end of the specimen was used as the gate, and this fact may be responsible for the absence of a core region.

The patterns of fiber distribution in the midplanes are shown schematically in Fig. 5. In PAI/C Type I specimens, most fibers are aligned in the mold fill direction and there is no evidence of a core region. Rather, the midplane appeared to consist of two regions welded along a curve, as shown in Fig. 6. The interregional boundary curve may be the result of an injection process where local welding took place between two streams of melt flow. The midplanes of PPS/C and PEEK/C, on the other hand, exhibit very specific patterns.

As for the midplane fiber orientations in flex bars, both PPS/C and PEEK/C show the same wavy patterns as in Type I specimens (Fig. 7), indicating that the flow behavior is similar in both Type I and flex specimens. PAI/C flex specimens also show a symmetric wavy pattern (Fig. 7C). The symmetric wavy pattern observed in PAI/C flex specimens is very close to the one computed by Gillespie et al. [4]. The modular nature of the wavy patterns is a manifestation of bands of fibers alternating in and out of the midplane [2].

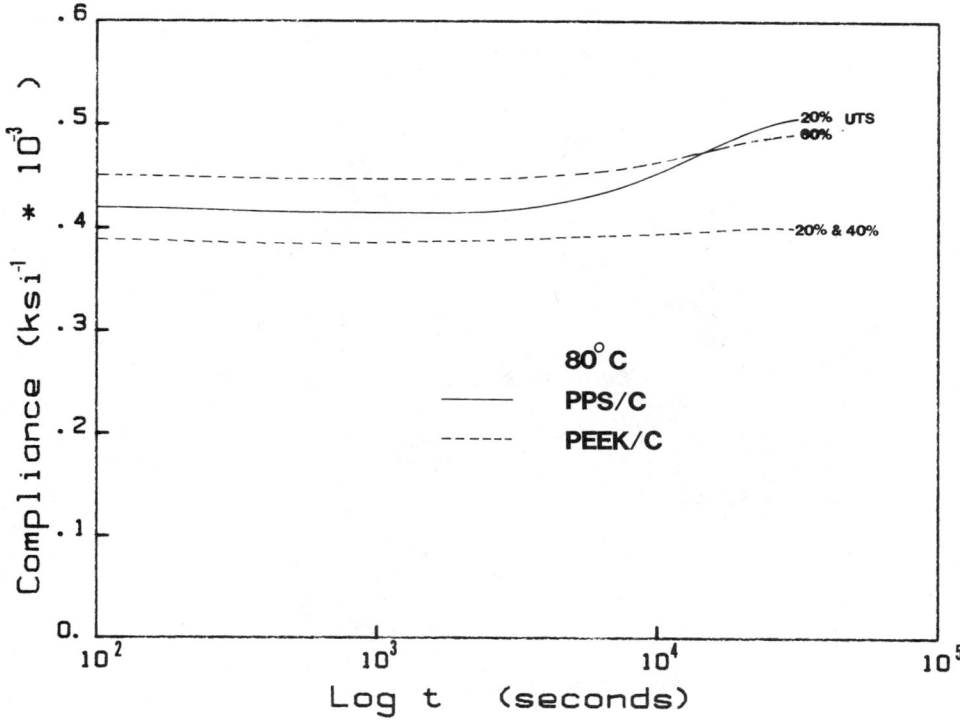

FIG. 3—*Creep compliances at 65°C.*

Both PPS/C and PEEK/C flex bars had their gates located at the center of an edge, while the gate for PAI/C bars was located at a corner. Yet, the corner gate on PAI/C specimens produced more symmetric wavy pattern.

The thicknesses of core regions can be seen on longitudinal cross sections (x-z planes). Figure 8 shows the core region in PPS/C being thinner than in PEEK/C.

Examinations of midplane as well as transverse and longitudinal cross sections yield a spatial distribution of fibers. The core is composed of a succession of wavy fronts where fibers are mostly transverse to the mold fill direction. On the other hand, the fibers in the skin are mostly aligned parallel to the mold-fill direction. The transversely oriented fibers render lower strength than the longitudinally oriented fibers. Therefore, failure would be initiated in the core upon tensile loading.

Being inhomogeneous, the distribution of fibers is rather difficult to characterize. Each fiber is represented by its orientation angle and length. Although these two parameters may in general be related to each other, they were assumed to be independent in the present study. Furthermore, since most fibers were lying parallel to the x-y plane, their orientation distribution was assumed to be two dimensional.

To determine the orientation distribution, the total thickness was divided into two regions: skin and core. In each region, fiber orientation angles were measured from the x axis on representative optical micrographs of planes parallel to the x-y plane. The average orientation distribution then was determined by multiplying the number of fibers at each angle by the relative

FIG. 4—*Fracture surface of a PEEK/C specimen showing a core layer (tensile fatigue at 70% UTS).*

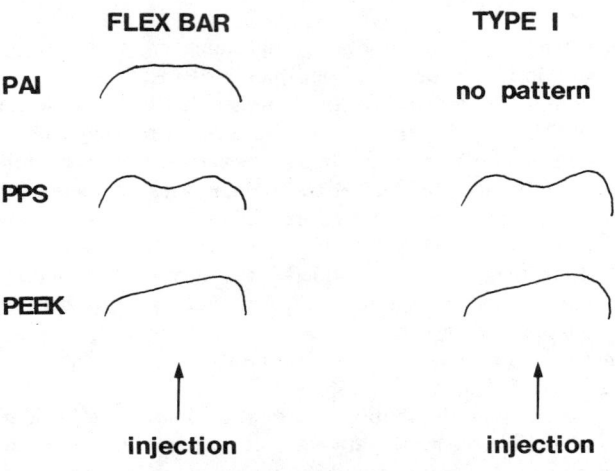

FIG. 5—*A schematic of wavy patterns observed on midplanes.*

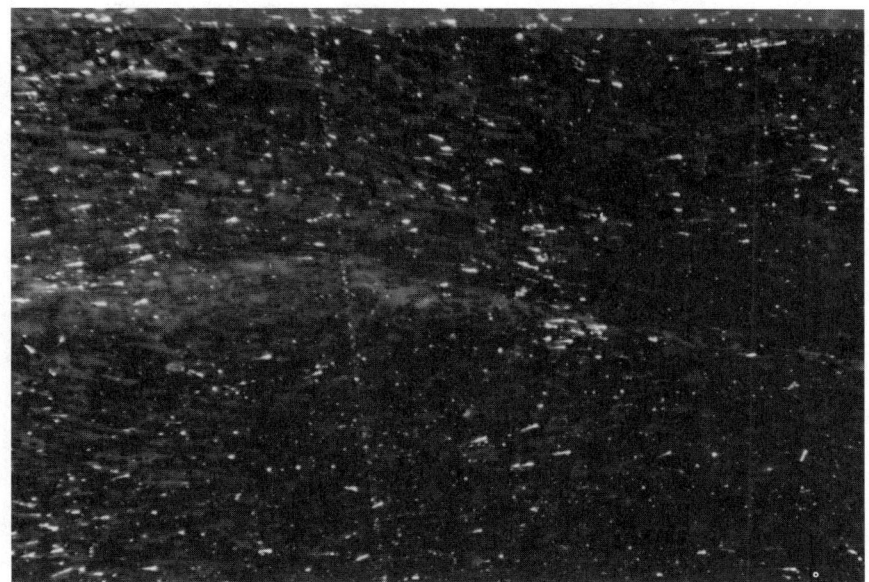

FIG. 6—*A weld line on midplane of a PAI/C Type I specimen.*

thickness of the corresponding region. The measured orientation distribution was fit by a cumulative distribution of the following form

$$F(\theta) = \frac{1 - \exp(-\lambda\phi)}{1 - \exp(-\lambda\pi/2)}$$

where

θ = radians.

Note that the same function but without the denominator was used in Ref 7.

The values of λ were found to be 5, 5.5, and 5, respectively, for PPS/C, PAI/C, and PEEK/C. A representative distribution is shown in Fig. 9. A higher value of λ indicates a better alignment of fibers. Although PAI/C shows a much better alignment in the midplane, its average fiber distribution throughout the volume is not much different from those of the other two composites. In all three composites, more than 80% of fibers have their orientation angles less than 22°.

Fiber length showed a substantial variation, ranging from 15 to 350 μm. However, there was little difference between the three composites. The average fiber length was found to be 250 μm with a coefficient of variation of 63%.

Micromechanics predictions for moduli and thermal expansion coefficients based on the foregoing fiber distribution compared well with experimental data, reinforcing the validity of the assumed two-dimensional distribution [7].

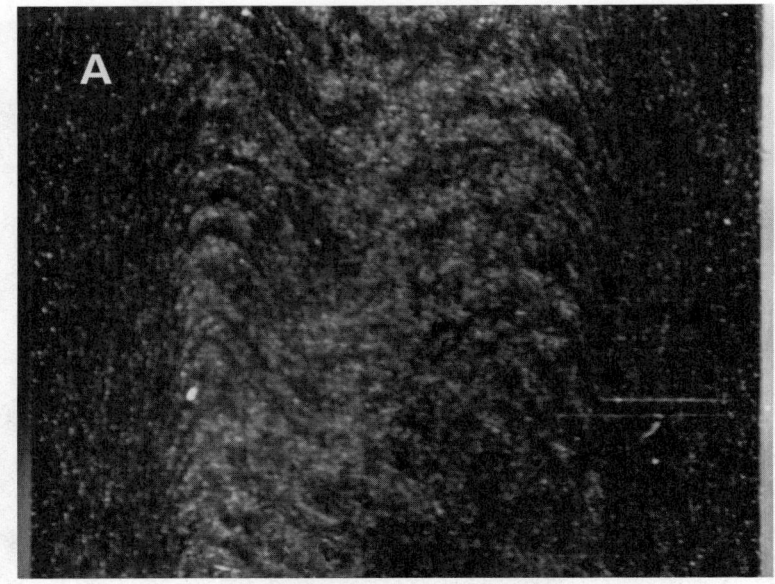

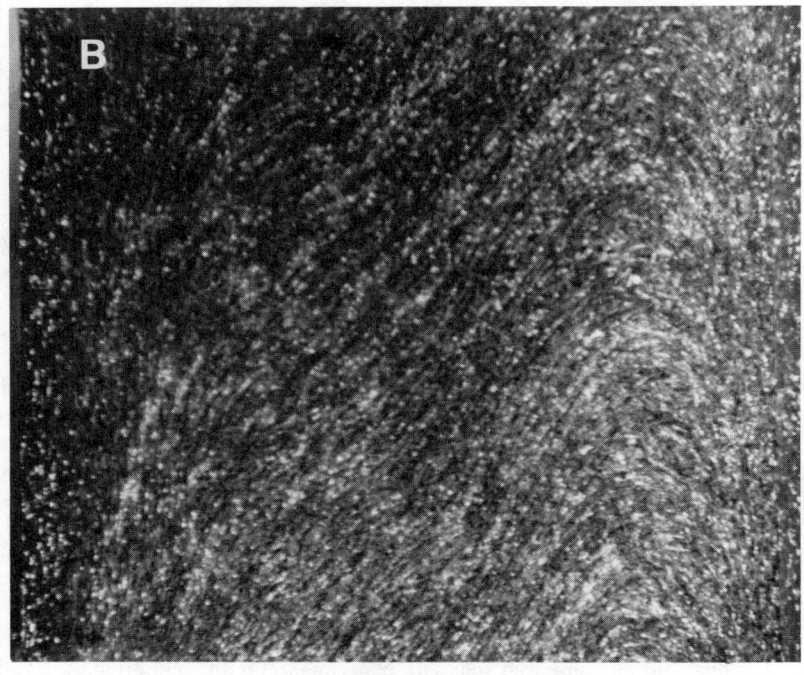

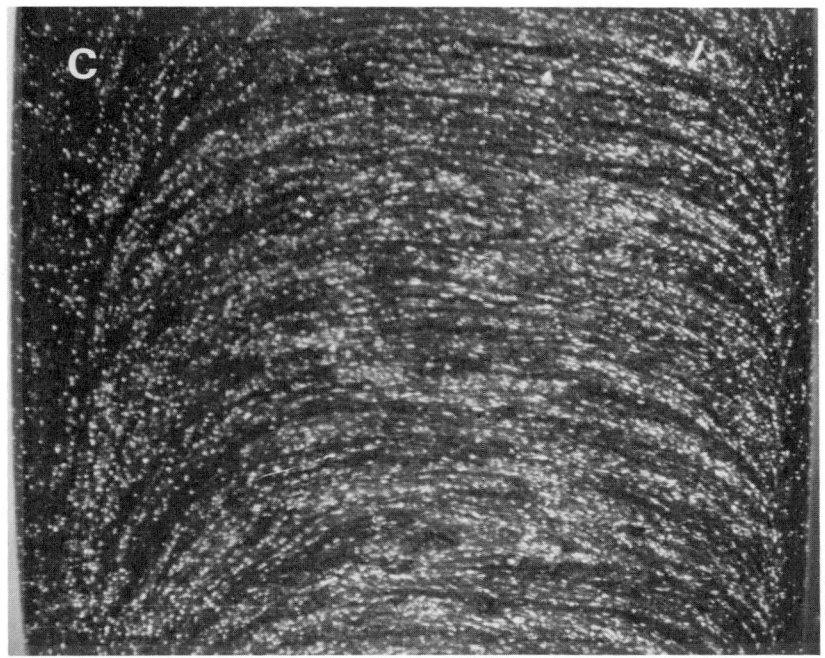

FIG. 7—*Fiber distributions on midplanes of flex bars:* (A) *PPS/C;* (B) *PEEK/C;* (C) *PAI/C.*

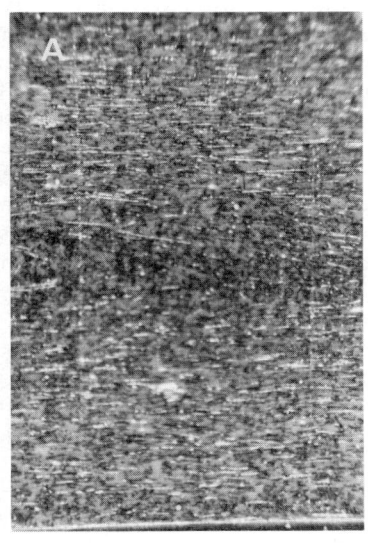
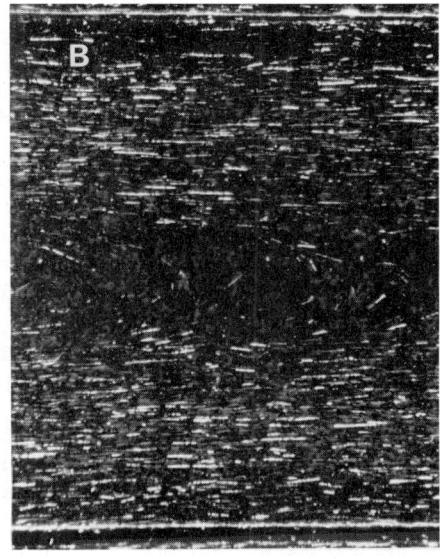

FIG. 8—*Longitudinal cross sections of Type I specimens:* (A) *PPS/C;* (B) *PEEK/C.*

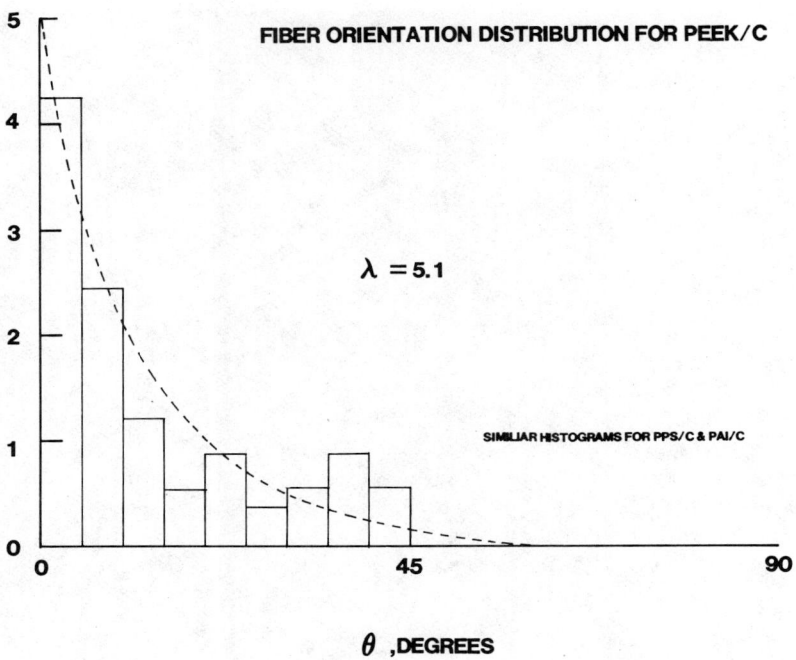

FIG. 9—A fiber orientation distribution for PEEK/C.

Fracture Morphology

Examinations of PPS/C and PEEK/C specimens on an optical microscope revealed that the fracture surfaces in the core region were concave toward the gate. Furthermore, the fibers in the core tended to be parallel to the fracture surface, indicating that the fracture in the core region was along the fibers. For example, Fig. 10 shows regions near the boundary of a core in a PPS/C specimen, which was tested in fatigue at 80% UTS. A side view of the core in the same specimen is also shown in Fig. 10C. Essentially the same patterns were observed on the fracture surfaces of PEEK/C specimens.

Since the strength parallel to the fibers is higher than that normal to the fibers, it may reasonably be assumed that failure was initiated in the core region in PPS/C and PEEK/C specimens. Unfortunately, no sign of failure initiation could be found on the fracture surfaces, perhaps because of the complex topography. Further details of fractographic examinations are described as follows:

PPS/C—Since failure in creep is gradual, more can be learned from creep fracture surfaces than static ones. To the naked eye, a creep fracture surface appears to consist of two regions: a white region and a dark region. Figure 11 shows micrographs taken from a white and a dark region, respectively. The white region is characterized by ductile failure of the resin accompanied by fiber separation from the matrix. The dark region, on the other hand, indicates brittle fracture in the resin with little fiber separation.

Because of the stress concentration, failure initiation is likely to be at the tip of a fiber. The resulting debonding will then spread along the fiber axis, allowing the matrix resin to deform plastically, pulling itself away from the fibers. Cracks eventually form in the resin and propagate into the neighboring regions. Failure of the resin due to propagation of cracks will appear

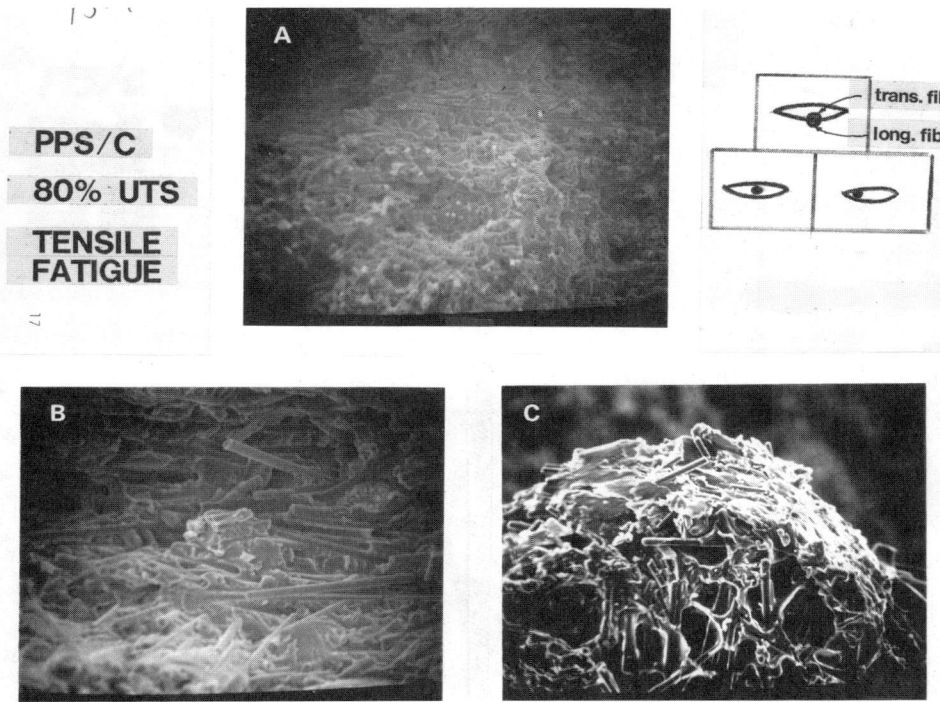

FIG. 10—*Fracture surfaces of a PPS/C specimen after fatigue failure at 80% UTS:* (A) *core-skin boundary;* (B) *core;* (C) *side view of core.*

brittle, and there will be little fiber separation during fast crack propagation. Thus, failure is believed to have been initiated in the white region where the matrix failure is ductile. It is noted that in both regions, fiber ends are quite free of resin. Therefore, the PPS/carbon fiber interface is judged to be rather weak.

In static tension, bonding appears better than in creep (Fig. 12), yet the resin failure looks rather brittle. The average fiber pull-out length was measured to be about 130 μm. Therefore, no significant amount of fibers are expected to have broken because the average initial fiber length is about 250 μm.

The failure processes in static tension are basically the same as in creep. The only difference is that in static tension the resin does not have enough time to yield and behave in a ductile fashion.

The fracture surfaces of fatigue specimens indicate fatigue failure to be a combination of creep and static failure. The resin failure on the fracture surfaces mostly appears brittle, but some areas reveal debonding and ductile fracture (Fig. 13). Since the fatigue test, especially at low stress levels, can be regarded as an interrupted creep test, the same failure mechanisms as in creep should apply. The only difference is that the resin does not have enough time to yield during a fatigue test.

PEEK/C—Fracture surfaces of creep specimens again reveal two distinct regions, similar to the PPS/C specimens. One region shows a ductile matrix failure with some fiber debonding, and the other shows a brittle matrix failure with no trace of fiber debonding (Fig. 14). Therefore, the failure mechanisms in PEEK/C are similar to those in PPS/C.

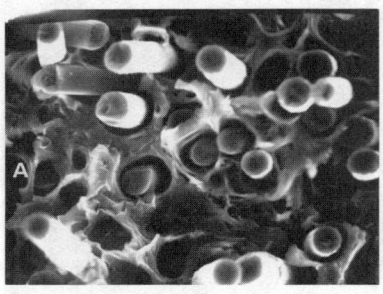

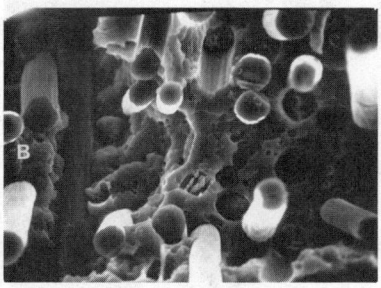

FIG. 11—*Fracture surfaces of a PPS/C specimen after creep rupture at 40% UTS and 80°C:* (A) *a white region showing ductile failure;* (B) *a dark region showing brittle failure.*

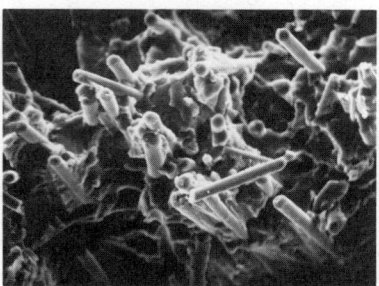

FIG. 12—*A static fracture surface of a PPS/C specimen.*

Static fracture surfaces indicate excellent fiber/matrix bonding because fibers including their ends are still coated with resin and the pulled-out fibers are shorter than in PPS/C (Fig. 15). Thus, fiber pullout occurs as a result of matrix failure rather than interfacial failure and without fibers breaking. Cracks are believed to be initiated inside the matrix and grow normal to the loading, yet avoid fibers. Fatigue fracture surfaces look similar to static fracture surfaces (Fig. 16). Good interfacial bonding is evident on the fatigue fracture surfaces also.

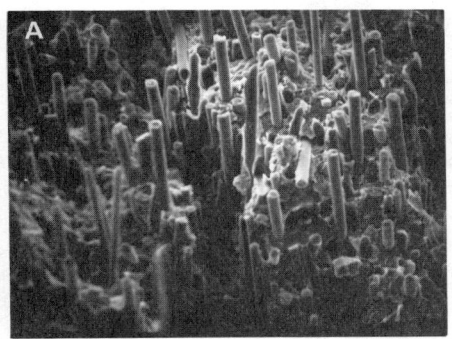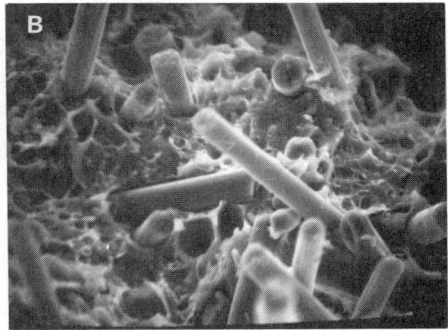

FIG. 13—*A fatigue fracture surface of a PPS/C specimen (70% UTS):* (A) *low magnification;* (B) *high magnification.*

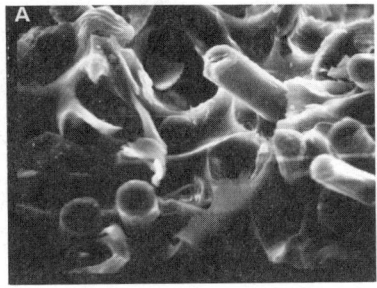

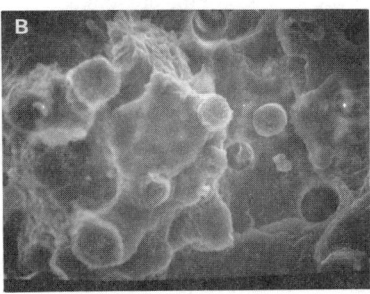

FIG. 14—*Creep rupture surfaces of a PEEK/C specimen (35°C, 60% UTS):* (A) *ductile spot;* (B) *brittle spot.*

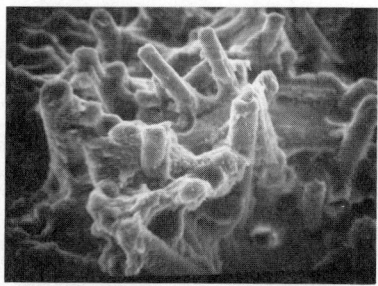

FIG. 15—*A static fracture surface of a PEEK/C specimen.*

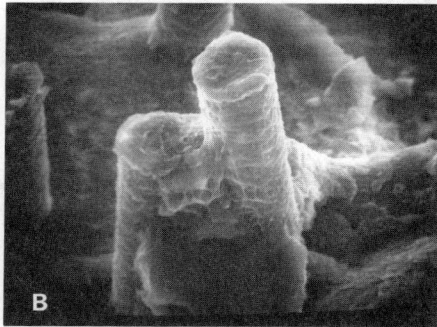

FIG. 16—*A fracture surface of a PEEK/C specimen after fatigue failure at 50% UTS:* (A) *low magnification;* (B) *high magnification.*

PAI/C—Fracture surfaces of the PAI/C composite look similar both in static tension and in tension-tension fatigue (Fig. 17). But they are different from the fracture surfaces of PPS/C and PEEK/C. One distinct feature of PAI/C fracture surfaces is that the matrix has the appearance of having been crushed. Yet the PAI/C shows good interfacial bonding. Of the three thermoplastic resins tested, PAI has the highest strength while being most brittle. Since interfacial bond is also strong, a crack approaching a fiber in PAI/C is more likely to break the fiber than grow around it. However, PAI still yields the highest strength for its composite.

Conclusions

The presence of a core region was confirmed in three different injection-molded high-performance thermoplastics reinforced with carbon fibers. What distinguishes the core from the skin region is that fibers in the former are transverse to the mold-fill direction. However, there was one exception. PAI/C Type I tension specimens did not show transverse fibers in the core; rather they exhibited a weld line with most fibers still parallel to the mold-fill direction.

In the core, the fracture surface followed the wavy pattern of fibers. It was difficult to locate the failure initiation sites. In creep tests, failure initiation involved ductile deformation in the

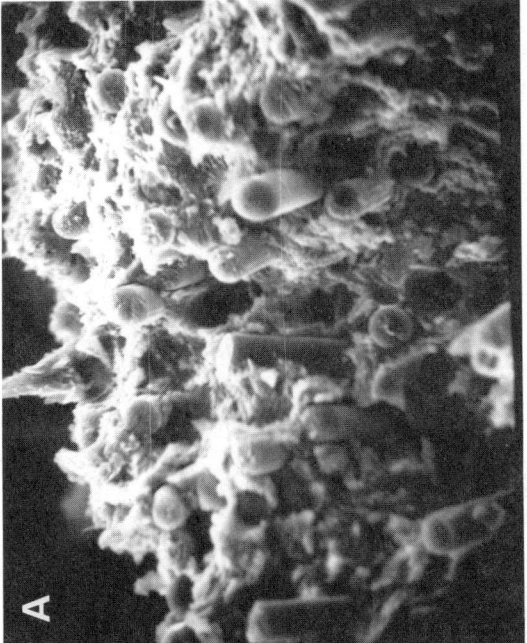

FIG. 17—*Fracture surfaces of PAI/C specimens: (A) static tension; (B) fatigue at 40% UTS.*

matrix. Regions of such ductile deformation appeared white to the naked eye. Ductile deformation of the matrix usually was accompanied by extensive fiber separation from the matrix. Among the three composites, the matrix failure was most ductile in PEEK/C and the most brittle in PAI/C.

PPS/C showed the most improvement over the matrix properties in both static and fatigue strengths in the presence of fibers. Such good improvement was possible in spite of the extensive fiber-matrix debonding observed on the fracture surfaces. The best interfacial bonding was observed in PEEK/C; however, failure was still in the matrix without much fiber fracture. Pulled-out fibers were coated with the matrix.

References

[1] Goettler, L. A., "Controlling Flow Orientation in Molding of Short-Fiber Compounds," *Modern Plastics,* April 1970, pp. 140–146.
[2] Fakirov, S. and Fakirova, C., "Direct Determination of the Orientation of Short Glass Fibers in an Injection-Molded Poly(ethylene terephthalate) System," *Polymer Composites,* Vol. 6, 1985, pp. 41–46.
[3] Bright, P. F., Crowson, R. J., and Folkes, M. S., "A Study of the Effect of Injection Speed on Fibre Orientation in Simple Mouldings of Short Glass Fibre-Filled Polypropylene," *Journal of Material Science,* Vol. 13, 1978, pp. 2497–2506.
[4] Gillespie, J. W., Jr., Vanderschwen, J. A., and Pipes, R. B., "Process Induced Fiber Orientation— Numerical Simulation with Experimental Verification," *Polymer Composites,* Vol. 6, 1985, pp. 155–159.
[5] Mandell, J. F., Huang, D. D., and McGarry, F. J., "Fatigue of Glass and Carbon Fiber Reinforced Engineering Thermoplastics," *Proceedings of the 35th SPI Reinforced Plastics/Composite Institute,* Society of Plastics Industry, New York, NY, 1980.
[6] Kacir, L., Narkis, M., and Ishai, O., "Oriented Short Glass Fiber Composites. II. Analysis of Parameters Controlling the Fiber/Glycerine Orientation Process," *Polymer Engineering Science,* Vol. 15, 1975, pp. 532–536.
[7] Hahn, H. T., Jerina, K. L., and Chiou, W., "Thermoelastic Behavior of Injection-Molded Thermoplastic Composites," *Proceedings of the International Conference on Composite Materials and Structures,* Technomic Publishing Co., Westport, CT, 1986, pp. 356–361.

David R. Carlile,[1] *David C. Leach,*[2] *D. Roy Moore,*[1] *and*
Nabil Zahlan[1]

Mechanical Properties of the Carbon Fiber/PEEK Composite APC-2/AS-4 for Structural Applications

REFERENCE: Carlile, D. R., Leach, D. C., Moore, D. R., and Zahlan, N., **"Mechanical Properties of the Carbon Fiber/PEEK Composite APC-2/AS-4 for Structural Applications,"** *Advances in Thermoplastic Matrix Composite Materials, ASTM STP 1044*, G. M. Newaz, Ed., American Society for Testing and Materials, Philadelphia, 1989, pp. 199–212.

ABSTRACT: The mechanical properties of carbon fiber/polyetheretherketone (CF/PEEK) composites have been characterized for structural applications. Stiffness and strength properties were measured on unidirectional laminates in tension, compression, and flexure, enabling the elastic constants for the material to be determined. Stiffness and strength were measured on multidirectional lay-ups in tension. The effect of temperature on strength was examined for fiber- and matrix-dominated lay-ups. Toughness was assessed using fracture mechanics techniques, instrumented falling weight impact tests, and postimpact compression tests to obtain both intrinsic and comparative toughness. Creep and fatigue properties were characterized on fiber- and matrix-dominated lay-ups. The evaluations provide a property data base for CF/PEEK composites.

KEY WORDS: PEEK, polyetheretherketone, carbon fiber composites, mechanical properties, toughness, creep, fatigue

Continuous fiber-reinforced thermoplastic composites have been developed in recent years to meet highly demanding engineering applications, particularly in the aerospace industry. One such composite is continuous carbon fiber (CF) reinforced polyetheretherketone (PEEK), APC-2. This composite has many similarities with existing thermosetting matrix composites in that it contains high levels of continuous fiber reinforcement (61% CF by volume). In other respects the materials differ in that PEEK is a tough, ductile, semicrystalline thermoplastic polymer with a glass-rubber transition temperature, T_g, of 143°C. The mechanical properties of PEEK for engineering applications have been examined previously [1], and the properties of CF/PEEK have been assessed in detail in previous papers [1–5]. These have included room temperature stiffness and strength [1,2], the effect of temperature on stiffness and strength [2,5], toughness [3,4], and long-term properties [5,6]. The purpose of this paper is to summarize the mechanical properties of CF/PEEK, drawing on the earlier, more detailed publications and on previously unpublished work to give an overview of the mechanical performance of APC-2. The properties are discussed in terms of stiffness and strength, toughness and long-term properties, and where appropriate the influence of temperature is also examined. The results are from our own experimental characterization on a consistent set of samples of the CF/PEEK composite.

[1]ICI Advanced Materials, P.O. Box 90, Wilton, Middlesbrough, Cleveland, United Kingdom TS6 8JE.
[2]ICI Composites, 2055 E. Technology Circle, Tempe, AZ 85284.

Experimental Information

Material

The material examined was the CF/PEEK composite APC-2, manufactured by ICI/Fiberite and reinforced with Hercules "Magnamite" AS-4 carbon fiber. The preimpregnated tape contained 61% by volume of fiber and was molded into laminates using recommended conditions [7]. The laminates were typically 0.5 by 0.5 m, though in some cases smaller panels (for example, 0.3 by 0.1 m) were made. The lay-ups will be discussed under each test. The quality of all laminates was verified by ultrasonic C-scan against an internal standard.

Experimental Methods

Stiffness and Strength Properties—The test methods used to characterize the stiffness and strength properties are summarized in Table 1. The tension and compression specimens, except for the ASTM Test Method for Compressive Properties of Rigid Plastics (D 695) specimens, were end-tabbed using tapered, glass fiber/epoxy resin tabs bonded to the CF/PEEK with an epoxy adhesive. Axial and transverse strains were measured using extensometers clipped to the specimen. Two methods were used to measure 0° compressive strength: the IITRI test [8] at 23°C and the ASTM D 695 test method at 23°C and at other temperatures. The ASTM D 695 test method was used because the end-tabs delaminated from the specimen prior to compressive failure of the specimen in the IITRI test at elevated temperature. This did not occur in the ASTM D 695 test, where a 1-mm-thick gage section was machined from the 5-mm-thick, 40-ply specimen. Tests were conducted at 23°C/50% relative humidity in a temperature- and humidity-controlled room. Some of the tests were also conducted at other temperatures using an environmental chamber.

Toughness Testing—The toughness of CF/PEEK was characterized using various techniques including the measurement of a critical stress-field intensity factor (K_{IC}), a critical strain energy

TABLE 1—*Test methods for stiffness and strength properties.*

Test	Lay-Up	Test Method	Comments
0° tension	$[0]_8$	ASTM D 3039	specimen width 12.7 mm
0° compression	$[0]_{16}$	IITRI	Ref 8
	$[0]_{40}$	ASTM D 695	gage section 1 mm thick by 6 mm machined from center of specimen
0° flexure	$[0]_{16}$	ASTM D 790	3-point bending, span: thickness ratio 50:1, loading roller diameter 20 mm
90° tension	$[90]_{16}$	ASTM D 3039	specimen width 12.7 mm
90° flexure	$[90]_{16}$	ASTM D 790	3-point bending, span: thickness ratio 25:1, loading roller diameter 6.35 mm
Short-beam shear	$[0]_{16}$	ASTM D 2344	span; thickness 5:1
±45 tension (in-plane shear)	$[\pm45]_{4S}$	ASTM D 3039	specimen width 19 mm
Multiaxial tension	$[0/90]_{4S}$ $[-45/0/+45/90]_{2S}$	ASTM D 3039	specimen width 19 mm
Open-hole tension	$[-45/0/+45/90]_{2S}$	ASTM D 3039	specimen width 38 mm, hole machined centrally in gage section, diameter 6.35 mm

release rate (G_{IC}), instrumented falling weight impact (IFWI), and postimpact compressive (PIC) properties.

The values of K_{IC} and G_{IC} were obtained from notched three-point bend tests on 40-ply, unidirectional laminates. The width and depth of the specimens varied from 5.6 to 12.0 mm, depending on orientation. A span of 40 mm was used for all testing. Further details of the specimen dimensions and the test method are given in Ref 9. Cracks were propagated in six directions relative to the reinforcing fibers to give interlaminar, intralaminar, and translaminar cracks. G_{IC} was also measured using the double cantilever beam (DCB) test to give an interlaminar fracture toughness. The laminates used were 40-ply, unidirectional, with a crack starter included between the center piles. The thickness of the laminates was 5.1 mm, and the beam width was 25 mm. Further experimental details are given in Ref 4. IFWI testing was conducted on $[-45/0/+45/90]_{2S}$ and $[0/90]_{4S}$ laminates. The support geometry was circular (50 mm diameter) with the specimen struck by a hemispherical impactor 12.7 mm in diameter. Both through-penetration and low-energy tests were conducted, and further details of the methods are given in relevant papers [9,10]. Postimpact compressive tests were conducted on $[-45/0/+45/90]_{5S}$ laminates using a specimen 152 by 102 mm [4,9]. The compressive test used a jig with antibuckling guides to give lateral support to the specimen [9].

Long-Term Properties—The long-term properties examined were creep and fatigue. Creep tests were conducted on $[0]_8$, $[(0/90)_2/0/90/0/(90,0)_2]$, and $[\pm45]_{4S}$ laminates, using lever loading tensile creep machines [11] and a servohydraulic Instron 8033.

Fatigue properties were measured on $[-45/0/+45/90]_{2S}$ and $[\pm45]_{4S}$ laminates. Most of the testing used square-wave loading with a frequency of 0.5 Hz, but other waveforms (sine wave) and other frequencies (5 Hz) were also examined [6].

Stiffness and Strength

The room-temperature mechanical properties of CF/PEEK, APC-2/AS-4, are summarized in Table 2. All the values in this table were measured experimentally. The 0° tensile properties

TABLE 2—*Room temperature mechanical properties of CF/PEEK, APC-2/AS-4.*

Property	Value
0° Tension	
Modulus	134 GN/m²
Strength	2130 MN/m²
Failure strain	1.45%
0° Compression[a]	
Modulus	119 GN/m²
Strength	1100 MN/m²
0° Flexure	
Modulus	121 GN/m²
Strength	1880 MN/m²
90° Tension	
Modulus	8.9 GN/m²
Strength	80 MN/m²
90° Flexure	
Modulus	8.9 GN/m²
Strength	137 MN/m²
0° Short-beam shear strength	105[b] MN/m²

[a]IITRI test.
[b]Specimens do not fail by interlaminar shear.

show a high utilization of the fiber (greater than 90%) for both modulus and strength. As with most carbon fiber composites, the 0° compressive modulus and strength are lower than the tensile values due to the influence of the matrix stiffness in compression. Failure was by local microbuckling of the fibers within or at the end of the gage section. The 0° flexural modulus and strength values lie between the tensile and compressive values, as would be expected. The predominant mode of failure in the flexural test is tensile with some compressive failure on the upper surface. The 90° properties show the greater influence of the matrix with relatively high values of transverse strength.

The short-beam test failure occurred in compression at the upper face of the specimen rather than in the interlaminar region. This invalidates the test, but the result is included because of the widespread use of the short-beam test. It previously has been observed that the initial mode of failure is by compressive buckling for ductile matrix composites in the short-beam shear test [12].

The experimentally measured elastic properties of CF/PEEK are summarized in Table 3. All the values were measured using a secant method at 0.25% axial strain. The in-plane shear modulus was obtained from tension tests on [±45] laminates. The Poisson ratios were obtained from the 0° and 90° tension tests. The values of ν_{12} and ν_{13} are very similar, indicating transverse isotropy. The results do not exactly give the expected equivalence of

$$\frac{E_{11}}{\nu_{12}} = \frac{E_{22}}{\nu_{21}}$$

almost certainly due to difficulty in accurately measuring ν_{21}. For practical purposes it is probably better to use the calculated value of ν_{21} from the above relationship rather than the experimentally measured value. The linear coefficients of thermal expansion at 0° and 90° in a unidirectional lay-up are summarized in Table 4. The full set of elastic constants (Table 3) and thermal expansion coefficients (Table 4) enable the thermoelastic properties of other lay-ups to be predicted.

The tensile stiffness and strength of various multidirectional CF/PEEK laminates are shown in Table 5. One feature of the CF/PEEK composite is the high tensile strength of the [±45] laminate, which is due to the large axial failure strain, typically in the range from 15 to 20%. This is accounted for by the high ductility of the matrix and a strong fiber-matrix interfacial adhesion.

The stiffness and strength also was examined over a wide range of temperatures. There is no change in tensile or flexural modulus at temperatures up to 143°C, the PEEK glass-rubber transition temperature, T_g, [2]. The effect of temperature on 0° strength properties is shown in

TABLE 3—*Elastic properties of CF/PEEK, APC-2/AS-4, at 23°C measured in tension using secant values at 0.25% axial strain.*

Property	Value, GN/m²
E_{11}	134
E_{22}	8.9
G_{12}	5.1
μ_{12}	0.33
μ_{13}	0.30
μ_{21}	0.04
μ_{23}	0.35

TABLE 4—*Thermal expansion coefficients for CF/PEEK APC-2/AS-4.*

Fiber Orientation	Temperature Range, °C	Linear Thermal Expansion Coefficient, /°C
0°	23–143	0.5×10^{-6}
0°	143–300	1.0×10^{-6}
90°	23–143	30×10^{-6}
90°	143–300	75×10^{-6}

TABLE 5—*Tensile properties of multidirectional CF/PEEK laminates (APC-2/AS-4). Modulus measured at 0.25% axial strain.*

Lay-Up	Modulus, GN/m²	Strength, MN/m²
$[\pm 45]_{4S}$	19.2	356
$[0/90]_{4S}$	73	1100
$[-45/0/+45/90]_{2S}$	47	910
Open-hole $[-45/0/+45/90]_{2S}$	...	411

Fig. 1. These results show that more than 75% of 0° tensile, flexural, compressive strength, and ±45 tensile strength is retained at 120°C when compared to room temperature. Above the PEEK T_g, the flexural and compressive strengths fall due to the reduction in the matrix stiffness leading to a microbuckling failure of the fibers in compression at lower stresses. However, strength is retained above T_g due to the semicrystalline nature of the polymer, and more than 40% of the room temperature compressive strength is retained at 180°C. PEEK matrix composites show high temperature stability and, for example, showed no change in room temperature transverse strength after being stored at 120°C for two months [13]. The strength properties show a slight increase at low temperatures, and, even at −269°C (4K), the 0° flexural strength is 2580 MN/m² [14]. Therefore, over a wide temperature range, there is little change in the stiffness or strength properties of CF/PEEK.

Toughness

The toughness of CF/PEEK has been widely characterized [3,4,9,15–20]. The main reason for this interest is the relatively high interlaminar toughness of the composite.

Values of K_{IC} and G_{IC} were obtained for crack propagation in various directions relative to the reinforcing fibers [9], and the six principal directions of crack propagation in a unidirectional laminate are illustrated in Fig. 2. Toughness tests were conducted to give crack propagation in these six directions using a notched three-point bend specimen. The results are given in Table 6 with a value of interlaminar fracture toughness measured using a DCB test. K_{IC} for the interlaminar and intralaminar cracks (Directions 1–4) measured in the notched three-point bend tests are all very similar because the fracture mechanism is predominantly intralaminar (transverse) cracking. The slight differences in G_{IC} within the three-point bend results, and between the three-point bend and DCB results, have been attributed to variations in the degree of fiber-bridging during the fracture process. Crack propagation which involves fiber breakage

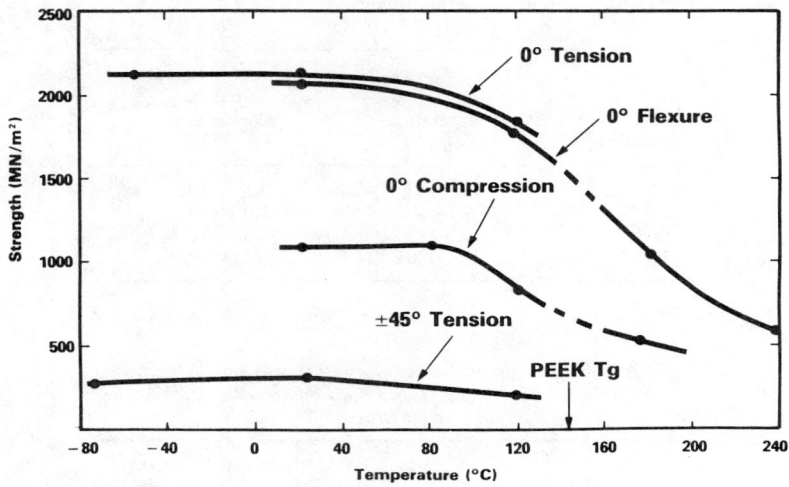

FIG. 1—*The effect of temperature on the strength properties of CF/PEEK.*

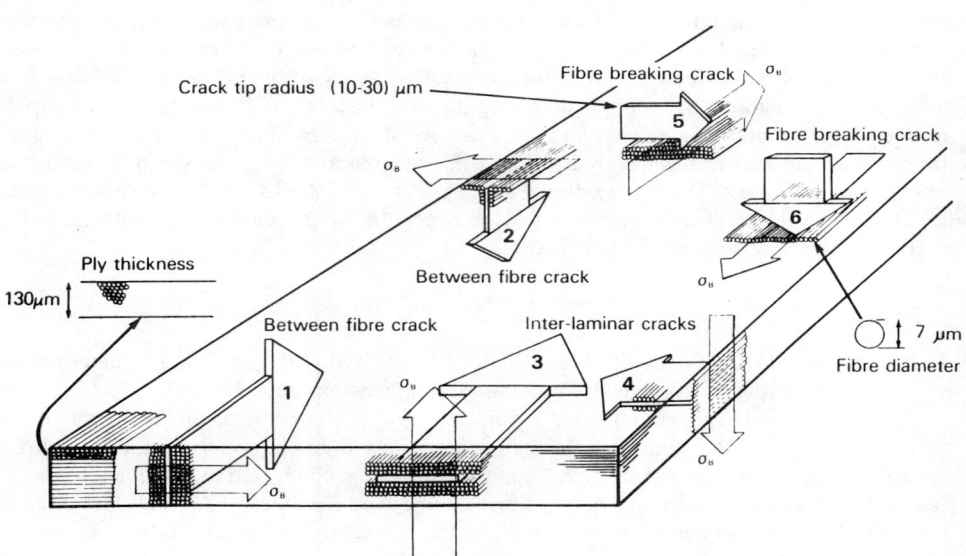

FIG. 2—*Principal directions of crack propagation in a unidirectional composite. σ_B indicates direction of bending stress for notched, three-point bend test.*

TABLE 6—*Fracture toughness results for CF/PEEK, APC-2/AS-4 measured at 23°C lay-up [0]$_{40}$, data from Ref 9, and this work.*

Crack Propagation Direction	Fracture Orientation[a]	K_{Ic}, MN/m$^{3/2}$	G_{Ic}, KJ/m^2	Test Method
Intralaminar (transverse fracture)	1 and 2	3.6	5.6	notched 3-point flexure
Interlaminar	3 and 4	3.7	3.4	notched 3-point flexure
Interlaminar	3	...	1.9	double cantilever beam (stable fracture)
Translaminar (fire breaking)	5 and 6	38	128	notched 3-point flexure

[a]See Fig. 2.

gives much higher values of K_{IC} and G_{IC} as demonstrated by the values for Crack Directions 5 and 6. The results from the DCB test show good agreement with results reported by other authors [16–20], and when tested at room temperature both stable and unstable fracture commonly have been seen. Even during unstable fracture, the toughness is high though the different methods of analysis for the unstable fracture will give different values of G_{IC}. The high fracture toughness is a consequence of the matrix surface ductility, the strong fiber-matrix adhesion, and subsurface failure processes in the composite [15].

Instrumented falling weight impact has been used to assess the energy absorbed in various impact failure processes from initial damage to complete penetration of the specimen [9,10]. The energy to puncture the specimens and the energy corresponding to the maximum force on the force-deflection curve is shown as a function of temperature in Fig. 3 for quasi-isotropic and cross-plied, 16-ply laminates. The energy absorbed up to the maximum force peak is similar for both lay-ups and does not change in the range from −55 to +80°C. The total energy absorbed is similar for both lay-ups and increases with increasing temperature. This is due to the increased ductility of the matrix at higher temperatures, resulting in greater energy absorption during the damage propagation phase of the IFWI test.

A photographic technique was used with the IFWI test and with low-applied energies in order to investigate the initial failure processes during impact loading [10]. This enabled notional G_{IC} values to be obtained for IFWI tests on cross-plied and quasi-isotropic laminates. These values are given in Table 7. These results show good agreement with predicted values [3]. The higher G_{IC} for the quasiisotropic lay-ups is due to the greater proportion of fiber fracture to intralaminar splitting in the quasi-isotropic compared to the cross-plied laminate.

The postimpact compression test frequently is used to simulate a service condition in which the composite is subjected to low-energy impact damage and subsequently is loaded in compression to fracture. Residual compressive strength is shown as a function of (applied energy)/ (thickness) in Fig. 4. As noted earlier, 40-ply, quasi-isotropic laminates were used in this testing with a thickness of 5.1 mm. Results on undamaged specimens have been omitted as these are highly dependent on the clamping of the antibuckling guides. The fall in residual compressive strength as a function of increasing impact energy is very shallow, and, even with impact energies which cause considerable surface damage, a high compressive strength is retained. It has been shown that the residual compressive properties are dependent on the extent of delamination damage as assessed by ultrasonic C-scan [4]. It also has been shown that, for CF/PEEK, there is no growth in the extent of delamination even at applied strains of 0.6% [4].

CF/PEEK shows relatively high toughness over a wide range of assessment techniques from intrinsic toughness (K_{IC} and G_{IC}) to service-related tests (IFWI, PIC). The material does not

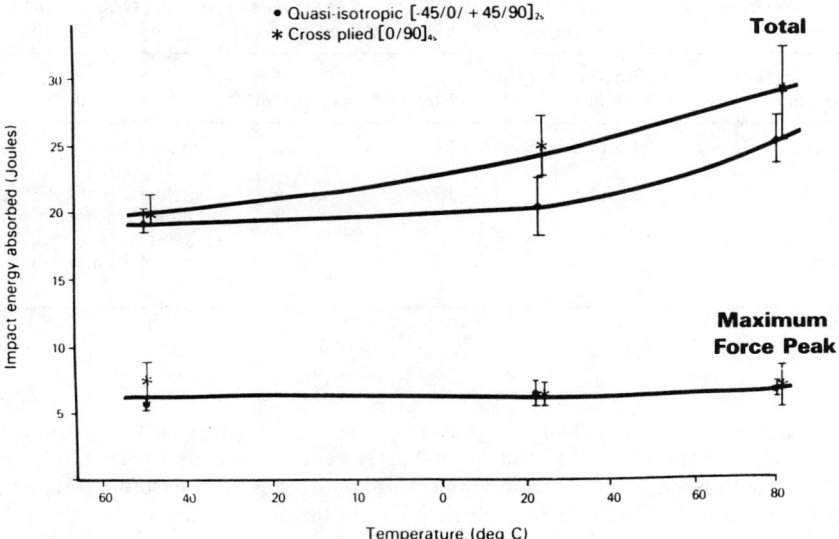

FIG. 3—*Instrument falling weight impact results versus temperature. Lay-ups [−45/0/+45/90]₂ₛ and [0/90]₄ₛ. Plaque size 75 by 75 mm, support diameter 50 mm, impactor nose diameter 12.7 mm, impact speed 5 m/s, 2.2 kHz analogue filter. Reproduced from Ref 9.*

TABLE 7—*Notional G_{IC} measurements for CF/PEEK, APC-2/AS-4, from photographed impact at 5 m/s and 23°C, data from Ref 3.*

Lay-Up	G_c, KJ/m²
[0/90]₄ₛ	66
[−45/0/+45/90]₂ₛ	80

show any transitions in toughness properties over a wide temperature range and shows high tolerance to a wide range of damage processes.

Long-Term Properties

It is important that composites retain their integrity and do not show large changes in properties as a function of time. In this respect the most critical properties are creep and fatigue.

Creep properties were characterized using an isochronous stress-strain plot together with constant stress, creep experiments in which tensile strain is measured as a function of time. Using these two functions, it is possible to obtain a full family of creep curves (strain versus time at constant stress) and isometric curves (stress versus time at constant strain) via a sequence of interpolation procedures. The creep of [0]₈ and [(0/90)₂/0/90/0/(90/0)₂] laminates is shown in Fig. 5 [21]. The creep curves at 23 and 120°C show little time dependence, indicating that the alignment of the fibers in the direction of applied stress ensures minimal reduction of modulus with time.

Creep tests also were conducted on matrix-dominated [±45]₄ₛ lay-ups. Interpolated creep

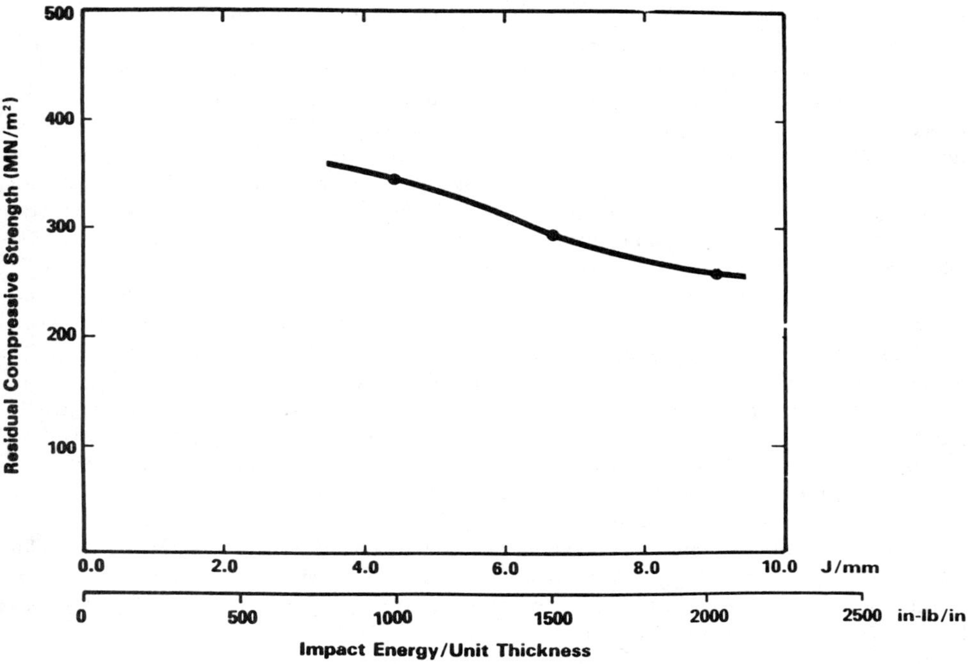

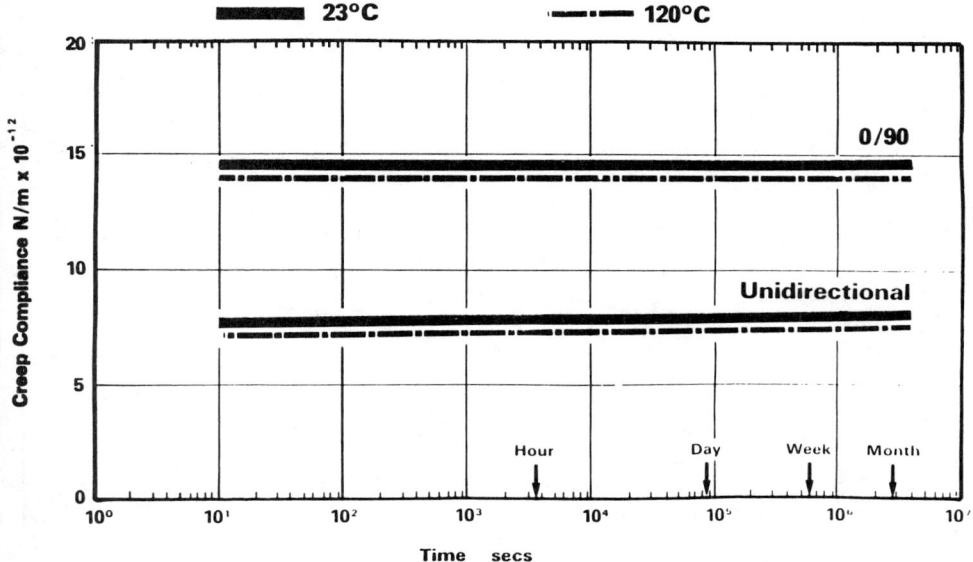

FIG. 4—*Postimpact compressive strength of CF/PEEK. Lay-up [−45/0/+45/90]ₛₛ, specimen size 152 to 102 mm.*

FIG. 5—*Creep of CF/PEEK in fiber-dominated lay-ups. Lay-ups [0]₈, [(0.90)₂/0/90/0/(90/0)₂]. Data from Ref 20.*

curves are shown for these laminates at 23°C (Fig. 6) and at 80°C (Fig. 7). For both of these tests, the applied stress in the creep test was set deliberately high, sufficient to cause yielding the matrix. Nevertheless, the CF/PEEK laminates were able to adequately sustain these stress levels for long loading periods (up to one year) without any indication of fracture. Creep would be expected in the [±45] lay-ups as, in common with all polymers, PEEK is a nonlinear viscoelastic material. The time dependence of modulus is accommodated in service by ensuring that, for engineering calculations which require a knowledge of modulus, a suitable value is used commensurate with the anticipated loading time. The family of creep curves at specific temperatures for a specific lay-up provide appropriate design data.

The behavior of the composite under cyclic loading is also important, and the fatigue of CF/PEEK has recently been examined in detail [6]. Quasi-isotropic [$-45/0/+45/90$]$_{2S}$ and [±45]$_{4S}$ laminates were examined using sine and square wave, waisted and parallel-sided specimens, and at 0.5 and 5-Hz loading frequency. If tests are conducted at high frequencies, then a temperature rise may occur in the specimen due to hysteresis effects. This would be expected to be greater for a matrix-dominated lay-up, and applied stress versus log (cycles to failure) (S/N) plots are shown for each lay-up at 0.5 Hz in Figs. 8 and 9. Both lay-ups show a sensitivity to frequency. This is attributed to temperature rise in the specimens. Temperature increases of 20°C above ambient were seen for the quasi-isotropic laminate at 5 Hz, and up to 150°C above ambient for the [±45] laminate [6] at the same frequency. Therefore, it was determined that 0.5 Hz was more appropriate for conducting fatigue tests on CF/PEEK. A summary S/N plot showing the fatigue response of quasi-isotropic and [±45] CF/PEEK is shown in Fig. 10. This provides a confident view of the fatigue performance of CF/PEEK laminates, particularly as lifetimes in excess of one year have been examined in the laboratory study.

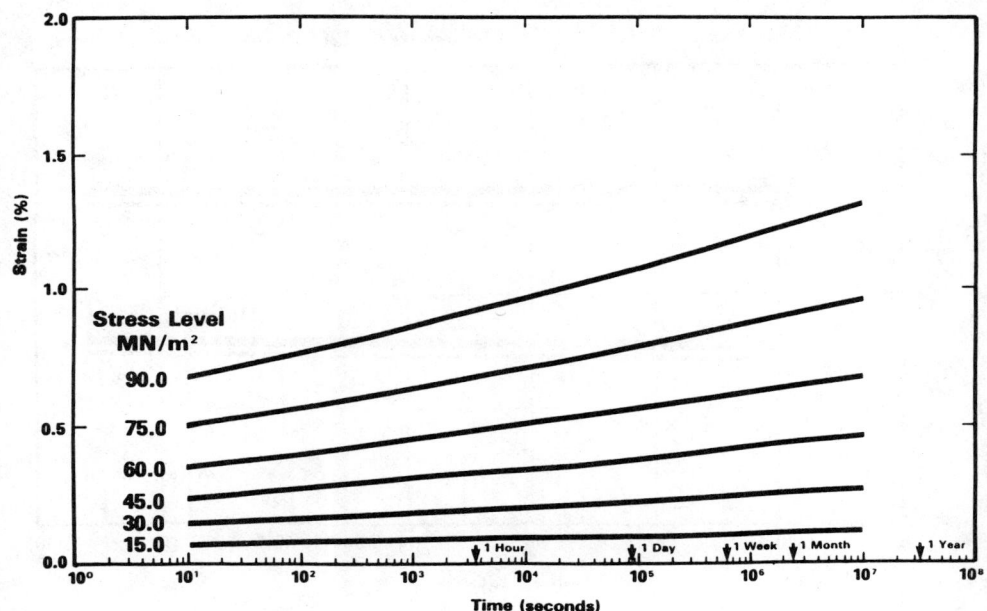

FIG. 6—*Interpolated creep curves for CF/PEEK at 23°C, lay-up [±45]$_{4S}$.*

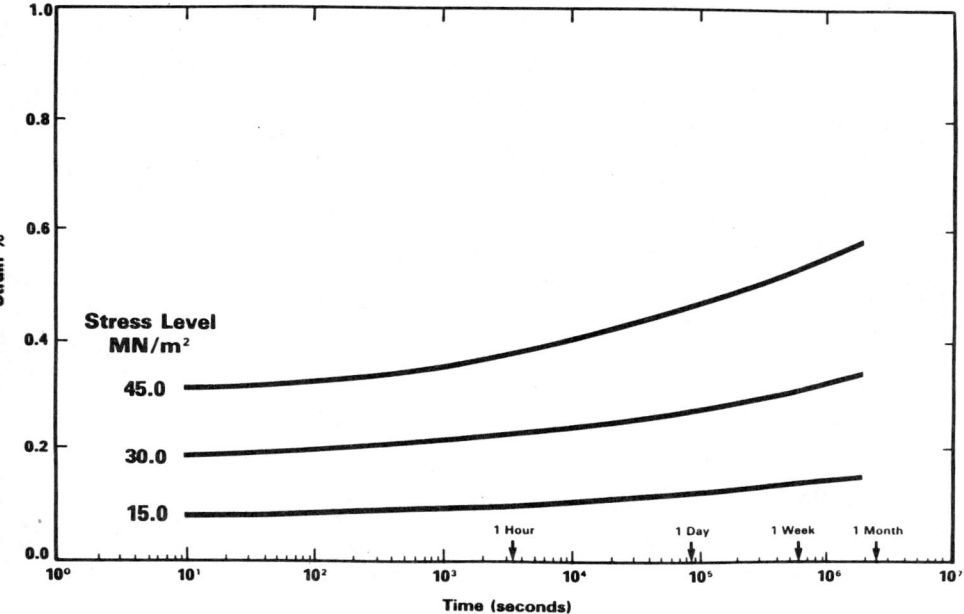

FIG. 7—*Interpolated creep curves for CF/PEEK at 80°C, lay-up [±45]₄ₛ.*

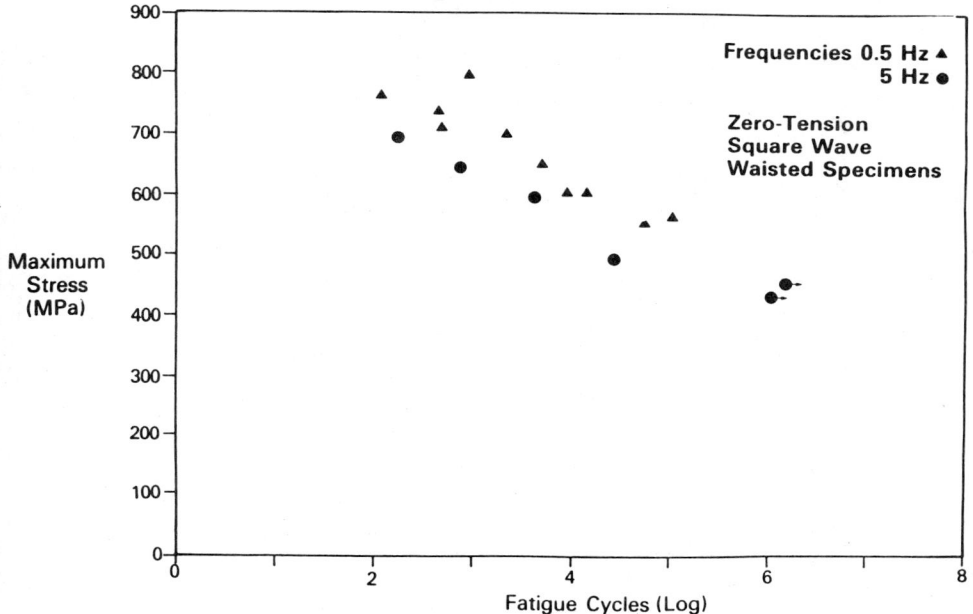

FIG. 8—*Influence of test frequency on fatigue performance of CF/PEEK, lay-up [−45/0/+45/90]₂ₛ.*

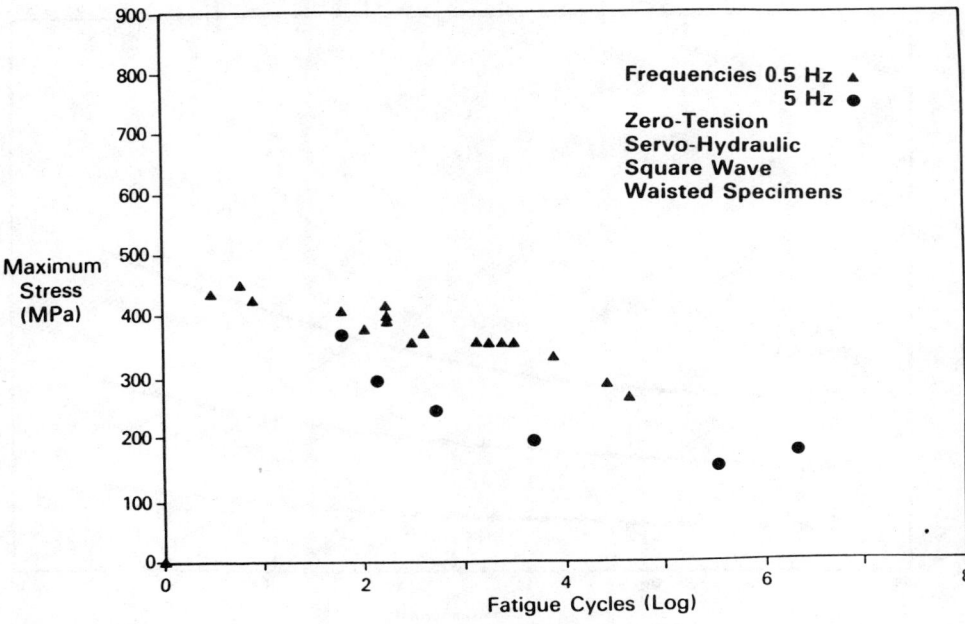

FIG. 9—*Influence of test frequency on fatigue performance of CF/PEEK, lay-up* $[\pm 45]_{4S}$.

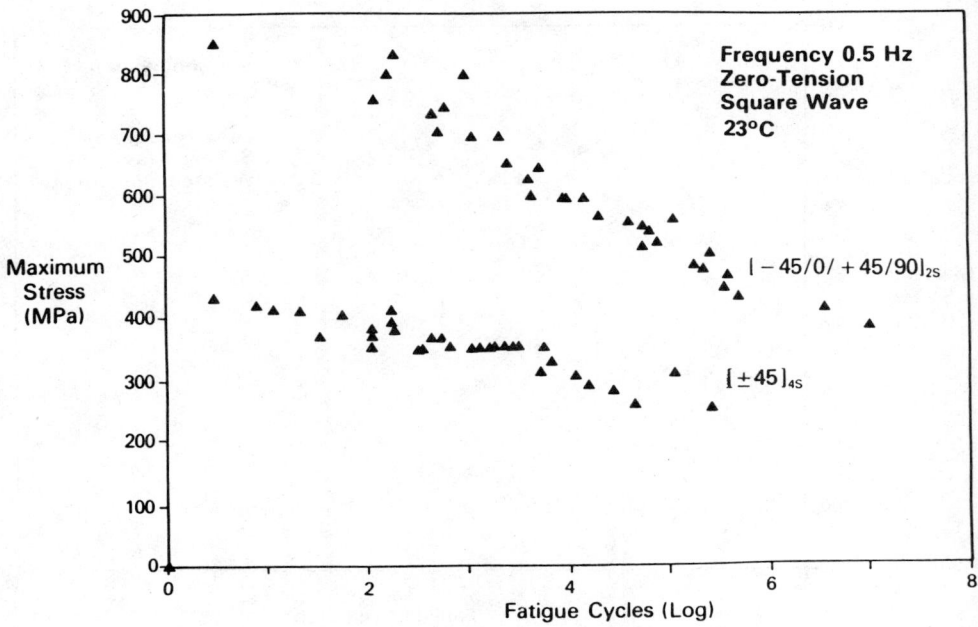

FIG. 10—*Maximum fatigue stress versus log (no. of cycles to failure) (S/N) plot for CF/PEEK.*

Conclusions

1. CF/PEEK composites, APC-2/AS-4, show a high utilization of the fiber properties in stiffness and strength, and similarly a high utilization of the matrix ductility in toughness.

2. A high proportion of room temperature stiffness and strength is retained at temperatures up to 120°C, and a slight increase occurs at cryogenic temperatures. Stiffness and strength are retained above the T_g of the matrix (143°C) due to the semicrystalline nature of the polymer.

3. CF/PEEK composites exhibit high toughness in a wide range of assessment techniques including fracture mechanics, instrumented falling weight impact, and postimpact compression.

4. The creep resistance of CF/PEEK in fiber- and matrix-dominated lay-ups is adequate at both room temperature and at elevated temperature.

5. The fatigue behavior of fiber- and matrix-dominated lay-ups have been studied for lifetimes in excess of one year (10^7 cycles), enabling cyclic loads to be accounted for in structural applications.

Acknowledgments

The authors acknowledge the experimental contributions from a number of colleagues in ICI, including D. C. Curtis, A. C. Lowe, R. S. Prediger, D. R. Tamblin, and B. Slater.

References

[1] Jones, D. P., Leach, D. C., and Moore, D. R., "Mechanical Properties of Poly(ether-ether-ketone) for Engineering Applications," *Polymer*, Vol. 26, 1985, pp. 1385–1393.

[2] Leach, D. C., Briggs, P. J., and Carlile, D. R., "Mechanical and Fire Properties of Aromatic Polymer Composites," *Proceedings of the Third Symposium on Spacecraft Materials in Space Environment, ESA SP-232*, European Space Agency, Noordwijk, Netherlands, 1985, pp. 153–160.

[3] Crick, R. A., Leach, D. C., and Moore, D. R., "Interpretation of Toughness in Aromatic Polymer Composites," *SAMPE Journal*, Vol. 22, No. 6, 1986, pp. 30–36.

[4] Leach, D. C., Curtis, D. C., and Tamblin, D. R., "Delamination Behavior of Carbon Fiber/Polyetheretherketone (PEEK) Composites, *Toughened Composites, ASTM STP 937*, N. J. Johnston, Ed., ASTM, Philadelphia, 1987, pp. 358–380.

[5] Leach, D. C., Cogswell, F. N., and Nield, E., "High Temperature Performance of Thermoplastic Aromatic Polymer Composites," *Proceedings of the 31st SAMPE Symposium*, April 1986, pp. 434–448.

[6] Curtis, D. C., Moore, D. R., Slater, B., and Zahlan, N., "Fatigue Testing of Multi-angle Laminates of CF/PEEK," *Composites*, Vol. 19, 1988, pp. 446–452.

[7] Aromatic Polymer Composites Data Sheet, ICI Fiberite, P.O. Box 787, Orange, CA.

[8] Hofer, K. E., Rao, N., and Larsen, D., "Development of Engineering Data on Mechanical Properties of Advanced Composite Materials," USAF Technical Report AFML-TR-72-205 Part 1, 1972.

[9] Leach, D. C. and Moore, D. R., "Toughness of Aromatic Polymer Composites Reinforced with Carbon Fibers," *Composites Science and Technology*, Vol. 23, 1985, pp. 131–161.

[10] Moore, D. R. and Prediger, R., "A Study of Low Energy Impact of Continuous Carbon Fiber Reinforced Composites," submitted to *Polymer Composites*.

[11] Turner, S., "Creep in Thermoplastic," *British Plastics*, July 1964.

[12] Whitney, J. M. and Browning, C. E., "On Short-Beam Shear Tests for Composite Materials," *Experimental Mechanics*, Sept. 1985, pp. 294–300.

[13] Cogswell, F. N. and Leach, D. C., "Continuous Fiber Reinforced Thermoplastic—a Change in the Rules for Composite Technology," *Plastics and Rubber Materials and Applications*, Vol. 4, No. 3, 1984, pp. 271–276.

[14] Evans, D., Morgan, J. T., Robertson, S. J., and Zahlan, N., "The Physical Properties of Carbon Fiber Reinforced PEEK Composites at Low Temperatures," *Proceedings of the 1987 Cryogenic Engineering/International Cryogenic Materials Conference*, St. Charles, Illinois, June 1987.

[15] Crick, R. A., Leach, D. C., Meakin, P. J., and Moore, D. R., "Interlaminar Fracture Morphology of Carbon Fiber/PEEK Composites," *Journal of Materials Science*, Vol. 22, 1987, pp. 2094–2104.

[16] Berglund, L. and Johannesson, T., "Mixed-mode Fracture of Carbon-Fiber/PEEK Composites,"

Proceedings of the First European Conference on Composite Materials, European Association for Composite Materials, Bordeaux, France, 1985, pp. 259-264.

[17] Davies, P. and de Charentenay, F. X., "The Effect of Temperature on the Interlaminar Fracture of Tough Composites," *Proceedings of the Sixth International Conference on Composite Materials,* Vol. 3, Elsevier, London, England, 1987, pp. 284-294.

[18] Hine, P. J., Brew, B., Duckett, R. A., and Ward, I. M., "Failure Mechanisms in Continuous Fiber Reinforced PEEK," *Proceedings of the Sixth International Conference on Composite Materials,* Vol. 3, Elsevier, London, England, 1987, pp. 397-404.

[19] Hashemi, S., Kinloch, A. J., and Williams, J. G., "Interlaminar Fracture of Composite Materials," *Proceedings of the Sixth International Conference on Composite Materials,* Vol. 3, Elsevier, London, England, 1987, pp. 254-264.

[20] Smiley, A. J. and Pipes, R. B., "Rate Sensitivity of Interlaminar Toughness in Composite Materials," *Proceedings of the First Technical Conference,* American Society for Composites, Technomic, Lancaster, PA, 1986, pp. 434-449.

[21] Dean, E. F. and Johnson, A. F., National Physical Laboratory, Teddington, UK, private communication.

Suong V. Hoa,[1] *Sui Lin,*[1] *and Jirui R. Chen*[1]

Effects of Moisture Content on the Mechanical Properties of Polyphenylene Sulfide Composite Materials

REFERENCE: Hoa, S. V., Lin, S., and Chen, J. R., "**Effects of Moisture Content on the Mechanical Properties of Polyphenylene Sulfide Composite Materials,**" *Advances in Thermoplastic Matrix Composite Materials, ASTM STP 1044*, G. M. Newaz, Ed., American Society for Testing and Materials, Philadelphia, 1989, pp. 213-230.

ABSTRACT: Effects of moisture absorption on the mechanical behavior of glass-reinforced polyphenylene sulfide (PPS) composites were examined. At 100°C and 95% relative humidity, significant moisture absorption into PPS composites took place within a period of one or two weeks. Large reduction in tensile strengths from 25 to 46% were observed. This reduction is significantly increased up to 60% if the thermoplastic material is not properly crystallized.

KEY WORDS: PPS composites, moisture content, crystallinity, tensile strength

Due to their good fracture toughness properties, high temperature performance, and ease in handling, thermoplastic matrix composites are gaining in popularity. Among many thermoplastic materials, polyphenylene sulfide (PPS) offers attractive properties in comparison with its cost and availability.

Polyphenylene sulfide is produced [1] by the reaction of *p*-dichlorobenzene and sodium sulfide at an elevated temperature in a polar solvent. The crystallinity of this polymer depends on thermal treatment. A PPS sample that is heated above the melting point and rapidly quenched yields a thermal analysis typical of an amorphous, but crystallizable thermoplastic. A glass transition temperature, T_g, of 80°C, a crystallization temperature, T_c, of 130°C, and a melting temperature, T_m, of 285°C can be observed on a differential scanning calorimetry (DSC) graph. Annealing the polymer above T_g (200°C/30 min) will produce a thermoplastic with approximately 65% crystallinity and with no sign of T_c on the DSC curve [2]. Good descriptions of the processing of the polymer can be found [3,4].

Polyphenylene sulfide can be used as a matrix for short or long fiber reinforcement [5]. The resistance of this material in harsh environments has also been studied by Vives et al. [6]. In Ref 6, PPS was subjected to a series of acids, bases, alcohols, aldehydes, ketones, amines, and esters for 24 h at 93°C. The percent retention of tensile strength was always more than 80% except in the case of 88% formic acid (75%), butylamine (49%), and ethylenediamine (65%).

Even though the thermoplastic material shows good resistance to many chemicals, the resistance of glass-reinforced PPS to the exposure to water is low. At 95°C, a 40% glass-reinforced PPS shows a reduction of more than 30% in flexural strength after a five-week exposure period. The same PPS material but unreinforced showed no change in flexural strength with exposure to water. In this paper, effects of exposure to 100°C and 95% relative humidity on the mechanical properties of glass-reinforced PPS (both short and long fibers) are examined.

[1]Professors and graduate student, respectively, Department of Mechanical Engineering, Concordia University, Montreal, Quebec, Canada H3G 1M8.

Materials

The materials used were supplied by Phillips Petroleum Company. Two types of samples, injection molded and advanced composites, were received. A summary of the characteristics of the as-received materials is shown in Table 1. The thicknesses of the mat-reinforced stampable sheets, AG10-20 and AG20-40, were usually about 1.65 mm. The thicknesses of the fabric-reinforced composites, AG31-60 and AG31-40 prepregs, were usually about 0.38 mm.

The as-received samples were subjected to differential scanning calorimetry (DSC) tests to determine their glass transition temperature and their crystallinity. These tests were carried out in accordance with the ASTM Test Method for Transition Temperatures of Polymers by Thermal Analysis (D 3418) on a Dupont 910 differential scanning calorimeter at a heating rate of 20°C/min in a nitrogen atmosphere with the temperature ranging from 30 to 330°C for the first run and from 30 to 180°C for the second run. The first run, the preliminary thermal cycle, was performed in order to erase any previous thermal history. Each sample was cooled with liquid nitrogen. The glass transition temperature, T_g, was determined after the first run and is shown in Table 1.

In the second run, the onset of the melting temperature on each curve was observed by extrapolation to be 259.5°C for the four advanced composite samples. The final melting temperature was determined to be 276.6°C. Two sets of these DSC curves are shown in Figs. 1 through 4. Figures 1 and 2 show the DSC curves for AG31-60 for the first run and second run, respectively, and Figs. 3 and 4 show the curves for AG31-40. It must be noted that for the AG31-60 sample, an exothermic peak was observed in the first run in the temperature range from 120 to 160°C. According to Kays and Hunter [7], this exothermic peak indicates that further crystallization was taking place and that the material was not as crystallized as possible.

Above their melting temperature, T_m, crystallizable resins are in an amorphous state. Below their glass transition temperature, T_g, crystallization occurs at a very slow rate. Most crystalline polymers will remain nearly completely amorphous when quenched in water from the melt; if cooled in air from the melt even at a rate as high as 20°C/min, however, they will attain near their maximum crystallinity.

Another method for achieving crystallinity in polymers is to anneal the amorphous film at temperatures between T_g and T_m. According to Mascia [9], as crystallization proceeds, the spherulites eventually will touch each other and, as a result, the uncrystallized polymer chains

TABLE 1—*Materials tested (as-received).*

Composite	E Glass, %	Fiber Length	Mineral Filled	Sheet Thickness, mm	T_g, °C
		INJECTION MOLDED			
R-4	40	Short	No	3.00	...
R-7	40	Short	Yes	3.05	...
A-100	40	Short	No	3.00	...
A-200	40	Short	No	3.00	...
		ADVANCED COMPOSITES			
AG31-60	60	Long satin weave fabric	N/A	0.378	90.5
AG31-40	40	Long satin weave fabric	N/A	0.378	93.0
AG11-20	20	25.4 mm chopped stampable sheet	N/A	1.650	90.1
AG20-40	40	25.4 mm chopped stampable sheet	N/A	1.670	91.5

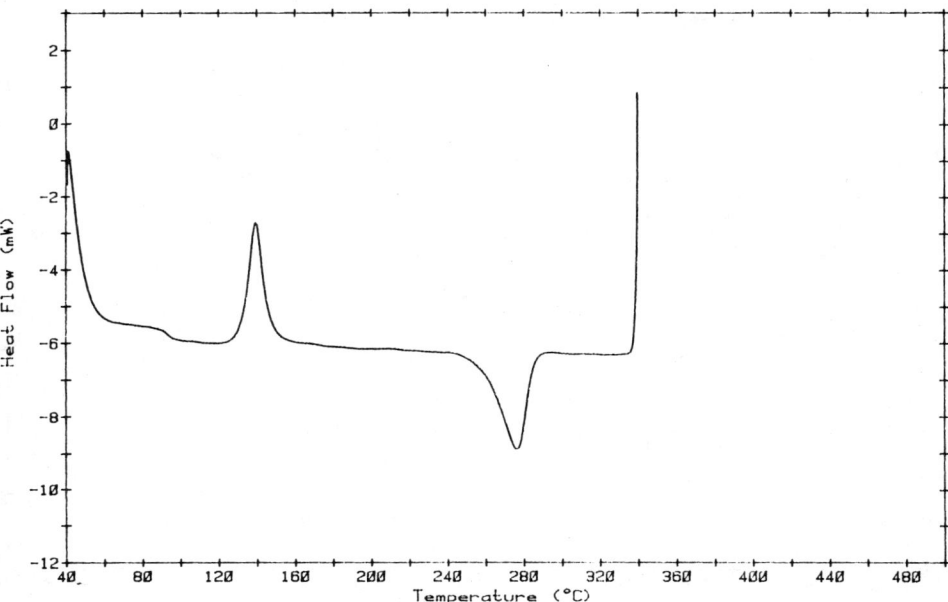

FIG. 1—*DSC curve, first run for AG31-60 (as-received).*

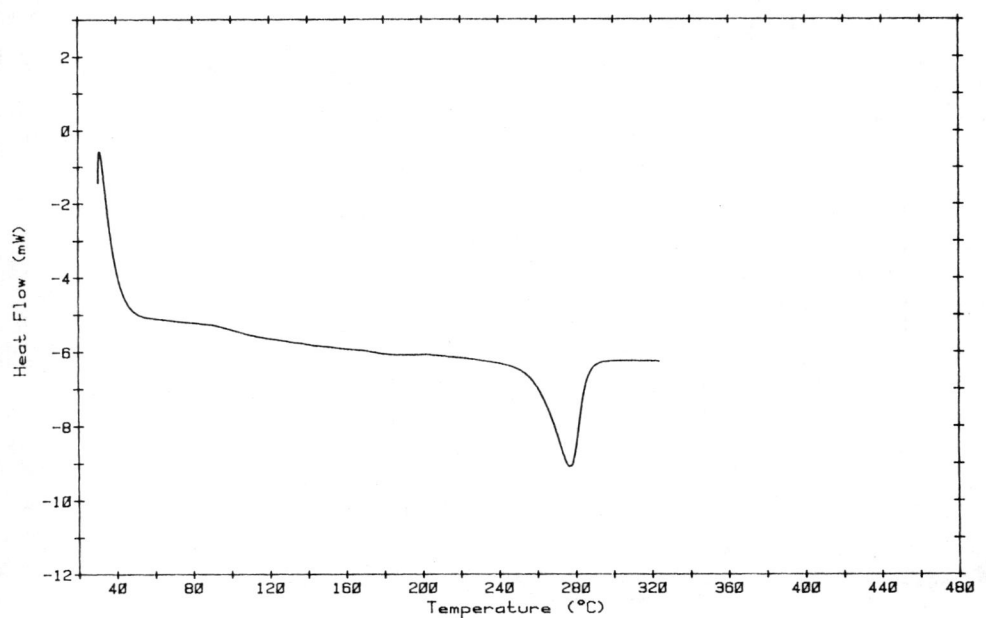

FIG. 2—*DSC curve, second run for AG31-60 (as-received).*

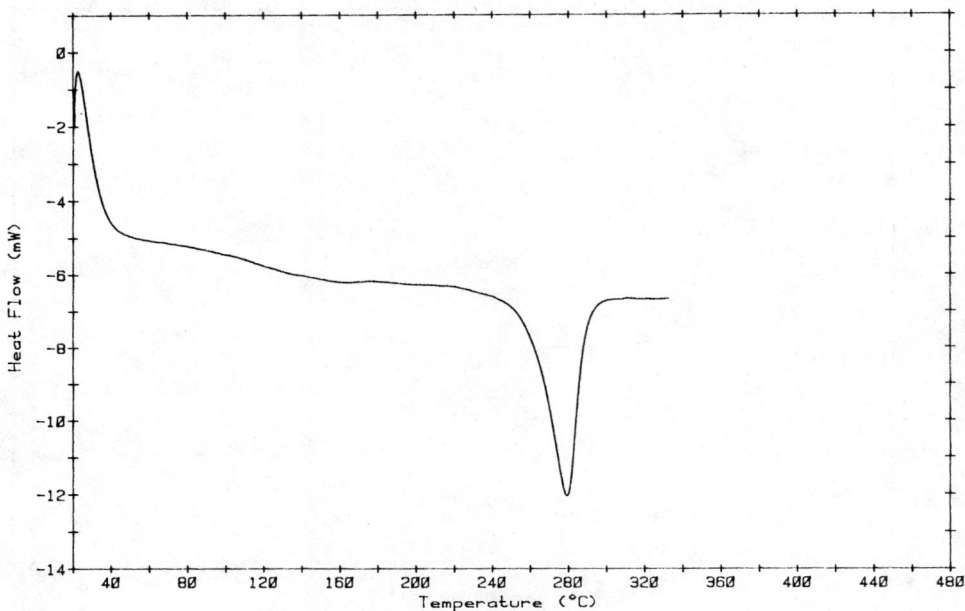

FIG. 3—*DSC curve, first run for AG31-40 (as-received).*

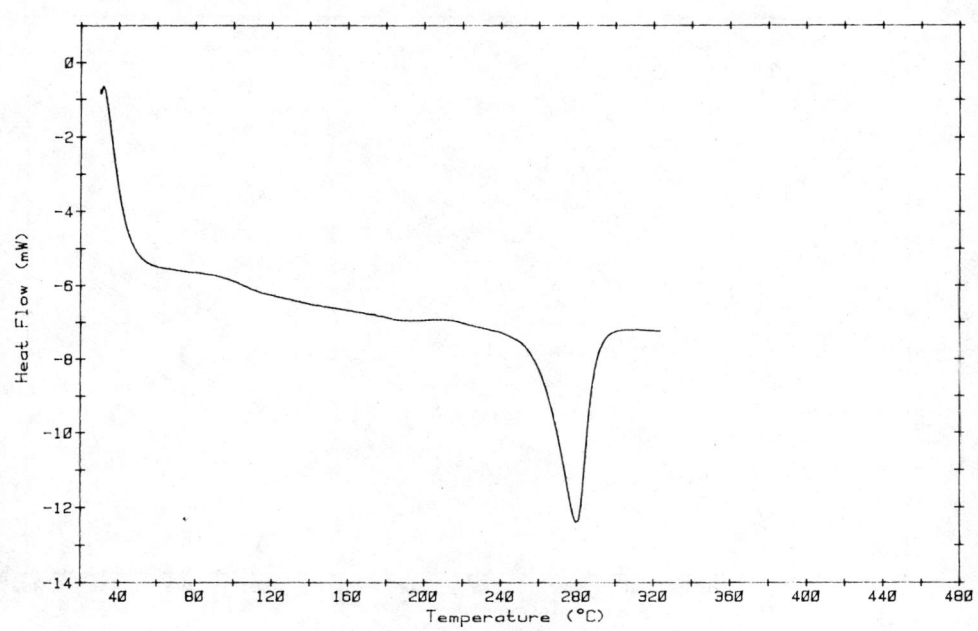

FIG. 4—*DSC curve, second run for AG31-40 (as-received).*

at the boundaries will find it increasingly more difficult to diffuse to the surface of existing crystals or to nucleate into new crystallites. If, after crystallization, the polymer is heated to temperatures just below the true melting point, some partial melting of the least perfect crystals takes place. The stable crystals, however, will continue to grow into thicker lamellae and a higher level of crystallinity is developed.

Crystallites have a marked influence on material properties. It is not unexpected, therefore, that the melting of the crystallites changes the properties abruptly and that the temperature at which this occurs sets an upper limit to the working range of most crystalline plastics. The crystallites of plastics materials vary in size and perfection, depending on local variations in molecular structure, such as chain branching and the previous thermal history of the sample. For these reasons, the melting of such crystallites is not a single catastrophic phenomenon occurring over a very narrow temperature range, as in the case for crystals of low molecular weight organic compounds, but a process that takes place progressively with increasing temperature. DSC curves for the same material in a subsequent run and DSC curves for other materials did not show this exothermic peak.

This lack of crystallinity in the AG31-60 samples caused a large reduction in tensile strength of samples exposed to hot-moist conditions, as will be shown later. In order to gain the maximum crystallinity in the AG31-60 samples, molding and annealing them are necessary. For this purpose, an infrared oven is used to preheat the samples and the molding is done on a press where the pressure can be controlled. It must be noted that the molding procedure of these materials differs because of the large differences in their thicknesses. The differences are as follows:

Molding Process of the Stampable Sheets—The as-received sheets were cut into required blanks and put into an infrared oven in which the temperature was controlled within the range from $320 + 10°C$ by a microcomputer-based controller. Heat by radiation was supplied from a single side. The smooth blanks became rough textured on one side in about 9 min and then were turned over to be heated from the other side for about 6 min. The totally rough textured blanks were taken out of the infrared oven and were quickly put together and transferred to a nearby press where the blanks were held under a hydraulic pressure of 200 bars as suggested by the company who supplied the materials. The molded part cooled sufficiently from about 300°C to the room temperature of 24°C in about 11 min (25°C/min) to retain its shape.

Molding Process of the Prepregs—The as-received prepregs were cut into required sizes and stacked one layer over another first before being put into the infrared oven. Since they were so thin that sufficient heat could penetrate several layers (four layers in this investigation) from one side, there was no need to turn them over. The temperature was controlled within the range from $330 + 10°C$. The prepregs were heated for 12 min and then taken out of the heating system and quickly transferred to the press, where they were held under a hydraulic pressure of 200 bars. The molded part cooled sufficiently from about 300°C to room temperature of 24°C in about 10 min (27°C/min) to retain its shape. It must be noted that the way to mold the stampable sheets is not applicable for the molding of the prepregs. Since the prepregs were so thin, they had to be stacked together before heating, whereas the stampable sheets did not have to be. Samples were cut from the molded parts. Dimensions of the new experimental samples are shown in Table 2.

Hot-Moist Condition and Tensile Properties

Injection molded samples as received from Phillips were exposed to the hot-moist condition of 100°C and 95% relative humidity for 13 days, and the advanced composites (both as received and molded) were exposed for 7 days. Percent weight uptakes for the injection-molded (as-received) samples are shown in Fig. 5. After 13 days, about 0.5% of water was absorbed and all samples did not show saturation of moisture uptake. The largest water uptake was observed in

TABLE 2—*Experimental results for PPS samples (molded)*.

Advanced Composite	Sample	Width, mm	Thickness, mm	Moisture Content, %	Tensile, MPa
AG31-60 (four layers)	1	13.56	1.47	0.00	133.44
	2	13.63	1.55	0.00	155.67
	3	13.63	1.45	0.00	154.33
	4	13.71	1.52	0.00	119.97
	5	13.67	1.37	11.27	90.77
	6	13.98	1.40	11.01	86.64
	7	13.82	1.50	10.92	87.54
	8	13.63	1.51	11.21	104.18
	9	13.60	1.52	15.23	74.98
	10	13.71	1.48	14.93	73.93
	11	13.64	1.40	14.62	78.55
	12	13.76	1.54	15.11	77.87
AG31-40 (four layers)	1	13.61	2.32	0.00	110.85
	2	13.69	2.18	0.00	122.30
	3	13.68	2.20	0.00	129.59
	4	13.63	2.12	0.00	115.94
	5	13.67	2.20	7.32	91.44
	6	13.64	2.23	7.07	103.56
	7	13.69	2.20	7.21	104.56
	8	13.68	2.31	7.78	88.61
	9	13.70	2.49	11.00	64.49
	10	13.65	2.22	10.66	77.55
	11	13.66	2.72	10.32	72.67
	12	13.65	2.42	10.03	63.57
AG11-20 (two layers)	1	13.52	3.66	0.00	97.80
	2	13.56	3.28	0.00	104.55
	3	13.59	3.37	0.00	89.70
	4	13.62	3.35	0.00	88.90
	5	13.97	3.23	3.72	74.24
	6	13.85	3.25	3.11	76.65
	7	14.00	3.24	3.63	71.65
	8	13.59	3.36	5.22	61.32
	9	13.69	3.28	5.36	55.68
	10	13.69	3.42	5.03	53.40
AG20-40 (two layers)	1	13.65	3.42	0.00	124.24
	2	13.65	3.02	0.00	135.85
	3	13.62	3.32	0.00	131.58
	4	13.60	3.30	0.00	122.52
	5	13.59	3.19	3.61	96.50
	6	13.59	3.17	3.06	96.33
	7	13.69	3.37	2.78	97.79
	8	13.68	3.13	4.15	93.42
	9	13.66	3.07	4.32	88.81
	10	13.65	3.05	3.73	91.87

Samples R-7, which were mineral filled. Figure 6 shows the percent weight uptake for advanced composite samples. In general, these samples showed much larger percent weight uptake than the injection-molded samples. After one week, the stampable chopped fiber mats, AG20-40 (molded), showed the least amount of weight uptake, 4%. The other stampable chopped fiber mats, AG11-20 (molded), showed about 5% weight uptake. The continuous-fiber satin weave fabric, AG31-40 (as received), showed 6% weight uptake after 7 days, while the molded AG31-40 samples showed 10% weight uptake after 7 days. The largest amount of weight uptake (25%)

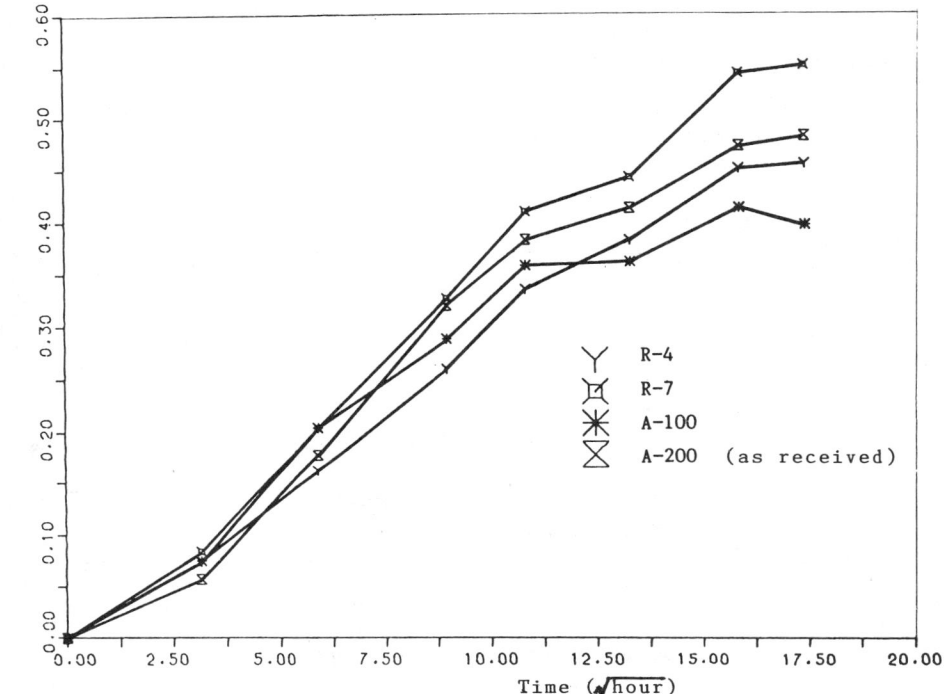

FIG. 5—*Moisture content versus square root of time.*

was exhibited in the continuous fiber satin weave fabric, AG31-60 (as-received), whereas the molded AG31-60 samples showed only about 15% weight uptake after 7 days. Compared with the AG31-40, this excessive amount of weight uptake in AG31-60 can be due to two factors: the larger amount of glass fibers, and the lower degree of crystallinity in the AG31-60 samples as discussed in the previous section. It is clear that the characteristics of moisture absorption in these samples were improved after molding the material, which erased the previous thermal history and annealed the amorphous samples.

Samples were taken out of the environmental chamber at different intervals of time and tested under tension in the MTS machine at a crosshead speed of 5 mm/min. Stress-strain curves for the injection-molded samples at typical amounts of moisture uptake are shown in Figs. 7 to 10. As expected, all samples showed a drop in both modulus and strength as the moisture content increased. The R-4 and R-7 samples showed smaller fracture strains in comparison with the A-100 and A-200 samples. All injection-molded samples showed a tensile strength of about 50 to 60 MPa. The drops in moduli of these samples due to the hot-moist conditions are shown in Table 3.

Stress-strain curves for the advanced composite samples at typical moisture content levels are shown in Figs. 11 to 14. There are significant reductions in both modulus and strength in all the samples, the largest being that of sample AG31-60 (as-received).

Tensile strengths of injection-molded samples (as-received) at different moisture contents are shown in Fig. 15. In general, there is a reduction of about 25% in tensile strength of all samples due to exposure to moist conditions for as short a time as two weeks.

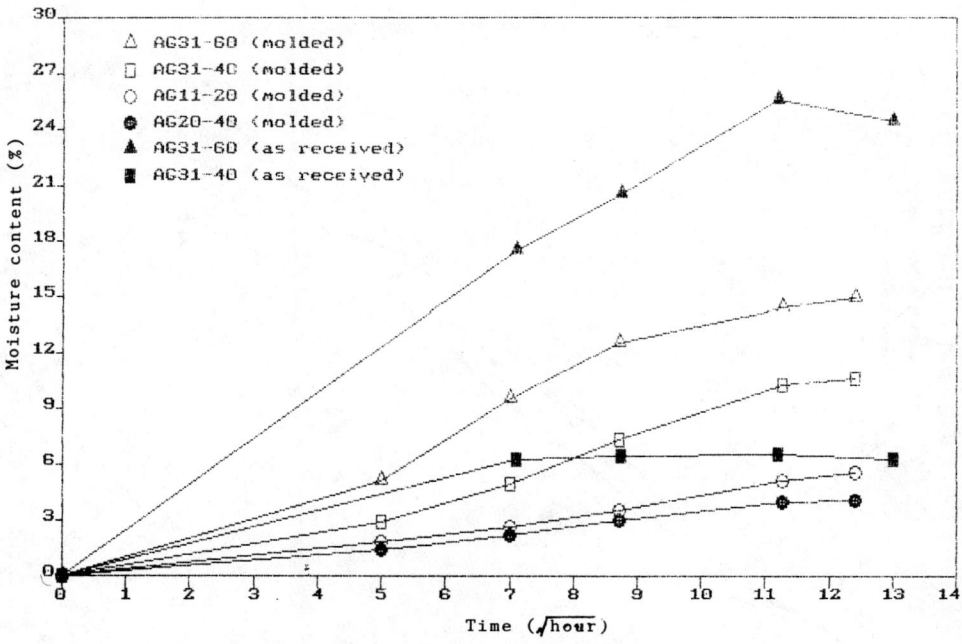

FIG. 6—*Moisture content versus square root of time.*

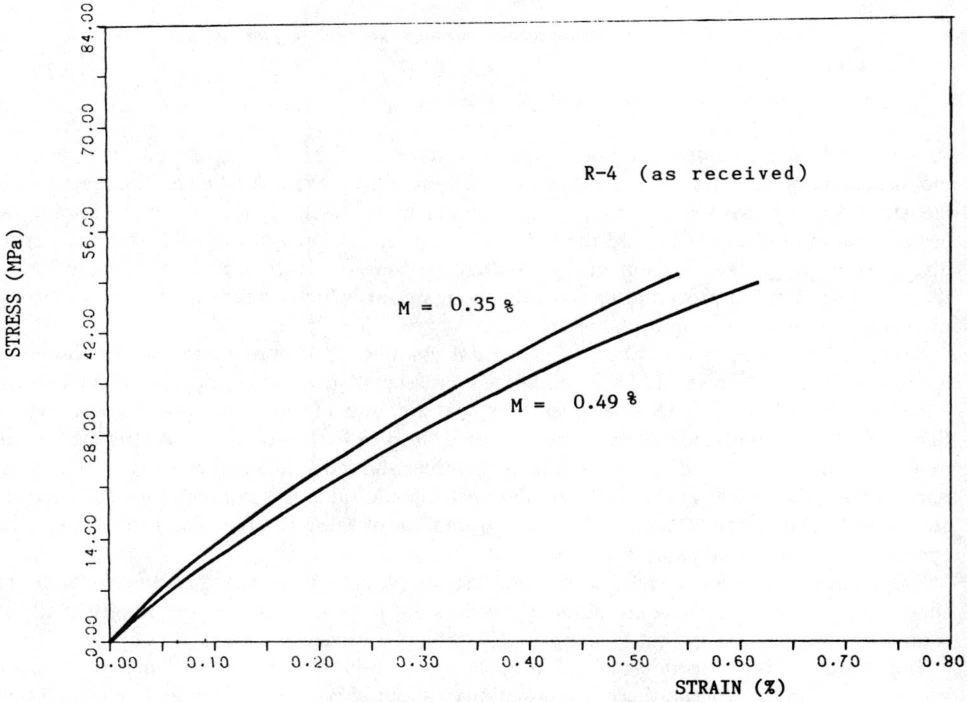

FIG. 7—*Stress-strain diagram with moisture content as a parameter.*

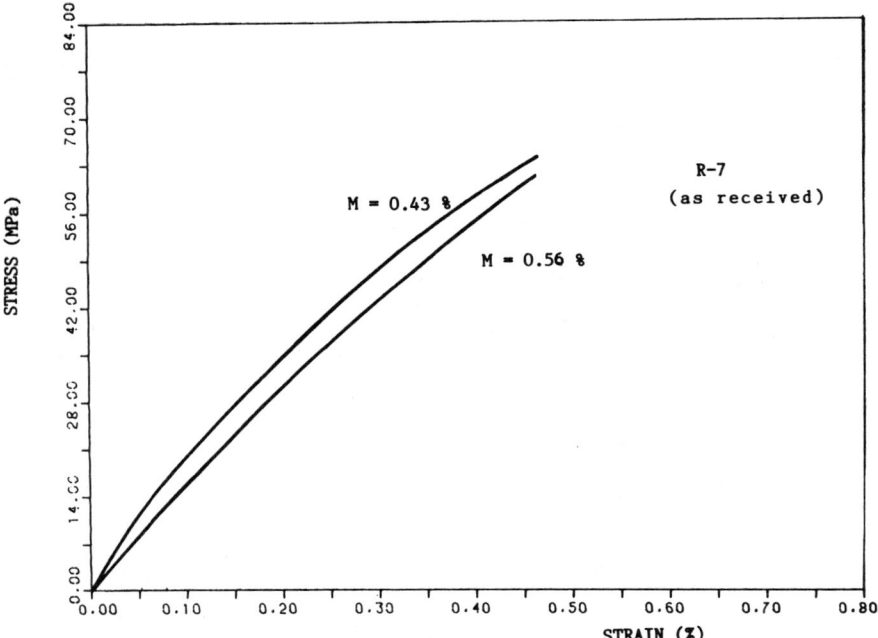

FIG. 8—*Stress-strain diagram with moisture content as a parameter.*

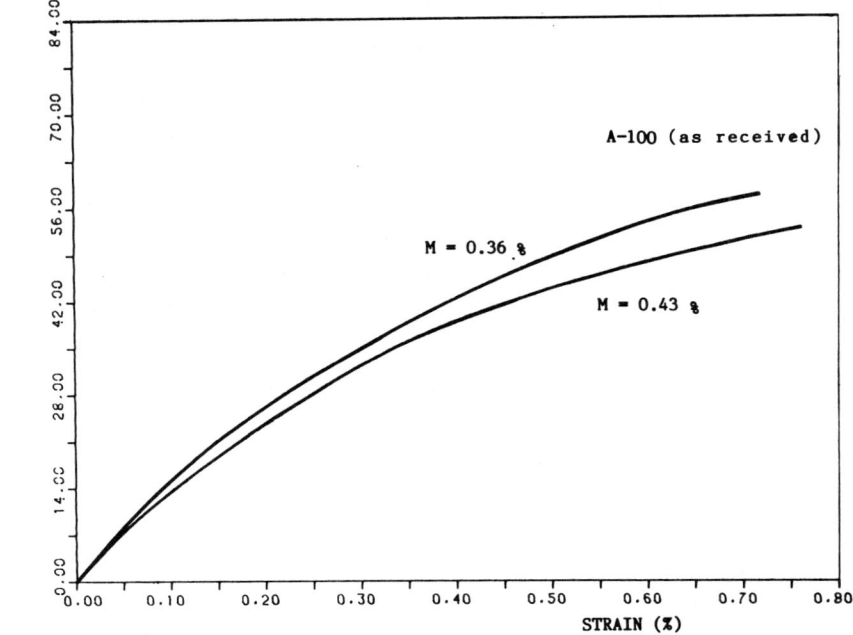

FIG. 9—*Stress-strain diagram with moisture content as a parameter.*

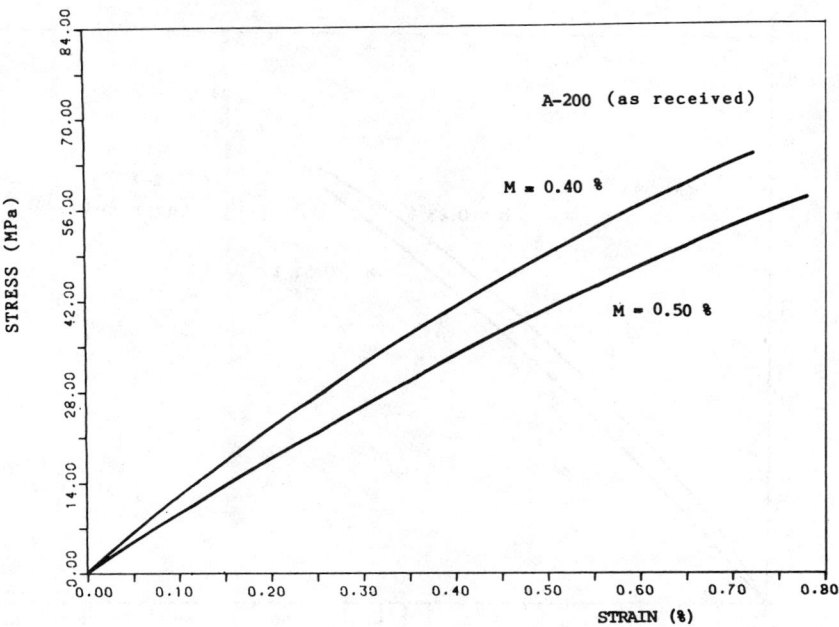

FIG. 10—*Stress-strain diagram with moisture content as a parameter.*

TABLE 3—*Elastic moduli of the composites calculated from the stress-strain curves with the change of moisture content.*

Composites	Moisture Content, %	Elastic Moduli, GPa
INJECTION MOLDED (AS-RECEIVED)		
R-4	0.35	15.4
	0.49	12.5
R-7	0.43	25.9
	0.56	16.8
A-100	0.36	16.5
	0.43	14.4
A-200	0.40	13.0
	0.50	9.5
ADVANCED COMPOSITES (MOLDED)		
AG31-60	0.00	42.1
	11.27	26.7
	15.11	19.0
AG31-40	0.00	26.6
	7.32	19.0
	10.32	14.9
AG11-20	0.00	24.5
	3.72	20.1
	5.36	17.3
AG20-40	0.00	27.1
	3.06	24.6
	4.15	23.1

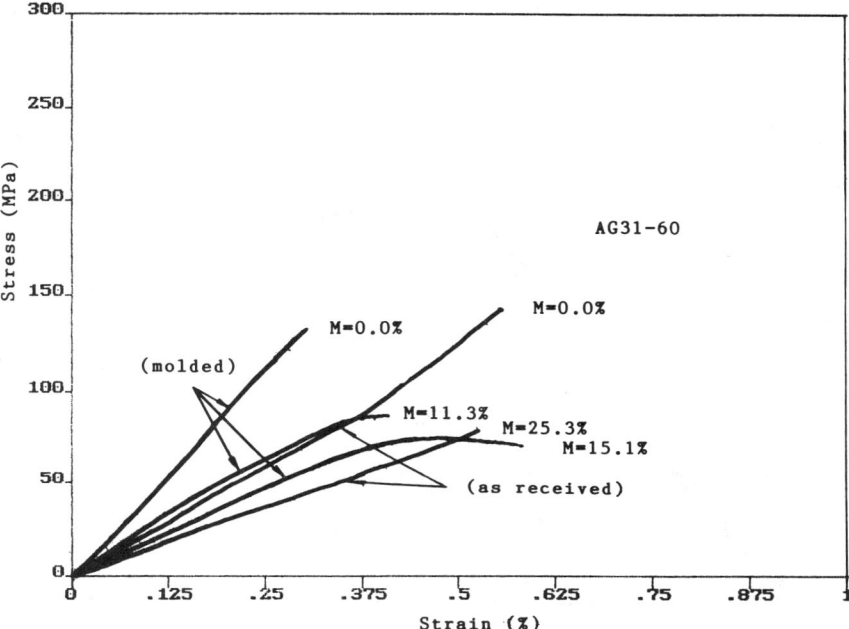

FIG. 11—*Stress-strain diagram with moisture content as a parameter.*

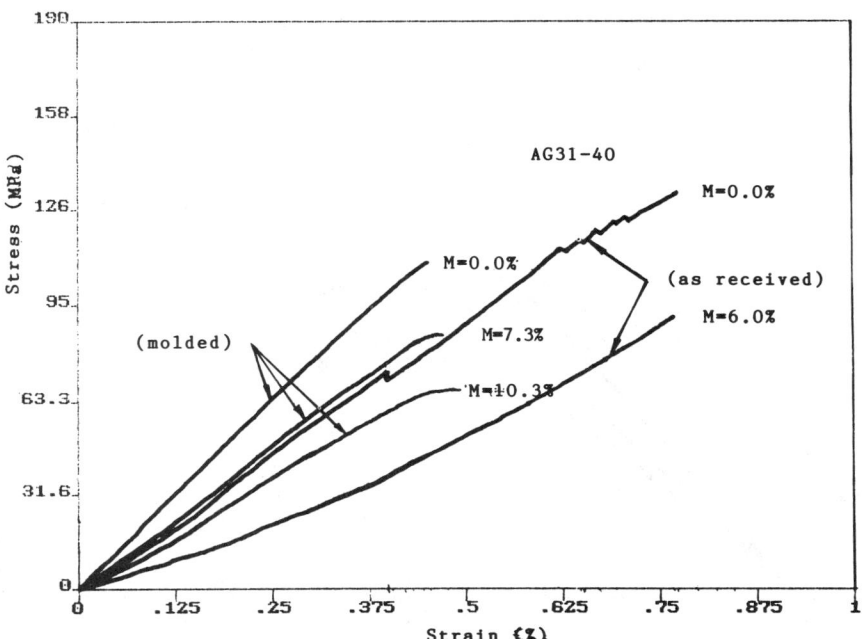

FIG. 12—*Stress-strain diagram with moisture content as a parameter.*

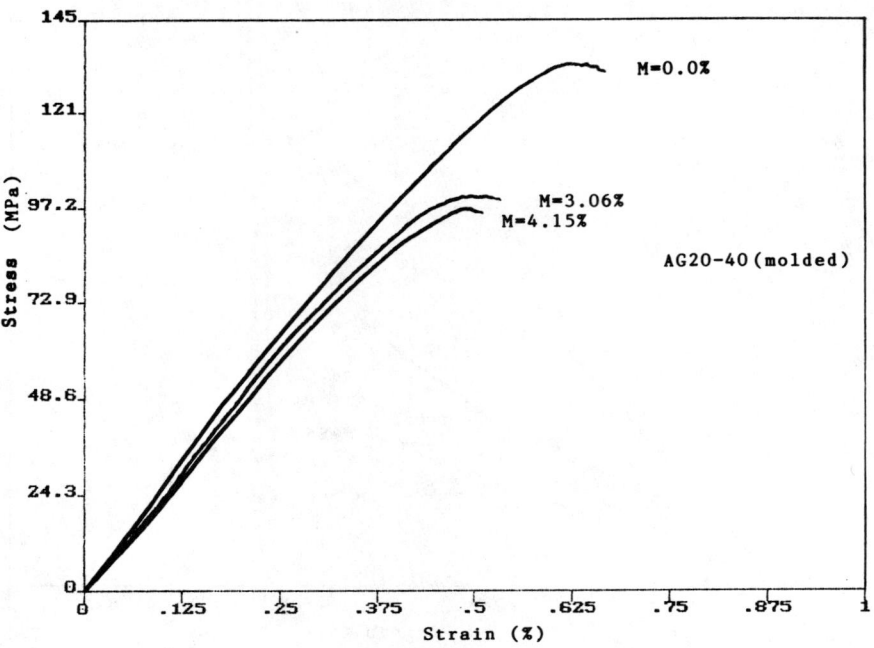

FIG. 13—*Stress-strain diagram with moisture content as a parameter.*

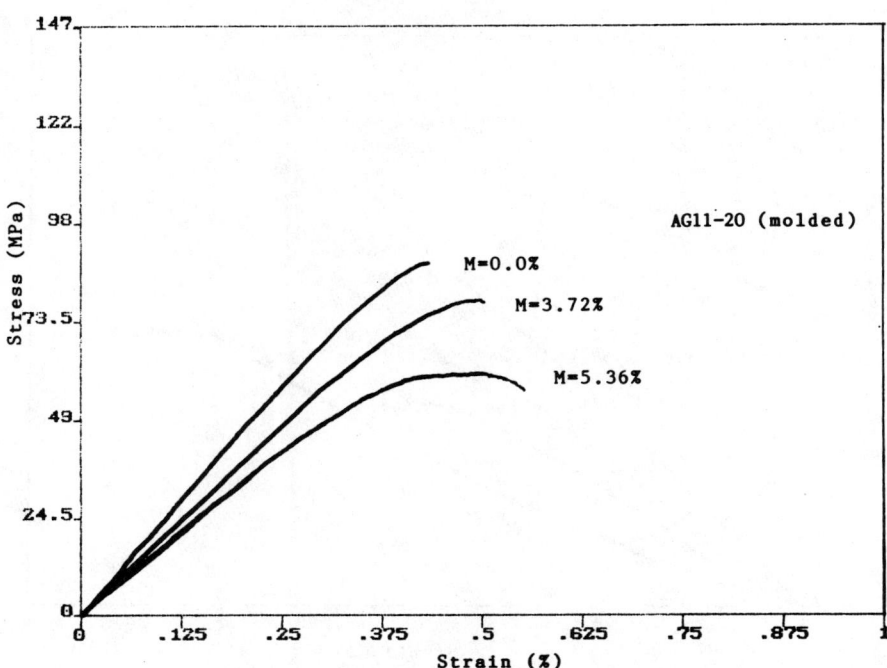

FIG. 14—*Stress-strain diagram with moisture content as a parameter.*

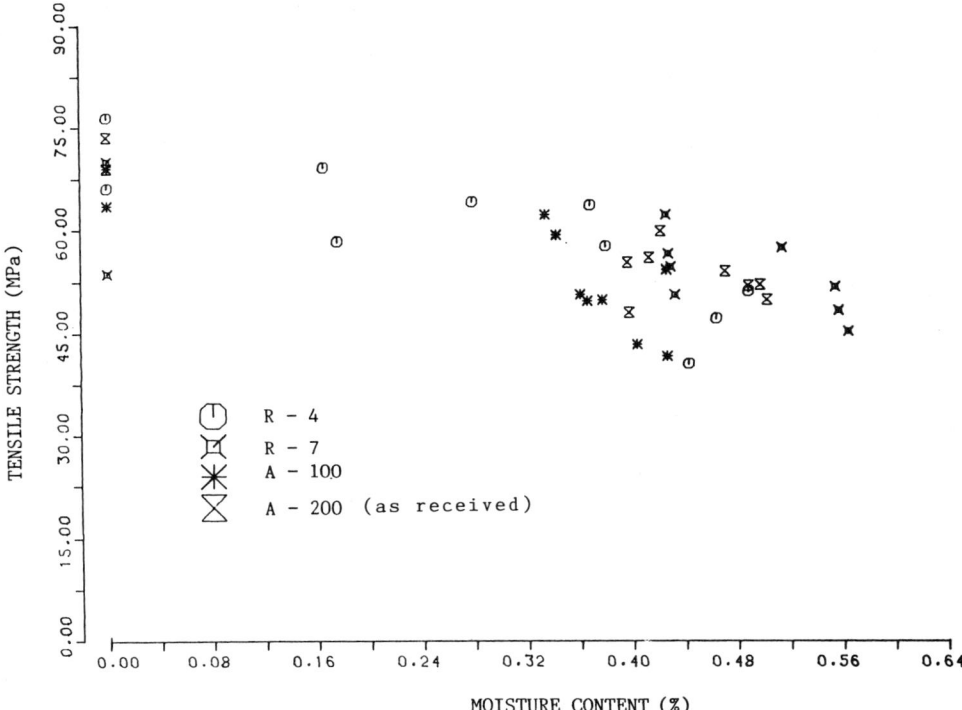

FIG. 15—*Change in tensile strength with moisture content.*

Figure 16 shows the changes in tensile strengths of advanced composite samples due to exposure to the hot-moist conditions. From an average virgin strength of about 150 MPa, the strength of AG31-60 samples (as-received) dropped to about 60 MPa (60% reduction) with the absorption of about 25% of moisture. This is the largest reduction in strength among the samples. From an average virgin strength of about 150 MPa, the strength of AG31-40 samples (as-received) dropped to about 105 MPa (30% reduction) after about 7% moisture was absorbed. From an average virgin strength of about 140 MPa, the strength of AG31-60 samples (molded) dropped to about 76 MPa (46% reduction) with the absorption of about 15% of moisture. From an average virgin strength of about 120 MPa, the strength of AG31-40 samples (molded) dropped to about 70 MPa (42% reduction) after about 10% moisture was absorbed. From an average virgin strength of about 95 MPa, the strength of AG11-20 samples (as-received) dropped to about 57 MPa (40% reduction) after about 5% moisture absorption. From a virgin strength of 129 MPa, AG20-40 samples dropped to 91 MPa (29% reduction) after 7 days of exposure.

It is indicated from Figs. 15 and 16 that the tensile strengths of the injection-molded samples (as-received) are usually lower than those of the advanced composites (either as-received or molded). It is noted that the larger the amount of glass fibers in the composites, the higher the virgin tensile strengths. It is shown from Figs. 5, 6, 15, and 16 that the larger the moisture content, the more the reduction in the tensile strength. The physical properties of glass-reinforced thermoplastics of all types generally deteriorate upon exposure to moisture at elevated temperatures such as the present 100°C [8]. Even if the resin itself is hydraulically stable, the

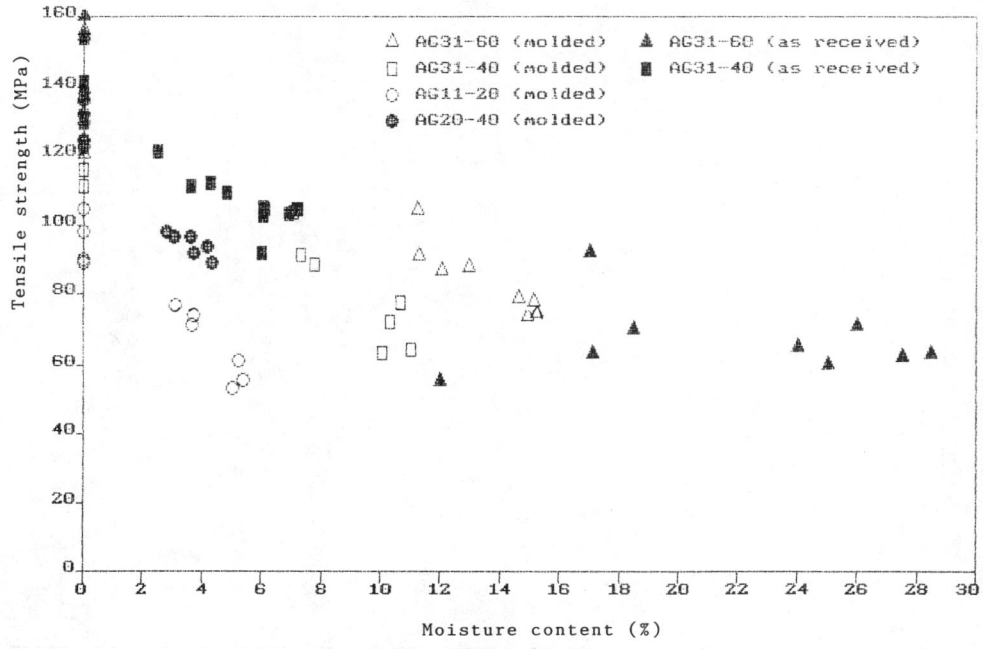

FIG. 16—*Change in tensile strength with moisture content.*

moisture penetrates the materials by a wicking action and attacks the glass-resin interfaces, causing delamination or debonding; reinforced PPS is no exception.

Electron Micrograph Observation

Typical electron micrographs for samples examined are shown in Figs. 17 to 22. Figures 17 and 18 show the microstructures of samples R-7 and A-100, respectively. These micrographs show that the fibers are short (about 17 μm long) and are covered with the thermoplastic matrix, with little exposure of the fiber-matrix interface. This is in contrast with the case of the advanced composites (Figs. 19 to 22). Figures 19 and 20 show the microstructures of the stampable chopped-fiber samples AG11-20. There are many fiber-matrix interfaces exposed and there are many capillaries which serve to suck in the moisture; fiber directions are also random. This lends the material to high transverse stresses and bending fracture of the fiber. Moisture tends to debond the glass fibers from the PPS resin as shown in Fig. 20.

Figures 21 and 22 show the microstructures of the continuous-fiber satin weave fabric samples. The number of crevices and capillaries magnifies. The material appears very dry and many fibers are exposed. Capillary action certainly is dominant here. There appears to be little difference between the microstructures of AG31-60 and AG31-40 in terms of the number of crevices and capillaries. The marked difference in tensile strength reduction shown in Fig. 16 can therefore be related to the lower degree of crystallinity in samples AG31-60 as shown in Figs. 1 through 4.

FIG. 17—*Sample R-7 at ×1500 (as-received).*

FIG. 18—*Sample A-100 at ×1490 (as-received).*

Conclusions

The effect of hot-moist conditions upon the mechanical properties of E-glass reinforced poly-phenylene sulfide has been investigated. The injection-molded samples showed less degradation due to exposure than the advanced composite samples (with the understanding that the advanced composite samples were stronger initially). Sealing of the fiber-matrix interface away from the moist environment prevented absorption and degradation. The advanced composite samples showed large areas of voids and crevices which tended to suck in water. Debonding between the glass fibers and the PPS resin happened because of the large amount of moisture absorption. The degree of crystallinity in the PPS matrix has a significant effect upon the degra-

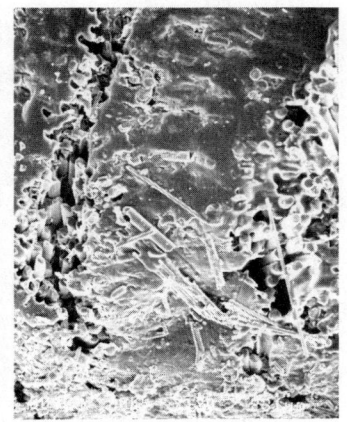

FIG. 19—*Sample AG11-20 at ×130 (as-received).*

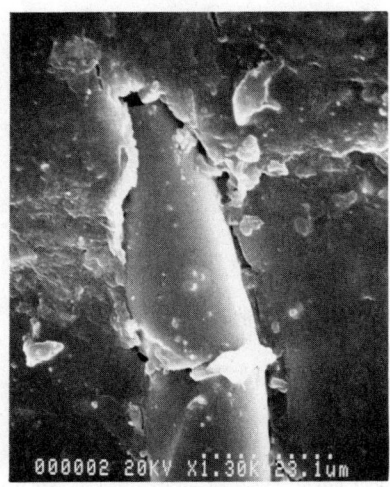

FIG. 20—*Sample AG11-20 at ×1300 (as-received).*

FIG. 21—*Sample AG31-60 at ×120 (as-received).*

FIG. 22—*Sample AG31-40 at ×300 (as-received).*

dation due to hot-moist conditions. The mechanical properties and the moisture absorption of the PPS composites can be improved by annealing the materials between the glass transition temperature and the melting temperature.

Acknowledgments

The financial support from Natural Sciences and Engineering Research Council of Canada under grant number A0413 and from the Quebec Ministere de l'Enseignement Superieure is appreciated.

References

[1] Edmonds, J. T., Jr. and Hill, H. W., Jr., "Production of Polymers from Aromatic Compounds," U.S. Patent 3,354,129, 1967.
[2] O'Connor, J. E., Ma, C. C. M., and Lou, A. L., "Polyphenylene Sulfide: Resin, Prepreg and High Performance Composites," *Proceedings of the 30th SPI Reinforced Plastics/Composites Conference,* Society of Plastics Industry, 1984, paper 11E.
[3] Brady, D. G. and Hill, H. W., Jr., "Polyphenylene Sulfide," *Engineering Thermoplastics: Properties and Applications,* Chap. 8, J. M. Margolis, Ed., Dekker, New York, 1985.
[4] Krone, J. R. and Walker, J. H., "Thermoforming Woven Fabric Reinforced Polyphenylene Sulfide Compounds," *Proceedings of Composites in Manufacturing,* Vol. 5, Los Angeles, CA, January 1986.
[5] O'Connor, J. E., Lou, A. Y., and Beever, W. H., "Polyphenylene Sulfide. A Thermoplastic Polymer Matrix for High Performance Composite," *Proceedings of the 5th International Conference on Composite Materials,* Metallurgical Society of AIME, San Diego, CA, July–August, 1985, pp. 963–970.
[6] Vives, V. C., Dix, J. S., and Brady, D. G., "Polyphenylene Sulfide (PPS) in Harsh Environments," *Effects of Hostile Environments on Coatings and Plastics,* D. P. Garner and G. A. Stahl, Eds., American Chemical Society, Washington, DC, 1983.
[7] Kays, A. O. and Hunter, J. D., "Characterization of Some Solvent Resistant Thermoplastic Matrix Composites," *Composite Materials: Quality Assurance and Processing, ASTM STP 797,* C. E. Browning, Ed., ASTM, Philadelphia, 1983, pp. 119–132.
[8] *Fillers for Plastics,* Wake, W. C., Ed., Butterworth, London, 1971, Chap. 1.
[9] Mascia, L., *Thermoplastics: Materials Engineering,* Applied Science Publishers Ltd., New York, 1982.

James P. Lucas[1] *and Ben C. Odegard*[1]

Moisture Effects on Mode I Interlaminar Fracture Toughness of a Graphite Fiber Thermoplastic Matrix Composite

REFERENCE: Lucas, J. P. and Odegard, B. C., "**Moisture Effects on Mode I Interlaminar Fracture Toughness of a Graphite Fiber Thermoplastic Matrix Composite,**" *Advances in Thermoplastic Matrix Composite Materials, ASTM STP 1044*, G. M. Newaz, Ed., American Society for Testing and Materials, Philadelphia, 1989, pp. 231–247.

ABSTRACT: Mode I interlaminar fracture toughness was determined for a high-temperature thermoplastic matrix, intermediate modulus graphite fiber composite laminate. Interlaminar fracture toughness and fracture energy were determined for a unidirectional layup, continuous-fiber reinforced thermoplastic composite by using a nonconventional specimen. A short bar hybrid test specimen (HTS) was developed and used in this investigation to determine the effect of moisture on mode I delamination fracture toughness and critical strain energy release rate. Exposure of the thermoplastic composite laminate in a water bath for over 400 h had an insignificant effect on delamination fracture toughness. Moreover, a fractographic examination did not reveal effects of absorbed moisture on fracture surface morphology and failure processes. Fracture was characterized primarily by matrix decohesion and fiber pullout. There was, however, clear evidence of fracture mechanisms that contributed significantly to mode I fracture energy of the composite. Crack bridging by isolated, single fibers and fiber bundles was prevalent, particularly after a considerable degree of crack extension in the laminate. Translaminar, secondary microcracking was observed in the vicinity of the primary crack tip.

KEY WORDS: fracture toughness, thermoplastic composites, moisture, hybrid test specimen, fracture energy, intermediate modulus fibers

Nomenclature

a	Crack length
A	Constant Eq 1
a_c	Critical crack length
a_1	Crack increment in V ligament
a_0	Chevron notch depth or initial crack length
b	Crackfront width
B	Specimen width
β	Correction factor
DCB	Double cantilever beam
E	Elastic modulus
E_{tr}	Transverse elastic modulus
G_{Ic}	Critical strain energy release rate
HTS	Hybrid test specimen
K_{IcSB}	Short bar stress intensity factor

[1]Members of technical staff, Sandia National Laboratories, Livermore, CA 94550.

P Applied load
P_c Critical load
P_{max} Maximum applied load
PEEK Polyetheretherketone
SB   Short bar
V   Chevron notch shape
r Distance from crack tip

Introduction

Usage of resin matrix continuous-fiber composites in aerospace applications and space technology is rising steadily. These materials are attractive to aerospace and space industries because of their unique combination of high specific strength and high specific stiffness. In addition, these composites can be designed to produce a stable, near-zero coefficient of thermal expansion. Such properties allow resin-matrix composite laminates to have extremely good dimensional stability compared to monolithic materials over a similar temperature range. Therefore, they are often used in applications where stiffness and precise tolerances of structures must be maintained, often times in aggressive environments.

However, prolonged exposure to certain environments may significantly affect both intrinsic material properties as well as engineered properties of resin matrix composites. For resin matrix composites, exposure to moisture may pose problems. Investigations have shown that moisture absorption in resin matrix composites leads to degradation of mechanical and physical properties [1-9], as well as reduced performance, for example, loss of dimensional stability [6,7,10]. Moisture tends to lower stiffness and strength of composites [3,5], particularly in the transverse direction. The fiber-matrix interface can also be damaged, causing fiber-matrix decohesion. Most of the sorbed moisture in the composite is absorbed by the matrix. The rate of moisture absorption (or desorption) in the composite matrix depends on resin chemical composition, processing, and environmental parameters [1-3]. Consequently, certain resin classes have a higher propensity for moisture absorption. The diffusion of moisture in composites is also influenced by fiber orientation and laminate ply layup [2]. The diffusivity of moisture of thermoset and thermoplastic resins is similar, but, the equilibrium moisture saturation level is much lower for thermoplastic resins [11].

The effect of moisture on mechanical properties has been well documented for thermoset matrix composites [3-7], but substantially few studies have addressed this issue for thermoplastic matrix composite materials. Perhaps the reason for the scarcity of moisture absorption data for thermoplastic matrix composites is attributable to the fact that moisture saturation levels are markedly lower for thermoplastics than for thermosets. Thus, moisture absorption effects in thermoplastic composites may be perceived as insignificant. Moisture effects on "static" mechanical properties, for example, tensile strength, of thermoset composites and thermoplastic (to a lesser extent) composites can be found in the technical literature [1,12]. In contrast, very little information and data are published on the effects of sorbed moisture on crack propagation and fracture phenomena of these composites [12]. This is certainly the case for mode I delamination fracture toughness of thermoplastic matrix composites. Since a high percentage of composite failures occurs by delamination fracture, it is important to determine whether exposure of resin-matrix composites to certain environments can influence fracture processes and, ultimately, failure. After all, delamination fracture toughness is the weak link for fracture and failure of continuous fiber reinforced resin matrix composites, particularly in brittle thermoset resin systems.

In this study, an investigation was conducted to determine the influence of absorbed moisture on delamination fracture toughness and fracture energy of a unidirectional, continuous fiber, thermoplastic matrix composite. We determined the interlaminar fracture toughness by designing and using an unconventional specimen. The newly designed, hybrid test specimen (HTS) evolved from the monolithic, short bar chevron notch fracture toughness specimen developed by Barker [13]. This new specimen consists of two metal adherends with a thermoplastic composite laminate in the center, through which a delaminating crack is initiated and propagated. Specimen design aspects are discussed along with fracture toughness and fracture energy results obtained by using the short bar hybrid test specimen.

Experimental Procedures

Materials

Fiber-reinforced thermoplastic material (IM6/PEEK) was obtained from Fiberite Corp. in the form of tape 229 mm wide. The prepreg (tape) thickness was 0.2 mm with a fiber volume fraction of 60%. The fiber modulus was 2.76×10^5 MPa and the fiber diameter was 6×10^{-3} mm. The matrix polymer was polyetheretherketone (PEEK). Laminates were fabricated by hot pressing. The stock materials (test specimen stock blanks) were fabricated by co-curing the metal to the composite laminate using a heated hydraulic laminating press. The stock blanks consisted of 24 plies of prepreg tape, oriented unidirectionally, ($[0]_{24}$), between grooved and anodized 6061-T6 aluminum plates (12.5 mm thick). Sheets of neat PEEK resin (0.25 mm thick) were placed between the aluminum plates and the composite tape on both sides of the configuration. The neat thermoplastic sheets filled the grooves in the metal plates, which aided mechanical locking between the metal adherends and composite laminate material in the center of the blanks, thus providing additional bonding strength. Figure 1 illustrates the assembly and configuration of the co-cured blank stock. Hybrid test specimens were subsequently machined from the co-cured stock blanks. The orientation of hybrid test specimens in relation to co-cured blank stock is also shown in Fig. 1. In order to properly cure the composite material of the blank stock, processing conditions were followed in accordance with those suggested by the composite tape manufacturer [14]. These conditions consisted of: first, heating the unconsolidated laminate material to 666 K (393°C) for 0.17 h under 0.49 MPa (70 psi) pressure. The second step involved raising the pressure to 1.4 MPa (200 psi) and maintaining it for 0.5 h at 666 K to complete consolidation. During processing, aluminum dams were used to confine resin flow and constrain excessive ply movement of the uncured laminate plies. After consolidation, the laminate was allowed to cool down slowly under pressure (~ 1.4 MPa) at a rate near 0.033 K/s.

Test Specimen Design

Hybrid test specimens (that is, metal adherend/resin matrix composite center) were fabricated by milling from the co-cured test blank stock. Figure 2 shows the geometric configuration and dimensions of the hybrid test specimen. The HTS has the dimensions of the chevron notched short bar (SB) fracture toughness specimen developed by Barker [13,15]. The original chevron notched short bar was used to determine plane strain fracture toughness of brittle monolithic materials [15]. However, the short bar test method has evolved in design and analysis to the point where the SB specimen can be used to determine fracture toughness in more ductile materials. The major difference between the HTS and monolithic SB test specimen is the composite laminate centered between metal adherends (Fig. 2). The metal adherends allow the appropriate geometric dimensions to be obtained as well as provide gripping points to load the specimen during testing.

In the original development of the SB method [14], a linear elastic fracture mechanics approach was used for monolithic materials possessing self-similar crack growth. For a inhomogeneous material, such as a composite laminate, self-similar crack growth is not strictly observed. Nonetheless, commonalities in the crack tip regime do exist. The ramification of such similari-

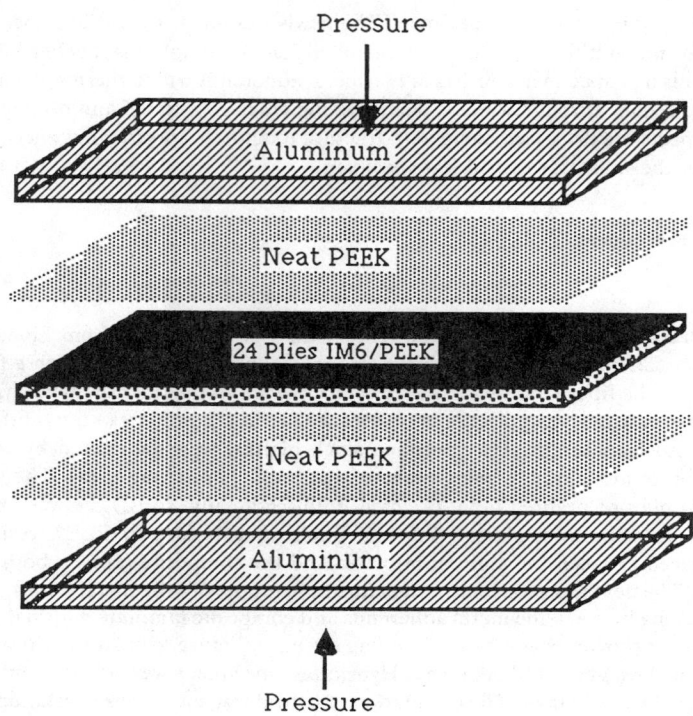

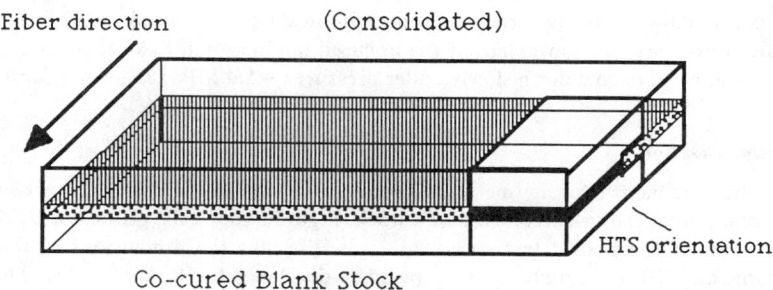

FIG. 1—*Configuration of the co-cured test blanks consisting of the composite laminate center and the aluminum adherend.*

ties is discussed regarding crack growth in the SB hybrid test specimen. As mentioned, in composite laminates [*16*], the crack tip elastic stress distribution does not strictly vary with a $1/\sqrt{r}$ (r is distance in front of the crack tip) relationship for large r as would be the case for a self-similar crack in monolithic materials. While a $1/\sqrt{r}$ stress distribution is not observed absolutely at the crack tip in HTS composite laminates, similarities exist in the crack tip stresses between composite laminates and monolithic materials which favors the application of linear

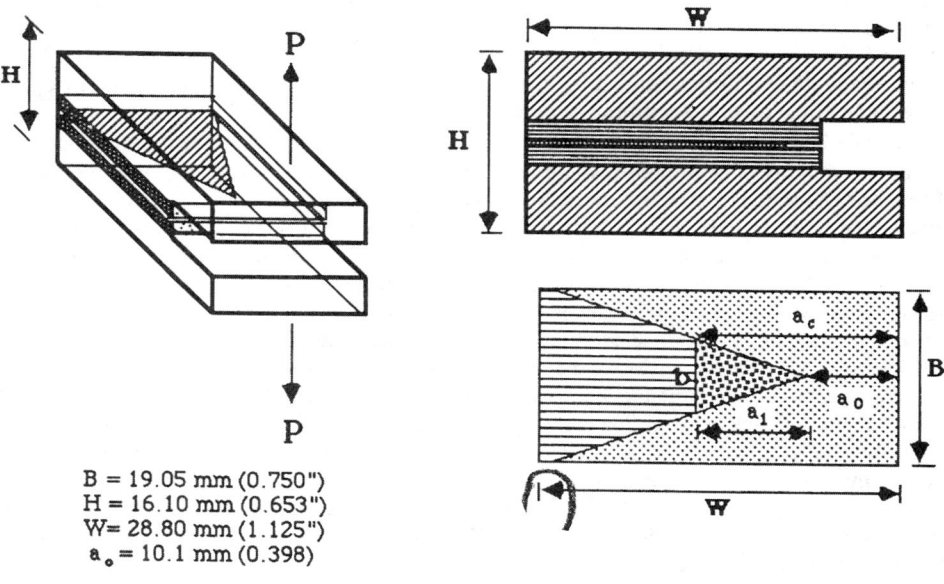

B = 19.05 mm (0.750")
H = 16.10 mm (0.653")
W= 28.80 mm (1.125")
a_o = 10.1 mm (0.398)

FIG. 2—*Geometry of the hybrid test specimen (HTS). The crack propagation area is indicated by the triangular chevron notched.*

elastic fracture mechanics (LEFM). First of all, small scale yielding exists in the matrix materials of the laminate at the crack tip due to constraint by high-modulus carbon fibers. Secondly, in the intermediate vicinity of the crack tip, the elastic stress distribution does vary as the inverse of $\sqrt{r}$ [16]. Hence, in a small regime in front of the crack tip, the laminate crack tip environment resembles a self-similar crack as in monolithic materials. Because of such similarities and agreement of experimental data with other test methods, we believe K_{IcSB} (HTS SB fracture toughness) as determined by the HTS is valid. Therefore, the fracture energy can be determined in accordance with the observed K_{IcSB} values.

In a number of other fracture mechanics test specimens used to determine interlaminar mode I fracture toughness in composites, a crack starter (for example, usually aluminum or Teflon film) is often used. The starter material is placed between lamina in the composite panel during processing in order to initiate a sharp crack necessary for valid test results. Crack starters are employed in double cantilever beam (DCB) specimens, which are commonly used to determine interlaminar fracture toughness of resin-matrix composites. Crack starters are not necessary for the HTS. A sharp interlaminar crack is initiated in the HTS at the triangular, chevron notch tip by applying a load, P, as illustrated in Fig. 2. The crack initiates at a low load level, but in a sufficiently high stress intensity field at the notch tip. As crack growth continues, the load required for propagation increases because of the broadening crack front, b, propagating in the chevron notch. Crack propagation is stable in the specimen until the crack length has reached a critical value, a_c. At the critical crack length, the competing effect of the increased specimen compliance with respect to incremental crack extension dominates and the crack tends to propagate in an unstable manner. The SB specimen is designed so that the critical crack length, a_c, occurs simultaneously with the maximum load, P_{max}, thus allowing the fracture toughness of brittle materials to be determined [13].

Moisture Absorption

Moisture was absorbed into the composite material by submersion of the hybrid test specimen in a heated water bath [3]. The water bath was gently agitated to prevent stagnation for the duration of the moisture-charging phase. Specimens were subjected to a constant water temperature 343 K (70°C). All accessible surfaces of the composite laminate of the HTS were exposed during charging. The absorbed moisture content of the composite laminate was determined by weighing the specimens before and after exposure. Since the entire HTS was submerged in the bath, part of the weight gained by the specimen was due to the formation of corrosion products on the metal adherends. Figure 3 shows the overall weight gain of the HTS and weight gain of the monolithic aluminum blanks due to oxide scale formation. To correct for weight gained due to scale formation, third-degree polynomials were fitted to weight gain versus time experimental data of hybrid test specimens and monolithic aluminum test blocks. Subtracting the appropriate polynomial expressions analytically then provided the "actual" weight gain (equivalent to change in moisture content) in the composite laminate. As shown by experiment and analysis [2,3], moisture absorption in the laminate varies linearly with the square root of the exposure time in the early stages of absorption. As stated previously, the sorbed moisture (that is, saturation level) is markedly lower for thermoplastics than for thermoset resin materials despite having quantitatively similar diffusion coefficients [11]. The moisture-time profile for a thermoset resin matrix composite laminate shown in Fig. 3 emphasizes the difference in moisture absorption characteristics for these two matrix resins.

Test Method

The theory of fracture toughness determination by the chevron notch short bar/rod method has been advanced by Barker, and others [13,17–22]. Therefore, we will not dwell on the theo-

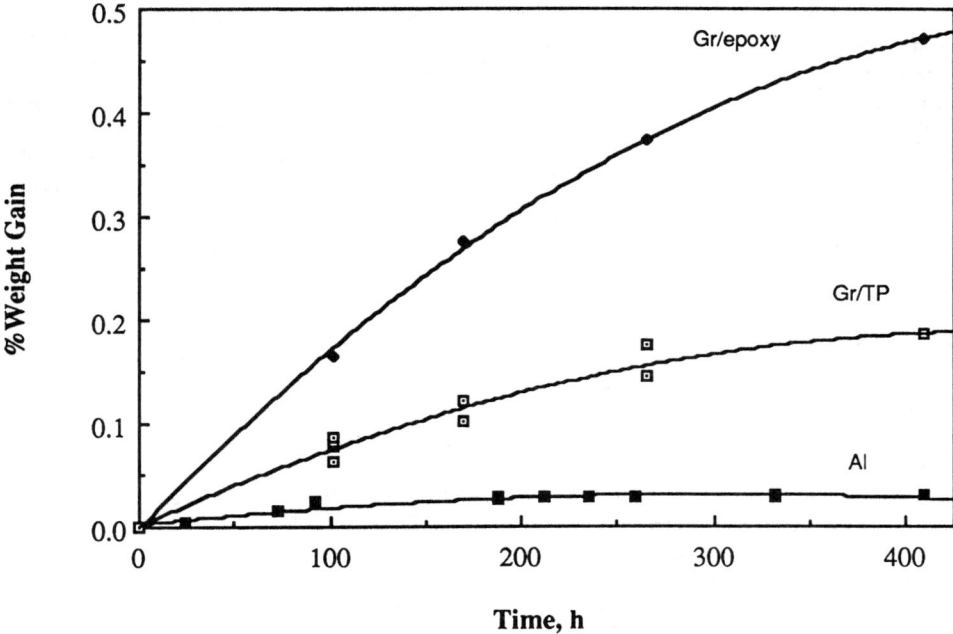

FIG. 3—*Weight gained during saturation phase for the graphite epoxy and graphite thermoplastic hybrid test specimens, as well as for the monolithic aluminum blanks.*

retical details here, but rather our focus will be to emphasize the advantages of this method. One advantage is simplicity. Unlike many other test methods for determining the fracture toughness of materials, a measure of incremental crack length is not required. Because a sharp crack is initiated at the tip of the chevron "V" notch, fatigue precracking and/or artificial crack starting by other means are not necessary. Once the compliance calibration is determined for a specific geometry, fracture toughness values can be obtained independent of material. Moreover, the specimen size required to obtain valid plane strain fracture toughness results is significantly smaller for the SB method than for most other methods [21–22]. Often, only the maximum load of a specimen is needed to determine the fracture toughness by the short bar method, provided that criteria for linear elastic fracture mechanics are satisfied. Barker has shown for the short bar geometry that a critical crack length, a_c, is reached under stable crack growth condition at the maximum load, P_{max}, which depends only on the geometry of the specimen and is independent of the material. From Barker's analysis, the short bar plane strain fracture toughness can be expressed as

$$K_{IcSB} = A\beta P_c/B^{3/2} \qquad (1)$$

where A is a constant for a fixed specimen geometry that includes terms such as b, the crack front width, B, the specimen thickness, and E the elastic modulus. P_c is the critical load and β is a specimen correction factor for slightly misdimensioned specimens. This factor, β, is always close to unity.

In the present tests, load was applied to the short bar HTS normal to the plane of the propagating crack front, (that is, opening mode, mode I) as indicated in Fig. 2. Testing was performed at room temperature in air at 37% relative humidity. Tests were run in displacement control on a Fractometer II chevron notched short bar test system. The mouth-opening displacement rate was 5.8×10^{-4} mm/s. Load was measured by the instrumented loading levers (load cell) of the Fractometer II test system. The mouth-opening displacement was measured with a clip gage positioned in the specimen grip slot. A schematic drawing of the load-displacement profile for an HTS is shown in Fig. 4. The response is linear on initial loading, corresponding to the compliance of the test specimen at crack length, a_0. As the load is increased, deviation from linearity occurs which correlates to the onset of crack growth at the chevron notch tip. The crack grows initially under a high stress field in a stable manner to a critical length. P_c occurs at the critical crack length and fracture toughness is then calculated by Eq 1.

Results and Discussion

HTS Laminate Moisture Absorption

Figure 5 shows that the weight gain is approximately 0.15% due to moisture absorption in the composite laminate. For our specimen, a 0.15% weight gain is higher than expected for the time exposed. A possible explanation for the higher-than-expected weight gain value is that moisture was also sorbed at the laminate/metal interface of the HTS in addition to sorption in the composite laminate matrix. Even though the hybrid test specimens were exposed by submerging in water bath for over 400 h, earlier sorption data of composite panels suggested that saturation of the HTS composite laminate was not reached through the thickness of the material. Moreover, no attempt was made to determine the moisture content-thickness profile for the HTS composite laminate. Therefore, the exact through-thickness moisture distribution of the composite laminate material was unknown.

Using the HTS, the fracture toughness is determined at a critical crack length after appreciable crack advancement in the laminate material. Because there is uncertainty in the moisture level as a function of laminate thickness, there will be uncertainty regarding the exact moisture content at the critical crack length where fracture toughness is determined. Instead of plotting fracture toughness as a function of the percent moisture absorbed (as would be the case for a saturated laminate), the fracture toughness is plotted in relationship to exposure time in the

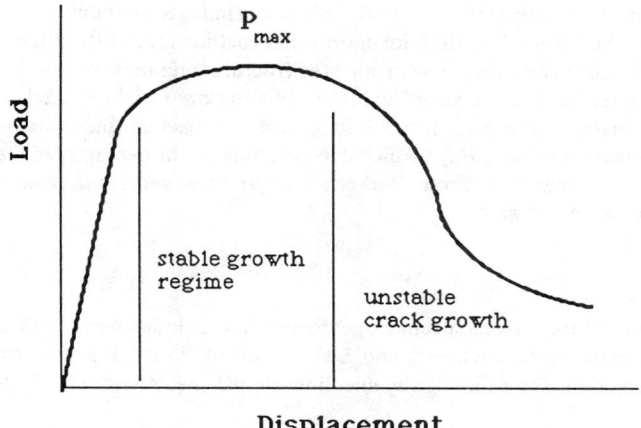

FIG. 4—*Typical load displacement profile for fracture toughness using an HTS.*

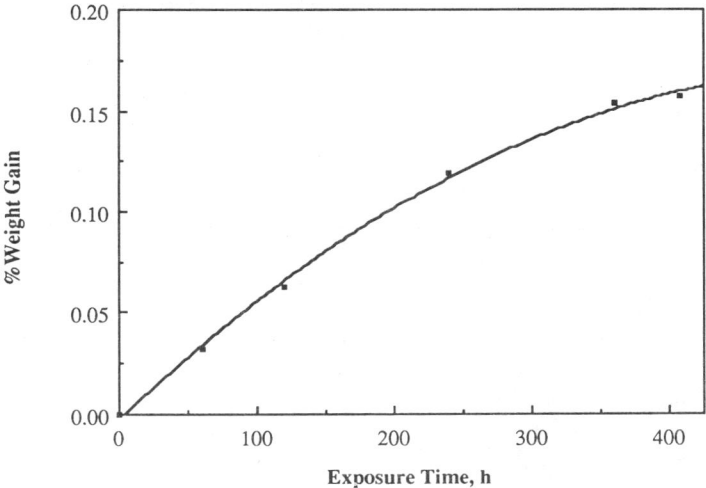

FIG. 5—*Moisture content absorption in the thermoplastic composite laminate versus time.*

moisture environment. Figure 6 shows the relationship of moisture on interlaminar fracture toughness of the composite after exposure for approximately 400 h. The results indicate that the interlaminar fracture toughness (or fracture energy) of PEEK composite is not affected by moisture for exposure up to 400 h.

Fracture

Using the short bar hybrid test specimen, we observed fracture toughness values comparable to those obtained by the commonly used double cantilever beam (DCB) method [23–26]. For the HTS, the average fracture toughness, K_{IcSB}, was 5.0 ± 0.1 MPa/m$^{1/2}$ (4.6 ± 0.09 ksi/in.$^{1/2}$) for IM6/PEEK thermoplastic composite. The HTS test method gave highly reproducible data as indicated by Table 1. The critical strain energy release rate, G_{Ic}, was calculated from the linear elastic fracture mechanics expression

$$G_{\mathrm{Ic}} = (K_{\mathrm{IcSB}})^2/E_{tr} \tag{2}$$

where E_{tr} is the transverse modulus of the composite laminate. The transverse modulus of the composite laminate dominates the compliance of the HTS and therefore is used in the calculation of the critical strain energy release rate. Since E_{tr} is equal to 1.04 × 10^4 MPa (1.5 × 10^6 psi), G_{Ic} was found to be 2469 J/m^2 (14.1 lb/in.). The fracture energy was in good agreement with results obtained elsewhere by DCB [24–28] for PEEK resin matrix composites.

As noted by other studies, true interlaminar fracture (that is, fracture only between single lamina) is difficult to achieve in testing because the crack front tends to wander between laminae. Crack wander occurs because of the crack tip stress distribution and also because the crack front interacts with inhomogeneities in the material [29–32]. For example, debonding of fiber-matrix interfaces, fiber intermingling (nesting) [30], fiber bridging [30,32–35], and transply microcrack nucleation are just a few of many failure processes and fracture mechanisms which may contribute with varying degrees to crack front deflection and translaminar crack propagation [31,32]. Mode I strain energy release rate, G_{Ic}, is usually increased as a result of

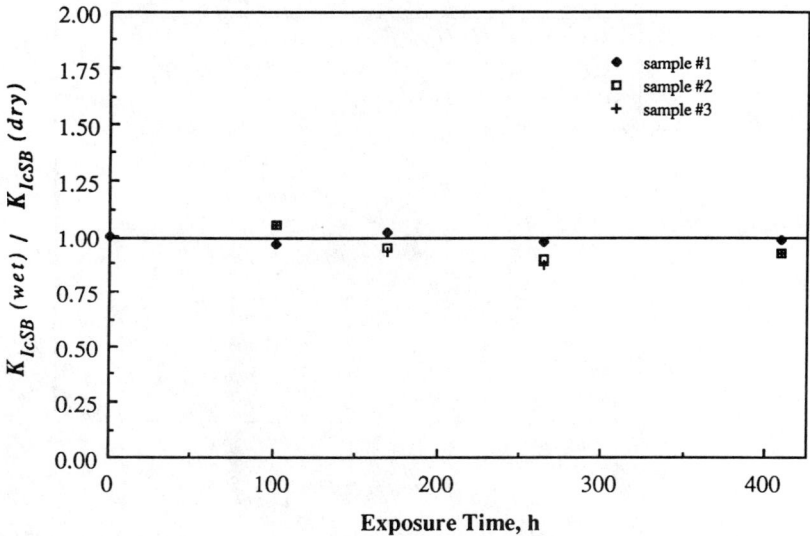

FIG. 6—*Effect of moisture on fracture toughness of thermoplastic composites.*

TABLE 1—*Fracture toughness and energy results.*

Specimen No.	Time, h	K_{Ic}, ksi/in.$^{1/2}$ (MPa/m$^{1/2}$)	G_{Ic}, lb/in. (J/m^2)
HTS 7-3	0.0	4.65 (5.07)	14.4 (2523)
HTS 7-4	0.0	4.65 (5.07)	14.4 (2523)
HTS 7-1	101.5	4.50 (4.90)	13.5 (2363)
HTS 7-2	169.4	4.72 (5.14)	14.8 (2599)
HTS 7-5	265.0	4.51 (4.92)	13.6 (2373)
HTS 7-6	409.0	4.58 (4.99)	14.0 (2447)

translaminar crack propagation because fracture processes contributing to translaminar crack propagation absorb energy. Table 2 lists fracture mechanisms and material parameters that influence G_{Ic}. The fracture features have been ranked qualitatively, indicating their contribution to increased G_{Ic}. For the HTS, fiber bridging of the crack surfaces does play a role along with other failure processes; however, design aspects of the HTS tend to mitigate the contribution of fiber bridging when compared to thin DCB specimens.

As mentioned earlier, the SB fracture toughness is determined after sufficient degree of stable crack growth that initiates at the machined-in chevron notch tip in the composite laminate. For the short bar hybrid test specimen used in this investigation, the fracture toughness was determined after an actual crack length, a_1, of approximately 4 mm. It has been shown by several investigations using DCB test specimens that fiber bridging of crack faces substantially increases G_{Ic} with increased crack advancement [34–37]. Moreover, it has been demonstrated [34,36–37] that fiber-bridging effects on delamination fracture energy is greatest when thin (for example, 3 to 5 mm thick) DCB test specimens are used. Apparently, the high degree of arm deflection of thin DCBs increases the fiber-bridged zone per unit crack length, making the

TABLE 2—*Factors influencing* G_{Ic} *[30–35].*

Feature	Influence on G_{Ic}	Comments
Fiber bridging	Very high	Very prevalent in tough-resin matrix composites; causes closure behind the crack front.
Secondary cracking	High	Secondary crack initiation in front of the primary crack causes crack deflection and translaminar crack propagation.
Fiber breakage	Moderate	Greater for high-fracture strain fibers.
Fiber pullout	Moderate	Fiber breaks during crack propagation.
Fiber volume	Low	May contribute to plastic zone constraint.
Fiber type	Low	High fracture strain fibers may influence fiber breakage and bridging effects.
Resin content	Low	Resin-rich areas promote higher toughness.

change in fracture energy with respect to crack increment, dG/da, significantly higher. However, an experiment by Phillips and Wells [37] showed that fiber bridging was severely reduced and dG/da approached zero when deflection of the DCB specimen arms was small, and consequently when the opposing crack faces of the laminate remained straight during crack extension. To maintain straight crack faces and a low degree of arm deflection with crack advancement, Phillips and Wells [37] utilized thick DCBs, and they indeed demonstrated the fracture energy to be less dependent on crack length, particularly for a small crack extension. Due to the HTS design, there is virtually no deflection of the composite laminate (due to low compliance of the metal adherends to which the laminate is bonded) and the crack faces remain straight during loading over crack extension range (~ 4 mm) through a triangular ligament at which G_{Ic} is determined. These features make the HTS more suitable than thin DCB specimens for determining delamination fracture energy because the crack length dependence on G_{Ic} will be less as the contribution from fiber bridging to fracture energy enhancement is reduced.

SEM fractographs revealed no discernible difference in the fracture surface morphology of unexposed and moisture-exposed hybrid test specimens. Fracture surface micrographs show that an interlaminar crack was initiated at the chevron notch tip. However, as the crack grows, there is evidence of fiber pullout, fiber breakage, and translaminar fracture in the crack tip region. In fact many of the features associated with crack propagation listed in Table 2 were observed to a certain degree on HTS fracture surfaces. Fiber/matrix debonding and fiber pullout are seen at the notch tip (Fig. 7). Debonded, isolated fibers also suggest fiber/matrix decohesion occurred during crack initiation. In Fig. 8a, cohesive matrix fracture is observed along with matrix failure which occurred in a resin-rich area. Fracture is characterized by a high degree of plastic deformation (ductile) locally in the resin-rich area. However, in resin-lean areas, the fracture appearance tends to be brittlelike, probably due to plastic constraint of the matrix material by high-modulus fibers, as suggested by other investigators [29,30,38]. Fiber breakage and nesting is observed in Fig. 8b. The energy-absorbing capacity of fiber nesting and breakage in the laminate is substantial and thus contributes to an increase in G_{Ic}.

Crick et al. [26] have attempted to make quantitative estimates of the energy absorption contributions of various failure processes. Their calculations suggested that fiber bridging alone contributed ~ 20% to G_{Ic} values for DCB specimens. Figure 9 shows several fracture features of an extended crack (~ 3 times a_c of the HTS) which occurred near the crack tip, and at considerable distance behind the advancing crack front. Fiber bridging of a long crack is clearly shown in Fig. 9a. Bridging of the crack surfaces by a single fiber (Fig. 9b) and also by fiber bundles (Fig. 9c) is quite evident. Necking of matrix material in the vicinity of fiber bridges suggests that good bonding occurred between fiber and resin, even though there was eventual fiber pullout by

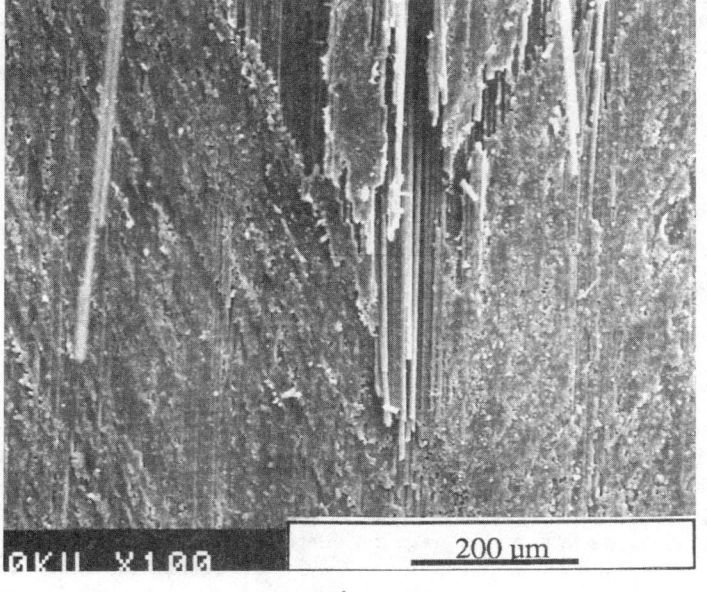

Crack Propagation ↑

top view

Crack Propagation ↑

side view

FIG. 7—*Fracture at the chevron notch tip is characterized by fiber/matrix debonding and fiber pullout.*

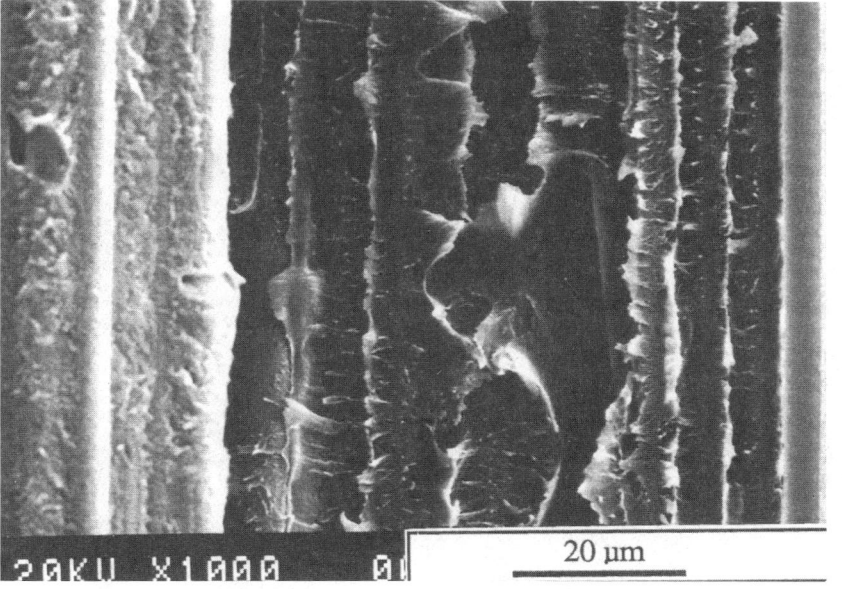

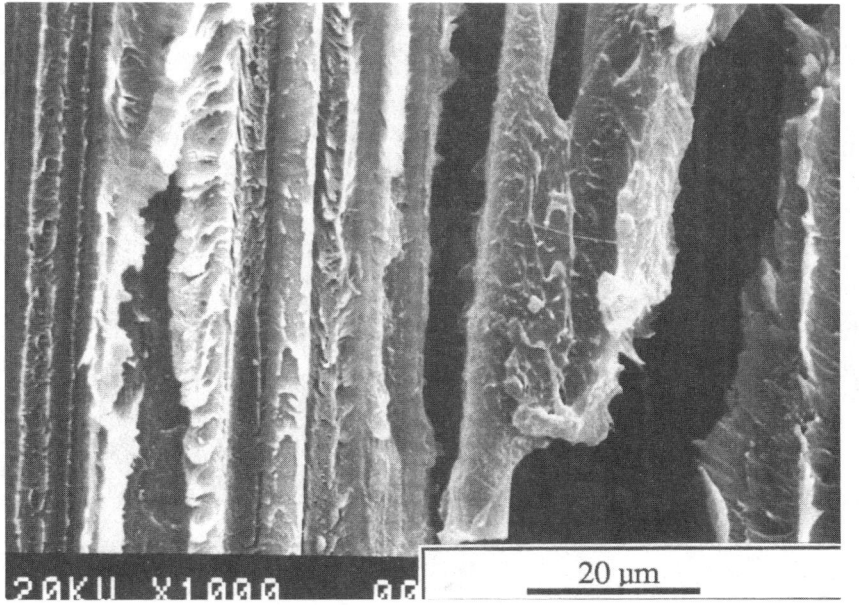

FIG. 8—*Variations in fracture morphology. Both ductile and brittle-like fracture modes are observed.*

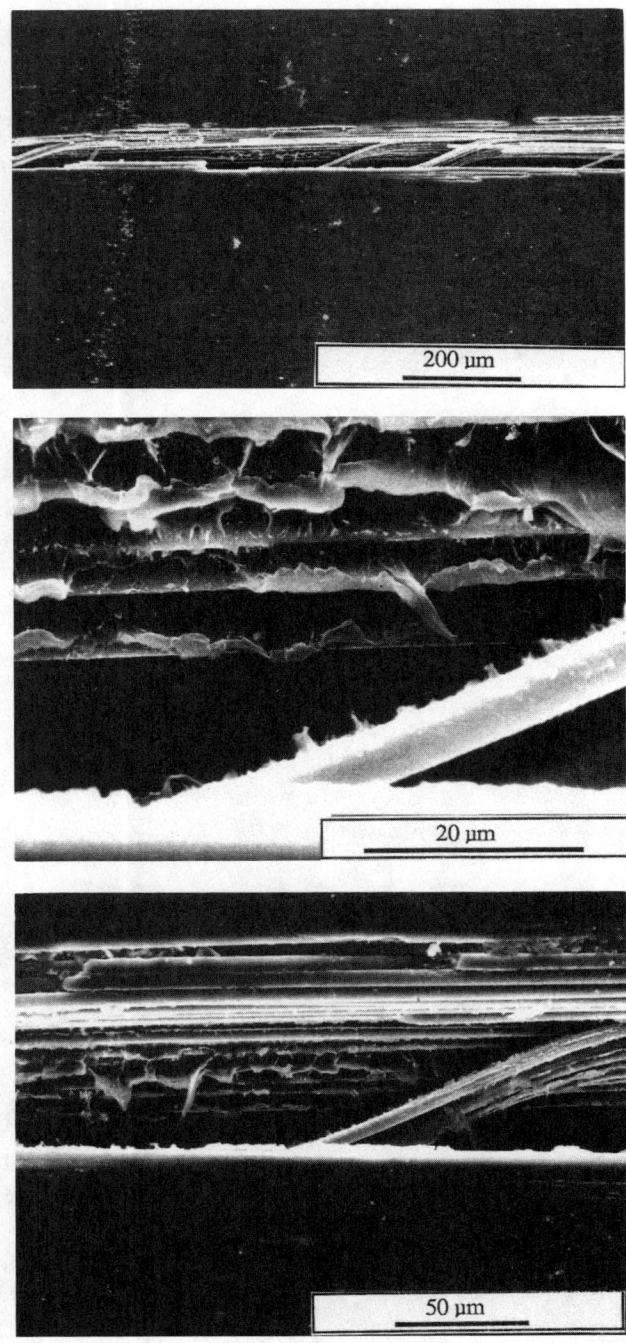

Crack Propagation

FIG. 9—*Evidence of fiber bridging of extended cracks. Single fiber and bundle fiber bridging is seen across the crack surface.*

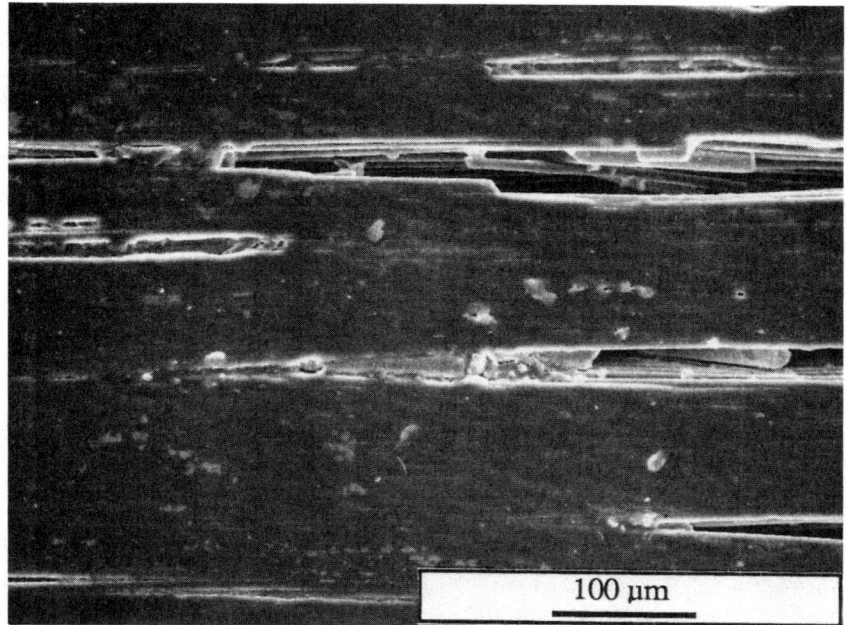

Crack Propagation

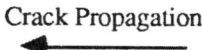

FIG. 10—*Secondary cracking in the vicinity of the primary crack front.*

debonding at the fiber matrix interface (Fig. 9*b*). Figure 10 shows microcracks that have nucleated in advance of the main crack as well as translaminar to the primary crack front. Both of these crack propagation features have been previously observed in graphite-epoxy composite laminates [*31,32,34*]. Microscopy examinations suggest that fiber bridging was involved in the fracture process, particularly at large crack extension. When fracture toughness and fracture energy are determined after minimal amount of crack extension, the influence of fiber bridging is considerably less.

Conclusion

The delamination fracture toughness and critical strain energy release rate have been determined for an IM6/PEEK thermoplastic composite laminate. Fracture toughness was determined with an unconventional short bar hybrid test specimen. Fracture toughness results determined by using the HTS were comparable to results obtained by conventional test methods, such as the double cantilever beam. After exposure to moisture for approximately 400 h, no significant effect of moisture was observed on the interlaminar fracture toughness. Fiber bridging was observed during post-test analysis of the hybrid test specimens with long cracks. A smaller fiber-bridging contribution to fracture toughness values is observed with hybrid test specimens than with typically thin double cantilever beam specimens.

Acknowledgment

We wish to gratefully acknowledge T. Sage for his support in fabrication of test specimens and conducting many other pertinent experiments. We also wish to thank A. Gardea and K. Davidson. This work was supported by the Department of Energy under Contract No. DE-AC04-76DP00789.

References

[1] Springer, G. S., "Environmental Effects on Composites," G. S. Springer, Ed., Technomic Publishing Co., Westport, CT, 1981, p. 1.

[2] Shen, C. H. and Springer, G. S., "Moisture Absorption and Desorption of Composites," *Environmental Effects on Composites*, G. S. Springer, Ed., Technomic Publishing Co., Westport, CT, 1981, p. 15.

[3] Loos, A. C. and Springer, G. S., "Moisture Absorption of Graphite Epoxy Composition Immersed in Liquid and in Humid Air," *Environmental Effects on Composites*, G. S. Springer, Ed., Technomic Publishing Co., Westport, CT, 1981, p. 34.

[4] Loos, A. C., Springer, G. S., Sanders, B. A., and Tung, R. W., "Moisture Absorption of Polyester-E Glass Composites," *Environmental Effects on Composites*, G. S. Springer, Ed., Technomic Publishing Co., Westport, CT, 1981, p. 51.

[5] Shen, C. H. and Springer, G. S., "Effects of Moisture and Temperature on the Tensile Strength of Composites," p. 79.

[6] Browning, C. E., Husman, G. E., and Whitney, J. M., "Moisture Effects in Epoxy Matrix Composites," *Composite Materials: Testing and Design, ASTM STP 617*, ASTM, Philadelphia, 1976.

[7] Shen, C. H. and Springer, G. S., "Environmental Effects in the Elastic Moduli of Composite Materials," *Environmental Effects on Composites*, G. S. Springer, Ed., Technomic Publishing Co., Westport, CT, 1981, p. 94.

[8] Odegard, B. C., "Moisture Absorption in Graphite/Organic Matrix Composites," SAND86-8248, Sandia Laboratory Report, Livermore, CA, 1987.

[9] Shirrell, C. D., "Diffusion of Water Vapor in Graphite/Epoxy Composites," *Advance Composite Materials: Environmental Effects, ASTM STP 658*, J. R. Vinson, Ed., ASTM, Philadelphia, 1978, p. 21.

[10] McKague, E. L., Reynolds, J. D., and Halkias, J. E., "Swelling and Glass Transition Relations for Epoxy Matrix Material in Humid Environment," *Journal of Applied Polymer Science*, Vol. 22, 1978, p. 1643.

[11] Grayson, M. A. and Wolf, C. J., "The Solubility and Diffusion of Water in Poly(Aryl-Ether-Ether-Ketone) (PEEK)," *Journal of Polymer Science: Part B: Polymer Physics*, Vol. 25, 1987, p. 31.

[12] Russell, A. J. and Street, K. N., "Moisture and Temperature Effects of the Mixed-Mode Delamination Fracture of Unidirectional Graphite Epoxy," *Delamination and Debonding of Materials, ASTM STP 876*, W. S. Johnson, Ed., 1985, p. 349.

[13] Barker, L. M., "A Simplified Method for Measuring Plane Strain Fracture Toughness," *Engineering Fracture Mechanics*, Vol. 9, 1977, p. 361.

[14] Fiberite: Technical Data Sheet No. 2, February 1985.

[15] Barker, L. M., "Short Rod K_{Ic} Measurements of Al_2O_3," Terra-Tek Research Report, TR 77-69, July 1977.

[16] Wang, S. S., Mandell, J. F., and McGarry, F. J., "An Analysis in Crack Tip Stress Field in DCB Adhesive Fracture Specimens," *International Journal of Fracture*, Vol. 14, No. 1, 1978, p. 39.

[17] Biner, S. B., Barnby, J. T., and Elwell, D. W. J., "On the Use of Short Rod/Bar Test Specimens to Determine the Fracture of Metallic Materials," *Journal of Fracture*, Vol. 26, 1984, p. 3.

[18] Munz, D., Busbey, R. T., and Srawley, J. E., "Compliance and Stress Intensity Coefficient for Short Bar Specimens with Chevron Notches," *International Journal of Fracture*, Vol. 16, 1980, p. 359.

[19] Barker, L. M. and Leslie, W. S., "Short Rod K_{Ic} Test of Several Steels at Temperatures to 700K," Fracture 1977, International Conference of Fracture, ICF4, Waterloo, Canada, Vol. 2, D. M. R. Taplin, Ed., 1977, p. 305.

[20] Pook, L. P., "An Approach to a Quality Control K_{Ic} Testpiece," *International Journal of Fracture Mechanics*, Vol. 8, 1972, p. 103.

[21] Barker, L. M., "Compliance Calibration of a Family of Short Rod and Short Bar Fracture Toughness Specimens," *Engineering Fracture Mechanics*, Vol. 17-4, 1983, p. 289.

[22] Barker, L. M., "Specimen Size Effects in Short-Rod Fracture Toughness Measurements," *Chevron-Notched Specimens: Testing and Stress Analysis, ASTM STP 855*, J. H. Underwood, S. W. Freiman, and F. I. Baratta, Eds., 1983.

[23] Donaldson, S. L., "Fracture Toughness Testing of Graphite/Epoxy and Graphite PEEK Composites," *Composites*, Vol. 16-2, 1985, p. 103.
[24] Leach, D. C. and Moore, D. R., "Toughness of Aromatic Polymer Composites Reinforced with Carbons Fibers," *Composite Science and Technology*, Vol. 21, No. 2, 1985, p. 131.
[25] Talbott, M. F., Springer, G. S., and Bergland, L. A., "The Effects of Crystallinity on the Mechanical Properties of PEEK Polymer and Graphite Reinforced PEEK," *Journal of Composite Materials*, Vol. 21, 1987, p. 1056.
[26] Crick, R. A., Leach, D. C., Meakin, P. J., and Moore, D. R., "Interlaminar Fracture Morphology of Carbon Fibre/PEEK Composite," *Journal of Materials Science*, Vol. 22, 1987, p. 2094.
[27] Glessner, A. L., Takemori, M. T., Vallance, M. A., and Gifford, S. K., "Mode I Interlaminar Fracture Toughness of Unidirectional Carbon Fiber Composites Using a Novel Wedge Driven Delamination Design," *Composite Materials: Fatigue and Fracture, Second Volume, ASTM STP 1012*, P. Lagasse, Ed., 1989, p. 181.
[28] Devitt, D. F., Schapery, R. A., and Bradley, W. L., "A Method of Determining Mode I Delamination Fracture Toughness of Elastic and Viscoelastic Composite Materials," *Journal of Composite Materials*, Vol. 14, 1980, p. 270.
[29] Scott, J. M. and Phillips, D. C., "Carbon Fiber Composites with Rubber Toughened Matrices," *Journal of Materials Science*, Vol. 10, 1975, p. 551.
[30] Hunston, D. L., Moulton, R. J., Johnston, N. J., and Bascom, W. D., "Matrix Resin Effects in Composite Delamination: Mode I Fracture Aspects in Toughened Composites," *Toughened Composites, ASTM STP 937*, N. J. Johnston, Ed., ASTM, Philadelphia, 1987, p. 24.
[31] Jordan, W. M. and Bradley, W. L., "Mechanisms of Fracture in Toughened Graphite-Epoxy Laminates," *Toughened Composites, ASTM STP 937*, N. J. Johnston, Ed., ASTM, Philadelphia, 1987, p. 95.
[32] Hibbs, M. F., Tse, M. K., and Bradley, W. L., "Interlaminar Fracture Toughness and Real-Time Mechanisms of Some Toughened Graphite/Epoxy Composites," *Toughened Composites, ASTM STP 937*, N. J. Johnston, Ed., ASTM, Philadelphia, 1987, p. 115.
[33] Schwartz, H. S. and Hartness, J. T., "Effects of Fiber Coating on Interlaminar Fracture Toughness of Composites," *Toughened Composites*, ASTM STP 937, N. J. Johnston, Ed., ASTM, Philadelphia, 1987, p. 151.
[34] Johnson, W. S. and Mangalgari, P. D., "Investigation of Fiber Bridging in Double Cantilever Beam Specimen," *Journal of Composite Technology and Research*, Vol. 9-1, 1987, p. 10.
[35] Russell, A. J., "Factors Affecting the Opening Mode Delamination of Graphite/Epoxy Laminates," Materials Report 82-Q, Defense Research Establishment Pacific, Victoria, B.C., Canada, December 1982.
[36] de Charentenay, F. X., Harry, J. M., Prel, Y. J., and Benzeggagh, M. L., "Characterizing the Effects of Delamination Defect by Mode I Delamination Test," *Effects of Defects in Composite Materials, ASTM 836*, 1984, p. 84.
[37] Phillips, D. C. and Wells, G. M., "The Stability of Transverse Cracks in Fibre Composites," *Journal of Materials Science Letters*, Vol. 1, 1982, p. 321.
[38] Bradley, W. L. and Cohen, R. N., "Matrix Deformation and Fracture in Graphite Reinforced Epoxies," *Delamination and Debonding of Materials, ASTM STP 876*, W. S. Johnson, Ed., 1985, p. 389.

Delamination

Jeffrey A. Hinkley,[1] *Norman J. Johnston,*[1] *and T. Kevin O'Brien*[2]

Interlaminar Fracture Toughness of Thermoplastic Composites

REFERENCE: Hinkley, J. A., Johnston, N. J., and O'Brien, T. K., **"Interlaminar Fracture Toughness of Thermoplastic Composites,"** *Advances in Thermoplastic Matrix Composite Materials, ASTM STP 1044,* G. M. Newaz, Ed., American Society for Testing and Materials, Philadelphia, 1989, pp. 251–263.

ABSTRACT: Edge delamination tension (EDT) and double cantilever beam (DCB) tests were used to characterize the interlaminar fracture toughness of continuous graphite-fiber composites made from experimental thermoplastic polyimides and a model thermoplastic. Residual thermal stresses, known to be significant in materials processed at high temperatures, were included in the edge delamination calculations. In the model thermoplastic system (polycarbonate matrix), surface properties of the graphite fiber were shown to be significant. Critical strain energy release rates for two different fibers having similar nominal tensile properties differed by 30 to 60%. The reason for the difference is not clear. Interlaminar toughness values for the thermoplastic polyimide composites (LARC-TPI and polyimidesulfone) were 400 to 700 J/m². Scanning electron micrographs of the EDT fracture surfaces suggest poor fiber/matrix bonding. Residual thermal stresses account for up to 32% of the strain energy release in composites made from these high-temperature resins.

KEY WORDS: thermoplastic-matrix composites, interlaminar fracture, fiber/matrix interface, edge delamination, double cantilever beam, polyimide composites, carbon fiber

Thermoplastics currently are being considered for use as matrix resins in high-performance composites. Besides the potential advantages of unlimited shelf life and rapid, inexpensive processing, a major reason for the interest in these materials is their much greater fracture toughness relative to typical 177°C (350°F) cure epoxy resins [1]. Improvements in resin toughness are desirable as one approach to improving delamination resistance and damage tolerance in composite structures.

Extensive testing using the double cantilever beam (DCB) test has shown a reasonable correlation between resin toughness, as measured by compact tension tests on neat resin castings, and composite interlaminar fracture toughness [1]. However, two complications arise in the DCB testing for interlaminar fracture toughness: G_{Ic} values as measured by the DCB test may be affected by fiber bridging; and the adhesion between carbon fibers and the thermoplastic matrix may be poor [1].

The present work examines the interlaminar failure process in four experimental composite materials using both the double cantilever beam (DCB) and the edge delamination tension (EDT) tests. The EDT test was chosen to give an alternate measure of the strain energy release for propagation of an interlaminar crack—one that would not be influenced by fiber bridging, and one that would quantify the influence of residual thermal stresses on delamination onset.

As outlined in Table 1, one set of experiments was conducted on composites with a model

[1]Chemical engineer and chief scientist—Materials Division, respectively, NASA Langley Research Center, Hampton, VA 23665-5225.
[2]Senior scientist, U.S. Army Aerostructures Directorate, Hampton, VA 23665-5225.

TABLE 1—*Test matrix.*

Material	Layup	Test Method	Number of Replicates
AS4/polycarbonate	(0_{24})	DCB	3
	$(+45_2-45_20_290_2)_s$	EDT	5
XAS/polycarbonate	(0_{24})	DCB	3
	$(+45_2-45_20_290_2)_s$	EDT	5
AS4/TPI blend	(0_{24})	DCB	2
	$(+45_2-45_20_290_2)_s$	EDT	3
AS4/polyimidesulfone	(0_{24})	DCB	2
blend	$(+45_2-45_20_290_2)_s$	EDT	3

thermoplastic resin matrix (polycarbonate). These experiments illustrate the effect of varying fiber surface properties on delamination toughness. The other set of experiments was conducted on composites with novel high-temperature thermoplastic polyimide matrices. These latter tests illustrate the significant effect of thermal stresses on delamination onset strain in composite laminates made at high consolidation temperatures.

Materials and Processing

The fibers, AS4 (Hercules, Inc.) and XAS (Hysol Grafil), were supplied unsized, but with the manufacturers' proprietary surface treatments. Previous work has shown that these two fibers behave very differently toward polycarbonate in single-fiber adhesion tests [2]. Polycarbonate resin (Lexan 101 from General Electric Co.) was selected as a model tough thermoplastic, largely because its behavior in neat resin form is well characterized. Six-foot-wide unidirectional prepreg was made by drum-winding where the resin was applied at room temperature as a 17% w/w solution in a 50:50 mixture of chloroform and methylene chloride. Before laminating, prepreg was cut and dried for 1 h at 204°C (400°F)[3] in a forced-air oven. Thermogravimetric analysis of the prepreg after drying showed less than 1% weight loss in heating to 300°C (570°F), indicating that the drying procedure was effective in removing the solvents.

Laminating was performed in a matched metal mold in a heated press at 1.4 MPa (200 psi). A 15-min hold at 260°C (500°F) was followed by a 2-h hold at 240°C (473°F). All crossplied laminates showed small areas of visible fiber waviness in the surface plies after consolidation, whereas unidirectional panels had relatively straight fibers (Figs. 1 and 2). This fiber waviness has been commonly observed to occur in other thermoplastic composites [3,4]. However, ultrasonic C-scan evaluations under conditions revealing microvoids in epoxy composites indicated that the panels were void free. The cross-plied XAS laminates were an exception in that some areas of voids were seen. As will be shown later, dye-penetrant-enhanced radiography of the EDT specimens of this XAS/polycarbonate material revealed extensive cracking parallel to the fibers. These cracks were present even in an untested specimen.

The two polyimide blends consisted of 1:1 mixtures of commercial LARC-TPI powder with a polyamic acid (either LARC-TPI or polyimidesulfone). They were prepregged onto unsized AS-4 fiber using a slurry process [5]. Each blend also contained 2.5 weight % of the diamic acid formed by the reaction of *p*-diaminobenzene and phthalic anhydride. This additive had been shown to improve the flow of thermoplastic polyimides [6]. For simplicity, the two blends will be referred to simply as TPI blend and polyimidesulfone blend. The prepreg was dried and imidized for 1 h at 260°C (500°F); then 7.6 by 15.2 cm (3 by 6 in.) laminates were press-molded in a 3 to 4½-h cycle with a maximum temperature of 349°C (660°F) and 6.9 MPa (1000 psi)

[3]Original measurements were in inch-pound units.

FIG. 1—*A S-4/polycarbonate panel* (±45)$_{2s}$ *illustrating regions of fiber waviness in surface ply.*

pressure. Laminates prepared at lower pressures had a high void content, as judged by C-scan. Repressing these panels at 349°C and 6.9 MPa gave acceptable laminates.

Test Procedures and Calculations

Specimens 2.5 cm wide by 15 cm long for the polycarbonate double cantilever beam test were cut from 24-ply unidirectional laminates containing a 12.5 μm by 3.8 cm (0.0005 in. thick by 1.5 in. wide) Kapton film at the midplane as a crack starter. Pin loading was introduced through aluminum blocks bonded to the beam ends. Results were analyzed using the compliance calibration method [7]. Crosshead speed was 1.27 mm/min (0.05 in./min), and specimen compliance was determined from the opening load versus time record. Crack propagation was steady. In the case of the polycarbonate matrix laminates, extensive fiber bridging caused the apparent G_{Ic} to increase rapidly as the crack advanced from the Kapton insert, typically reaching a steady-state value after a crack extension of only 1 cm. The steady-state values were typically 30 to 100% larger than the initial values. Initial values only are reported, since they are regarded as more characteristic of the toughness that would be obtained in the absence of fiber bridging. For all of the tests, three to five specimens were tested, and the results were averaged. Procedures for the polyimide matrix DCB tests were identical, except the specimens were only twelve plies thick. The data reduction took into account geometric nonlinearity by the method of Devitt et al. [8].

Edge delamination tests were performed on $(+45_2/-45_2/0_2/90_2)_s$ specimens that were 1.27 cm (0.5 in.) or 2.54 cm (1 in.) wide with a 7.62 cm (3 in.) section between the grip tabs. No differences in delamination strain were seen between the two different widths. A layup containing 0° plies was chosen to minimize nonlinearity in the stress-strain curve before onset of delam-

FIG. 2—*AS-4/TPI blend (0_{24}) showing relatively straight fibers in surface plies.*

ination, and sixteen plies were used to ensure that the delamination strain was below the failure strain of the 0° fibers [4, 9, 10]. Delamination was detected visually on the free edges, which had been coated with water-based typewriter correction fluid. Total critical strain-energy release rates are calculated by the following formula

$$G_c = \frac{(\epsilon_c^2 \, t)}{2} (E_{\text{lam}} - E^*)$$ (1)

where

ϵ_c = delamination strain,
t = laminate thickness,
E_{lam} = original laminate modulus, and
E^* = modulus of a completely delaminated specimen.

The modulus of a completely delaminated specimen, E^*, may be calculated using the rule of mixtures equation as follows:

$$E^* = \frac{1}{t} \sum_i E_i t_i$$ (2)

where

E_i = modulus of the sublaminates formed by the delamination [3,9,11], and
t_i = thickness of the sublaminates formed by the delamination.

The sublaminate moduli may be calculated from ply properties, or from the measured stiffness of symmetric sublaminates [3]. For the materials used in the present study, these customary procedures had to be modified slightly. For each material, ply properties were obtained as before from tensile measurements on (0_8), (90_{12}), and $(\pm45)_{2s}$ laminates [11]. The results are shown in Table 2.

These properties were used with classical laminated plate theory to predict the modulus of the quasiisotropic layup employed in the EDT test. When this prediction was compared with the measured values E_{lam}, however, it was found that the actual specimen moduli fell below the predictions by 5 to 30%. Since it was known that the specimens contained wavy fibers, the discrepancy between the measured moduli and the laminate theory prediction was attributed to fiber waviness in the quasiisotropic laminates. Such waviness was also noted in Ref 4 for AS-4/PEEK. The rather large variability in E_{lam} among specimens cut from the same plate was assumed to be due to variations in the degree of fiber waviness across the plate, primarily in the 0° plies, since they dominate the tensile modulus. It follows, then, that a proper value of E^* for each specimen should take account of this variability. Therefore, the following procedure was adopted. Various values of E_{11} were used in the laminate theory until the measured value of E_{lam} was obtained. This adjusted ply modulus, E_{11}, was then used to calculate the corresponding sublaminate modulus E^* for that specimen. Typically, the adjustment amounted to about 15% of E_{11}. This procedure gave the most accurate values of $(E_{lam} - E^*)$ for each test specimen, and yielded the most realistic values of G_c from Eq 1.

The effects of thermal stresses introduced during cool down from the processing temperature were evaluated using a finite-element (FEM) calculation [12]. Total G could have been calculated using plate theory analysis [13], but FEM was used for convenience since it was available. The mesh used is shown in Fig. 2 in Ref 12. The rectangular mesh had 367 nodes with 102 eight-noded parabolic elements. This element size at the delamination tip has been shown to yield accurate strain energy release rate components [14].

The stress-free temperature for each material was determined by fabricating a $(0_2/90_6)$ laminate, which bowed upon cool-down. A strip of this laminate was heated in an oven at approximately 4.4°C/min and the height of the arc was monitored with a direct current displacement transducer (DCDT) (Fig. 3). The temperature at which the strip flattened completely was taken as the stress-free temperature (Fig. 4). As expected, this temperature closely matched the glass transition temperature of the matrix resin.

TABLE 2—*Ply properties.*

Material	Msi[a]				Fiber Volume, %	Average Ply Thickness, h, $\times10^3$ in.[b]
	E_{11}	E_{22}	G_{12}	ν_{12}		
AS4/polycarbonate	17.7	1.10	0.71	0.37	55.8	6.08
XAS/polycarbonate	19.0	1.37	0.44	0.37	55.1	5.63
AS4/LARC-TPI plus LARC-TPI powder	20.8	1.64	0.91	0.34	56.5	5.21
AS4/polyimidesulfone plus LARC-TPI powder	18.4	1.29	0.78	0.34	52.0	6.55

[a]1 Msi = 6.9 GPa.
[b]1 in. = 2.54 cm.

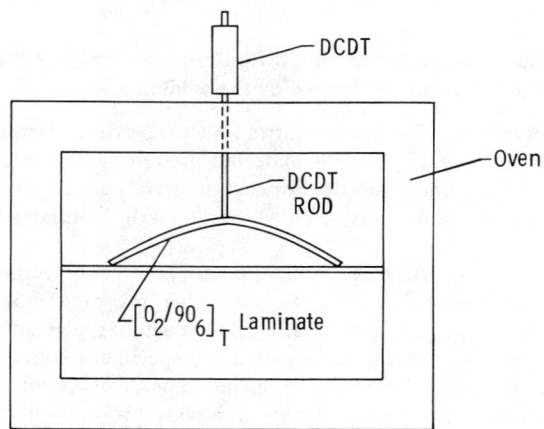

FIG. 3—*Schematic of apparatus for determining stress-free temperature.*

DETERMINATION OF STRESS-FREE TEMPERATURE

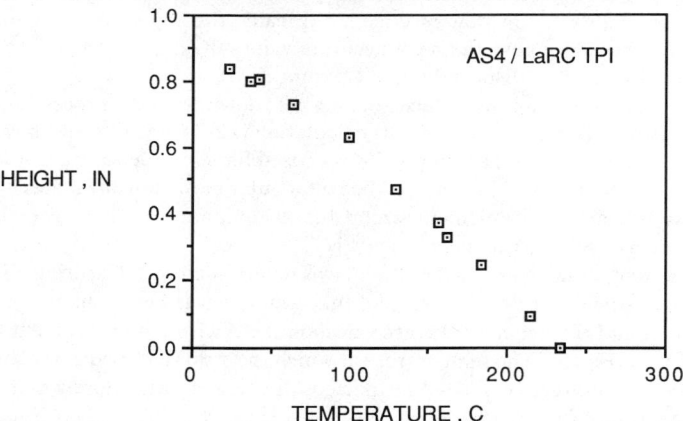

FIG. 4—*Arc height of $(0_2 90_6)_T$ laminate as a function of temperature (1 in. = 2.54 cm).*

Thermal expansion coefficients of the laminae were assumed to be zero in the fiber direction. Transverse expansion coefficients were measured near room temperature using an interferometric technique, or calculated from the resin and fiber properties using the concentric cylinder model of Hashin [15]. In the two cases where both were available, the measured and calculated values showed fairly good agreement, indicating the analytic technique was reasonably accurate. Results are shown in Table 3.

Results and Discussion

Initial G_{Ic} values for the AS-4/polycarbonate and XAS/polycarbonate, obtained from the first increment of crack growth beyond the insert in the DCB test, were 1.3 and 0.89 J/m^2,

TABLE 3—*Thermal expansion properties.*

Laminate	Measured Stress-Free Temperature, °C	Linear Coefficient of Thermal Expansion, μm/°C		
		Resin	Calculated Composite, α_2	Measured Composite, α_2
AS4/Polycarbonate	130	67.5	41.2	36.7
XAS/Polycarbonate	a	67.5	51.5	a
AS4/TPI blend	229	45	29.3	a
AS4/Polyimidesulfone blend	243	45b	31.9	27.9

aNot determined.
bAssumed.

respectively. Scanning electron micrographs (SEM) of the fracture surfaces show, in both cases, extensive matrix deformation (Fig. 5). There does seem to be a qualitative difference in fiber/matrix adhesion in the two cases, however. On the AS-4/polycarbonate (PC) fracture surface (Fig. 5), many fibers are judged by SEM examination to be completely bare, stripped of resin, and threads of drawn polymer litter the surface. In the XAS composite, by contrast, most fibers seem to retain shreds of polymer (Fig. 6).

These observations are consistent with the degrees of adhesion measured for the two systems using embedded-fiber fragmentation tests [2]. Table 4 reproduces some results from Ref 2. The critical fragment length, l_c, which provides a quantitative measure of fiber/matrix adhesion, seems to be quite large for AS-4/PC, indicating very poor adhesion. As a point of reference, the critical length for AS-4 in an epoxy resin is about 0.4 mm. The much smaller critical length for XAS/PC is indicative of better adhesion, although one would need to know the tensile strengths of the two fibers at submillimetre gage lengths to calculate the relative strengths of the interfacial bonds. A second, independent line of evidence, however, tends to confirm the difference in adhesion to the two fibers. This is the birefringence pattern that arises around the broken fiber ends. It is quite distinctive and shows characteristic features of poor adhesion with AS-4 and good adhesion with XAS. Based on the data above, therefore, poor fiber/matrix bonding, such as that between AS-4 and polycarbonate, appears to yield higher composite interlaminar frac-

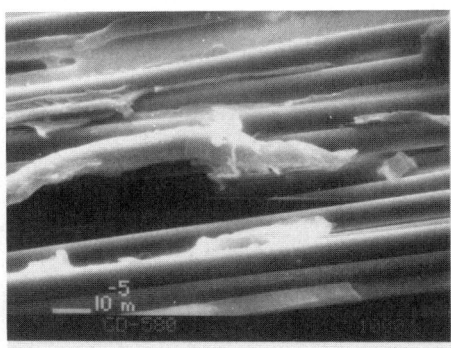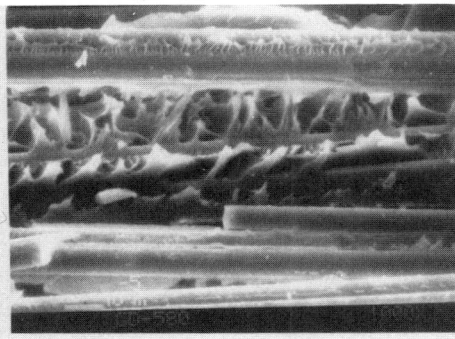

FIG. 5—*Scanning electron micrographs of DCB fracture surfaces, AS-4/Lexan polycarbonate.*

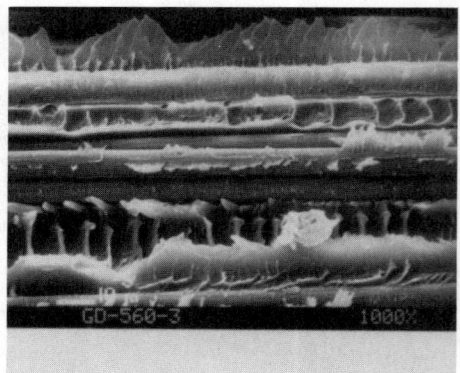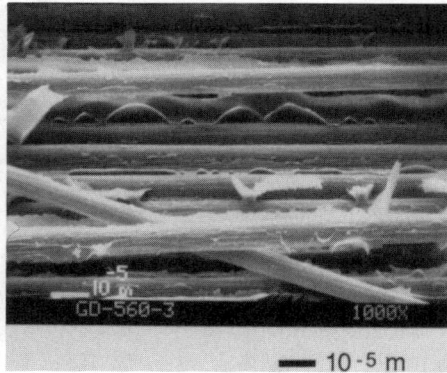

— 10⁻⁵ m

FIG. 6—*Scanning electron micrographs of DCB fracture surfaces, XAS unsized/Lexan polycarbonate.*

TABLE 4—*Embedded fiber fragmentation results.*

Fiber/Matrix	Critical Fragment Length,[a] mm	Birefringence Pattern
AS-4/PC	0.74	Poor adhesion
XAS/PC	0.36	Good adhesion

[a]The critical fragment length is inversely proportional to the fiber/matrix interfacial shear strength if fiber breaking strength is constant.

ture toughness as measured by the DCB test. This is a surprising result and should be regarded with caution.

In Ref 16, fiber/matrix debonding in DCB tests was said to contribute to increased G_{Ic} by increasing fracture surface area, by promoting fiber bridging, and by relieving stress triaxiality, thereby allowing more resin deformation. These effects, and especially fiber bridging, normally are observed after the delamination has grown some distance from the insert [17]. However, in the present study, increased toughness was found for initiation values. A possible explanation would be that in this case fiber bridging began to develop immediately at the insert. This conclusion must be regarded as tentative, though, because the hypothesized bridging was not detected directly.

In the edge delamination tests, delamination occurred, as expected, at the 0/90 interfaces with the failure plane wandering back and forth through the central 90° plies [3,9,11]. Dye-penetrant X-ray photos (Fig. 7) show that in the AS-4/PC composites, the 90° cracks do not extend much beyond the delaminated region, indicating that they form after the delamination [11]. In contrast, dye penetrant-enhanced radiographs of the XAS/PC showed extensive matrix or interfacial cracks throughout the laminate width in all plies (Fig. 8). As mentioned earlier, this cracking was present before any mechanical load was applied. Micrographs of the 0/90 delamination fracture surfaces of both the AS-4/PC and XAS/PC EDT specimens are dominated by what appear to be bare fibers and resin tracks left by fibers that peeled out cleanly (Figs. 9 and 10).

The resin in the XAS/PC composite appears to have fractured brittlely (Fig. 10). This apparent matrix brittleness was completely unexpected, and two possible explanations for it have

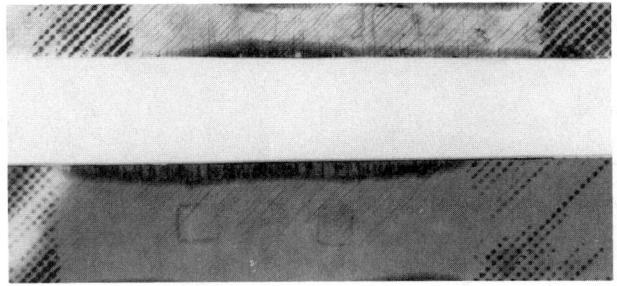

FIG. 7—*Dye-penetrant enhanced X-ray photographs of EDT specimens after testing. AS-4/polycarbonate.*

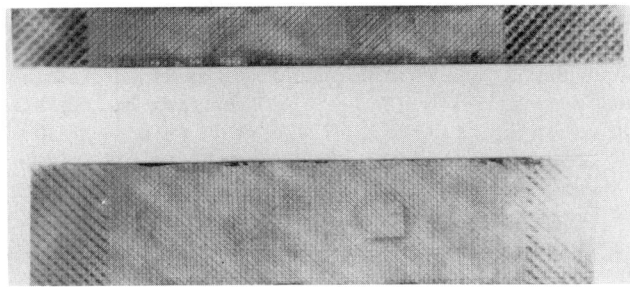

FIG. 8—*Dye-penetrant enhanced X-ray photographs of EDT specimens after testing. XAS/polycarbonate.*

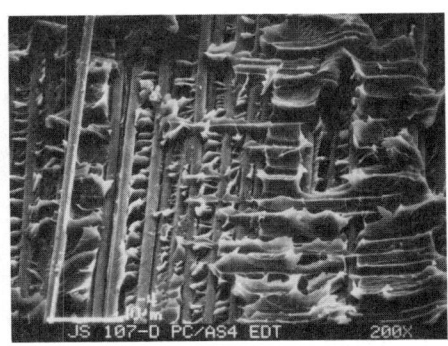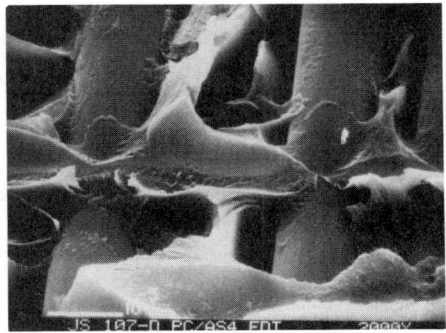

FIG. 9—*SEM photomicrographs of 0/90 delamination surfaces in EDT specimens, AS-4/polycarbonate,* $(+45_2, -45_2, 0_2, 90_2)_s$ *laminate.*

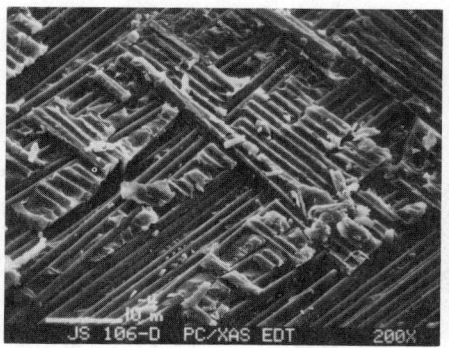

FIG. 10—*SEM photomicrographs of 0/90 delamination surfaces in EDT specimens, XAS/polycarbonate, $(+45_2, -45_2, 0_2, 90_2)_s$ laminate.*

been considered. If polymer crystallinity was nucleated by the fiber surface [18], the apparent brittle behavior might have been understandable, but no evidence of matrix crystallinity was found by differential scanning calorimetry or by wide-angle X-ray scattering. Another possibility is that trace impurities on the fiber surface could catalyze resin degradation. This possibility was ruled out also, since resin dissolved from the XAS composite had not decreased in inherent viscosity. Thus the reason for the brittle appearance of the resin in Fig. 10 is still not clear.

The EDT results are compared with those from the DCB test in Fig. 11, in which the superscripts M and T stand for mechanical and thermal stress contributions to G_c. Both tests indicate that the AS-4 composite possesses the higher interlaminar toughness. Note, however, that the total G_c as measured by the EDT is lower than the DCB result for G_{Ic}. This trend has been observed in other materials [3], but with polycarbonate the difference is striking. Including

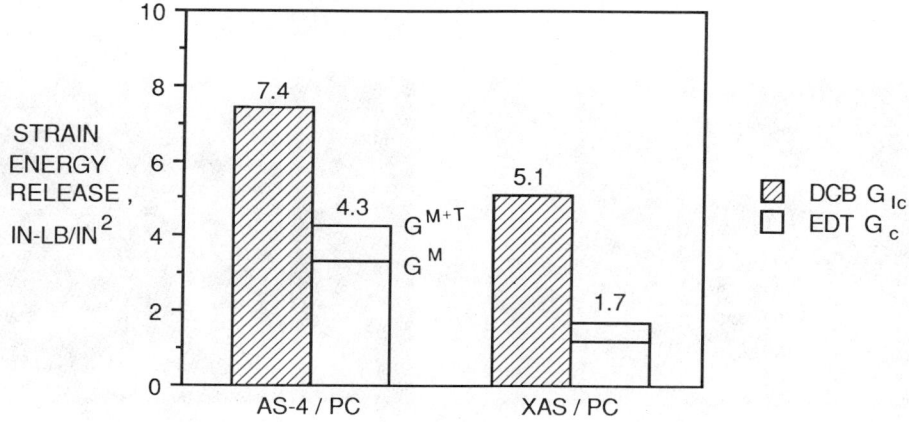

FIG. 11—*Comparison of G_{Ic} (from DCB test) with total G_c^{M+T} (from EDT) for polycarbonate laminates (1 in. · lb/in.² = 175 J/m²).*

thermal stresses in the EDT calculations raises the EDT results, but only by less than 20% because the ΔT between the stress-free temperature and room temperature is not very large. The large differences between the DCB and the EDT values in this case are tentatively attributed to the effects of fiber bridging in the DCB test, although this has not been shown conclusively. The processing [1] and other factors that lead to bridging in unidirectional specimens are not yet fully understood.

AS-4/Polyimide Blends

Double cantilever beam specimens of these materials showed much less obvious fiber bridging than the polycarbonate, although an increase in G_{Ic} with crack length was noted. Because there was no resin squeeze out during fabrication of these panels, the fibers probably could not nest effectively, and the fracture surfaces were quite flat.

In the edge delamination tests, these two materials behaved similarly, and their fracture surfaces again show regions of bare fibers and resin troughs where fibers have peeled out (Figs. 12 and 13). The SEM of the $PISO_2$/LaRC-TPI surface in Fig. 13 indicates some fine-scale hetero-

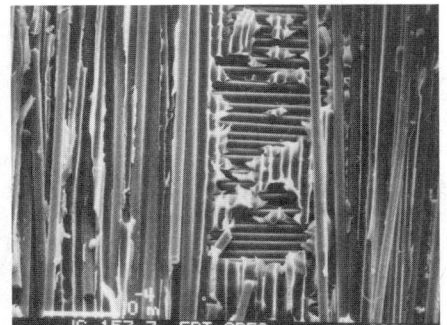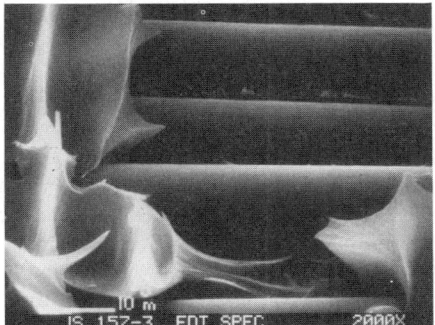

FIG. 12—*SEM photomicrographs of fracture surfaces of polyimide matrix EDT specimens, 1:1 LaRC TPI:LaRC TPI/AS-4, $(+45_2,-45_2,0_2,90_2)_s$ laminate.*

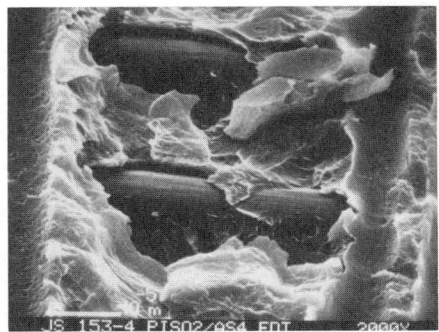

FIG. 13—*SEM photomicrographs of fracture surfaces of polyimide matrix EDT specimens, 1:1 $PISO_2$:LaRC TPI/AS-4, $(+45_2,-45_2,0_2,90_2)_s$ laminate.*

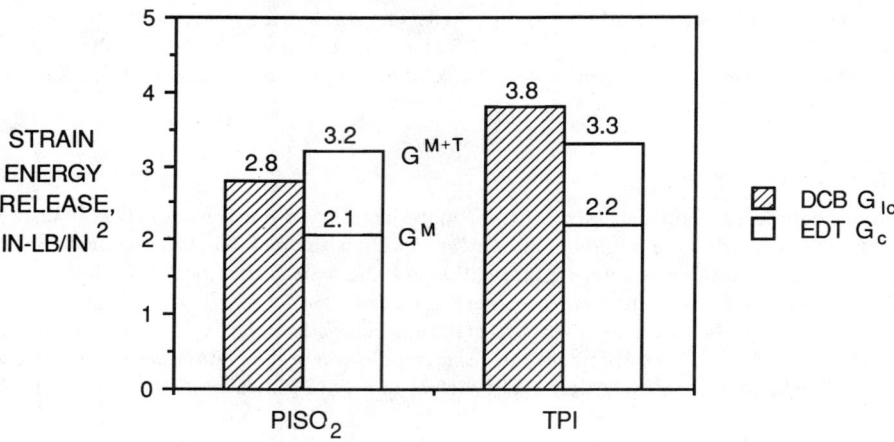

FIG. 14—G_{Ic} *(from DCB) and* G_c^{M+T} *(from EDT) for polyimide blends (1 in.* $\cdot$ *lb/in.^{2} = 175 J/m^2).*

geneity, which may result from incomplete blending of the LaRC powder with the PISO$_2$ binder. Whether better mechanical mixing is needed, or whether these two polymers are thermodynamically incompatible, is not known.

With the higher softening temperatures of these polyimides relative to conventional epoxy matrices, the issue of thermal stresses assumes much greater importance, as will be clear from an examination of the EDT results in Fig. 14. The strain-energy release including thermal stresses G^{M+T} is substantially larger than that due to the mechanical strain only, G^M. As mentioned earlier, one reason for using the EDT test is to illustrate this role of residual stresses in delamination. The EDT test is performed with a realistic laminate which inherently contains residual thermal stresses [4,10,13]. In the case of these thermoplastic polyimides, the contribution of residual stresses to delamination is significant, amounting to 28 and 32% of G_c^{M+T} for the TPI and polyimidesulfone blends, respectively.

Figure 8 also shows the relative magnitudes of G_{Ic} determined from the DCB test and total G_c^{M+T} calculated from the EDT test. For the polyimide materials, unlike the polycarbonate, one sees that total G_c is comparable in magnitude to G_{Ic}. The layup chosen for the EDT test in this work produces almost a pure Mode I delamination, so this would be expected. Furthermore, Johnson and Mangalgiri [19] have shown that for other tough resin composite systems, including the thermoplastic PEEK APC2, $G_{Ic} = G_{IIc} = G_c$ at delamination onset.

Summary and Conclusions

The edge delamination test and DCB test gave similar toughness rankings for the fiber/resin combinations studied here. Fiber bridging seems to increase the DCB results typically by at least a factor of two. For the tough model thermoplastic it appears that even at onset from the insert, the measured toughness may be high, although this effect is not fully understood.

A procedure based on adjusting the modulus of 0° plies in a laminate is a reasonable way to account for the effects of fiber waviness on composite laminate stiffness in thermoplastic materials. This procedure was needed to accurately measure G_c from the EDT test.

Thermal residual stresses must be considered when analyzing results of EDT tests on higher-temperature matrix materials. In one graphite/polyimide system, the thermal contribution amounted to 32% of the total G_c.

References

[1] Hunston, D. L., Moulton, R. J., Johnston, N. J., and Bascom, W. D., "Matrix Resin Effects in Composite Delamination: Mode I Fracture Aspects," *Toughened Composites, ASTM STP 937*, Norman J. Johnston, Ed., American Society for Testing and Materials, Philadelphia, 1987, pp. 74–94.

[2] Bascom, W. D., Jensen, R. M., and Cordner, L. W., "The Adhesion of Carbon Fibers to Thermoplastic Polymers," *Proceedings of the Sixth International Conference on Composite Materials*, F. L. Matthews, Ed., Elsevier, London, 1987, p. 5.424.

[3] Johnston, N. J., O'Brien, T. K., Morris, D. H., and Simonds, R. A., "Interlaminar Fracture Toughness of Composites. II—Refinement of the Edge Delamination Test and Application to Thermoplastics," *28th National SAMPE Symposium and Exhibition*, 1983, Vol. 28, p. 502.

[4] O'Brien, T. K., "Fatigue Delamination Behavior of PEEK Thermoplastic Composite Laminates," *Journal of Reinforced Plastics*, Vol. 7, No. 4, 1988, pp. 341–359.

[5] Johnston, N. J. and St. Clair, T. L., "Thermoplastic Matrix Composites: LARC-TPI, Polyimidesulfone and Their Blends," *18th SAMPE International Technical Conference*, October 1986.

[6] Pratt, J. R., St. Clair, T. L., Burks, H. D., and Stoakley, D. M., "Polyimide Processing Additives," *32nd International SAMPE Symposium*, April 1987, Vol. 32, p. 1036.

[7] Wilkins, D. J., Eisenmann, J. R., Camin, R. A., Margolis, W. S., and Bensen, R. A., "Characterizing Delamination Growth in Graphite-Epoxy," *Damage in Composite Materials, ASTM STP 775*, K. L. Reifsnider, Ed., American Society for Testing and Materials, 1982, pp. 168–183.

[8] DeVitt, D. F., Schapery, R. A., and Bradley, W. L., "A Method for Determining the Mode I Delamination Fracture Toughness of Elastic and Viscoelastic Composite Materials," *Journal of Composite Materials*, Vol. 14, 1980, p. 270.

[9] O'Brien, T. K., "Mixed-Mode Strain-Energy Release Rate Effects on Edge Delamination of Composites," *Effects of Defects in Composite Materials, ASTM STP 836*, American Society for Testing and Materials, Philadelphia, 1984, pp. 125–142.

[10] O'Brien, T. K., Johnston, N. J., Raju, I. S., Morris, D. H., and Simonds, R. A., "Comparisons of Various Configurations of the Edge Delamination Test for Interlaminar Fracture Toughness," *Toughened Composites, ASTM STP 937*, N. J. Johnston, Ed., American Society for Testing and Materials, Philadelphia, 1987.

[11] O'Brien, T. K., Johnson, N. J., Morris, D. H., and Simonds, R. A., "A Simple Test for the Interlaminar Fracture Toughness of Composites," *SAMPE Journal*, Vol. 18, No. 4, 1982, p. 8.

[12] Raju, I. S., "Q3DG—A Computer Program for Strain-Energy Release Rates for Delamination Growth in Composite Laminates," NASA CR 178205, October 1986 (computer program is available through COSMIC).

[13] O'Brien, T. K., Raju, I. S., and Garber, D. P., "Residual Thermal and Moisture Influences on the Strain Energy Release Rate Analysis of Edge Delamination," *Journal of Composites Technology and Research*, Vol. 8, No. 2, Summer 1986, pp. 37–47.

[14] Raju, I. S., Crews, J. H., Jr., and Aminpov, M. A., "Convergence of Strain-Energy Release Rate Components for Edge-Delaminated Composite Laminates," NASA TM 8935, April 1987.

[15] Hashin, Z., "Analysis of Properties of Fiber Composites with Anisotropic Constituents," *Journal of Applied Mechanics*, Vol. 46, 1979, pp. 543–550.

[16] Hibbs, M. F., Tse, M. K., and Bradley, W. L., "Interlaminar Fracture Toughness and Real-Time Fracture Mechanism of Some Toughened Graphite/Epoxy Composites," *Toughened Composites, ASTM STP 937*, N. J. Johnston, Ed., American Society for Testing and Materials, Philadelphia, 1987, pp. 115–130.

[17] Russell, A. J. and Street, K. N., "Moisture and Temperature Effects on the Mixed Mode Delamination Fracture of Unidirectional Graphite-Epoxy," *Delamination and Debonding of Materials, ASTM STP 876*, W. S. Johnson, Ed., American Society for Testing and Materials, Philadelphia, 1985.

[18] Kardos, J. L., Cheng, F. S., and Tolbert, T. L., "Tailoring the Interface in Graphite-Reinforced Polycarbonate," *Polymer Engineering and Science*, Vol. 13, 1973, p. 455.

[19] Johnson, W. S. and Mangalgiri, P. D., "Influence of the Resin on Interlaminar Mixed-Mode Fracture," *Toughened Composites, ASTM STP 937*, N. J. Johnston, Ed., American Society for Testing and Materials, Philadelphia, 1987.

Golam M. Newaz,[1] *Arnold Lustiger,*[2] *and Jong-Yeong Yung*[3]

Delamination Growth Under Cyclic Loading at Elevated Temperature in Thermoplastic Composites

REFERENCE: Newaz, G. M., Lustiger, A., and Yung, J-Y., **"Delamination Growth Under Cyclic Loading at Elevated Temperature in Thermoplastic Composites,"** *Advances in Thermoplastic Matrix Composite Materials, ASTM STP 1044,* G. M. Newaz, Ed., American Society for Testing and Materials, Philadelphia, 1989, pp. 264–278.

ABSTRACT: The cyclic Mode I delamination growth characteristics of unidirectional carbon/polyetheretherketone (PEEK) composites at ambient and elevated temperatures was studied. At room temperature, crack growth rate was found to increase as a function of cycles, a common characteristic for such materials under load-controlled conditions. In contrast, at elevated temperature 93°C (200°F), early delamination growth was characterized by decreasing growth rate (stable growth). With continued cycling, there was a breakdown of the process zone (craze-plastic zone) leading to rapid crack propagation. Increased temperature and lower rates of deformation tend to enhance ductility in PEEK-based thermoplastic composites through fibril formation resulting from the breakup of crystalline lamellae. This ductility tends to stabilize the crack front, most likely through craze formation ahead of the crack tip, resulting in an anomalous decrease in growth rate with increasing cycles. This behavior is termed as slow crack growth (SCG). The process zone (craze-plastic zone) eventually gives way to crack growth after a critical crack-tip opening displacement is reached. The crack growth process is associated with less ductility, even at elevated temperature 93°C (200°F). When crack growth occurs without any attendant process zone ahead of the crack, as in the case of the specimens tested at room temperature, less ductility is evident. These fracture surfaces display evidence of interlamellar failure. As a result of ductility and the active role of the process zone at elevated temperature, neither the strain energy release rate, G, nor the stress intensity factor, K, were found to be appropriate parameters to characterize the earlier part of delamination growth at elevated temperature. Instead, the process zone (craze-plastic zone) growth at elevated temperature was shown to be best described by a relaxation controlled growth model.

KEY WORDS: Mode I delamination, thermoplastic composite, elevated temperature, process zone, cyclic loading, fracture morphology

Compared with conventional thermosetting composites, thermoplastic composites offer some unique advantages. An important advantage of thermoplastic composites is the potential for fast and easy part fabrication. In addition, the damage tolerance characteristics of advanced thermoplastics generally are found to be greater than their thermoset counterparts. However, the thermomechanical response of these composites under long-term loading is not well known.

Delamination is a common failure mode in advanced composites. Much attention has been paid to studying delamination problems in recent years [1]. Resistance to initiation and growth

[1]Principal research scientist, Battelle Columbus Division, Columbus, OH 43201.
[2]Member, Technical staff, AT&T Bell Laboratories, Murray Hill, NJ 07974.
[3]Research scientist, Battelle Columbus Division, Columbus, OH 43201; currently at Sunstrand Corp., Rockford, IL 61107.

of delamination cracks has a direct bearing on the service life of many laminated composite airframe structures. It has been found that thermoplastic composites such as carbon fiber rein-forced PEEK exhibit almost an order of magnitude greater Mode I delamination toughness compared with traditional carbon/epoxy composites. Studies have been conducted to investi-gate and compare the delamination resistance of thermoplastic composites with traditional ad-vanced composites [2–5]. Also, long-term behavior such as delamination growth under fatigue loading has received attention [6]. However, research to date has been confined to room temper-ature studies only.

The influence of elevated temperature on delamination growth characteristics is an impor-tant consideration in applications where the composite material is exposed to service tempera-tures above ambient conditions. Thermoplastic composite response at elevated temperature must be understood well to utilize reliably these materials for future aerospace applications, especially in primary structures of advanced fighters.

In this investigation, Mode I delamination growth characteristics at elevated temperature in carbon/PEEK (APC-2) composite were studied. The major goal of the study was to identify failure mechanisms at elevated temperature and relate them to observed cyclic delamination growth response of the composite. An additional objective was to model the delamination growth behavior at elevated temperature.

Materials and Testing

Unidirectional carbon/PEEK 24-ply laminates were fabricated from APC-2 prepregs with AS-4 carbon fibers obtained from ICI (Imperial Chemical Industries). The laminates were made using compression molding in a flat press with appropriate temperature controls. The consolidation temperature and pressure cycle was 5 min at 385°C (725°F) at 0.69-MPa pressure and another 5 min at 149°C (300°F) at 2.07-MPa pressure. The fiber volume fraction in the laminates was 0.63. A 0.127-mm Teflon insert was embedded from one edge of the laminate for the starter delamination crack. The dimensions and configuration of the specimen are shown in Fig. 1.

All fatigue tests were conducted in a Universal test machine. For elevated temperature tests, a special inexpensive test chamber was fabricated from Styrofoam with laminated wood as an outer wall. A hot air gun with a blower fan was used to circulate hot air. Thermocouples on the

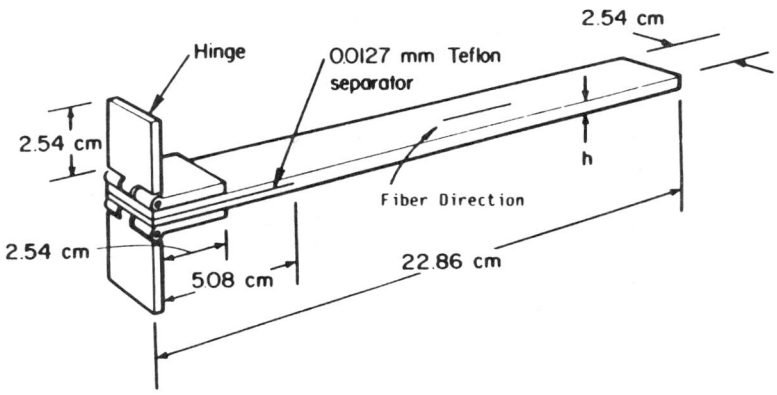

FIG. 1—*Double cantilever beam specimen.*

specimen were used to verify the uniformity and constancy of the test chamber temperature. Delamination fatigue tests were conducted at 93°C (200°F).

All fatigue tests were conducted in the load-controlled mode with a triangular waveform with an R ratio of 0.01. The test frequency was 0.5 Hz. The peak load was set at 138 N, which was 85% of the load level that caused instantaneous delamination growth. The load-deflection response was monitored using an x-y plotter. Deflection was measured at the load-line and represented crosshead displacement.

Crack and process zone lengths were recorded at regular intervals as a function of fatigue cycles by observing the specimen through a glass window. An optical eyepiece with a ×10 magnification was used to monitor the crack and the process zone length.

Scanning electron microscopy was used to study the mechanisms associated with Mode I delamination growth.

Results and Discussion

Typical responses of delamination crack growth at room temperature are shown in Figs. 2 and 3. The delamination process zone (craze-plastic zone) growth at elevated temperature 93°C (200°F) is shown in Fig. 4. The crack and process zone lengths are represented by a and α, respectively. There are significant differences between room and elevated temperature response as seen in Figs. 2 and 4. The slow crack growth (SCG) behavior up to 750 000 cycles in Fig. 4 is attributed to the development of the process zone while the original delamination crack remained stationary. In reality, some crack growth may also have taken place. A clear understanding of the interaction of the original delamination crack and process zone will require an extensive study. Existence of a typical process zone due to the influence of elevated temperature is shown in Fig. 5. This sample was subjected to 100 cycles at 200°F. It was then examined under SEM. A porosity-like feature can be seen at the trailing edge of the process zone. This appears to be a defect which occurred during SEM sample preparation.

The behavior at room temperature is similar to what we observe for homogeneous materials for the load-controlled condition, that is, the delamination crack growth rate continues to increase as a function of fatigue cycles until final failure occurs. On the other hand, the behavior

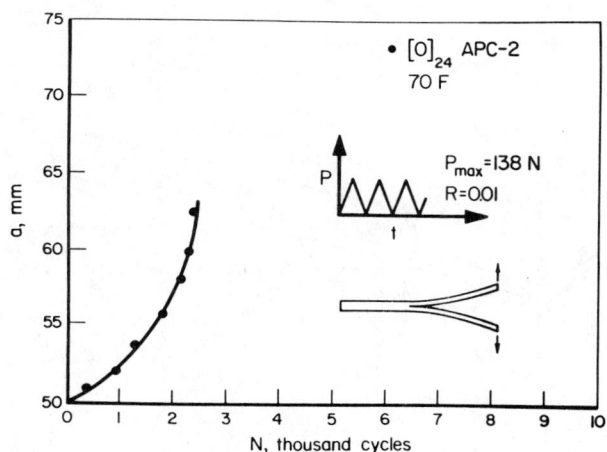

FIG. 2—*Delamination growth as a function of fatigue cycles in carbon/PEEK composite at 23°C (73°F).*

DIRECTION OF CRACK PROPAGATION ➡

FIG. 3—*Step cracking ahead of original delamination crack at 23°C (70°F).*

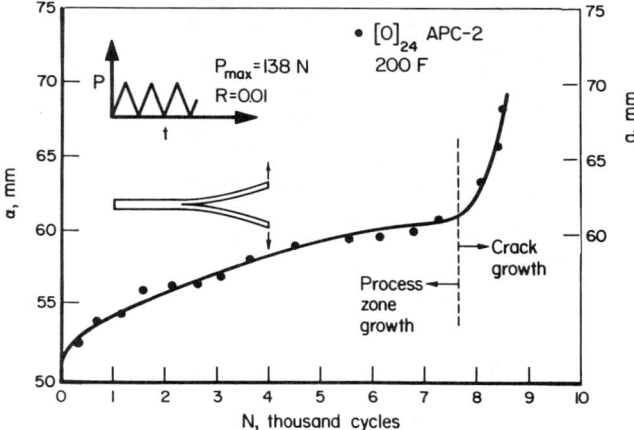

FIG. 4—*Delamination growth as a function of fatigue cycles in carbon/PEEK composite at 93°C (200°F).*

at elevated temperature is much more complex. For load-controlled fatigue loading, SCG behavior is an indication that some toughening mechanism is operative ahead of the original crack. Contribution to this toughening is offered by the process zone.

At early life, the delamination process zone grows at an anomalously decreasing growth rate. Eventually, there is a breakdown of the process zone. Consequently, delamination crack starts to propagate fairly rapidly with increasing growth rate. The early decreasing growth rate behavior at elevated temperature is attributed to a toughening mechanism, that is, ductility in the form of process zone formation and growth ahead of the original delamination crack. This aspect is examined in detail below.

Conventional analysis of fatigue crack growth data is accomplished using the Paris-type law

$$\frac{da}{dN} = A(\Delta G)^n \tag{1}$$

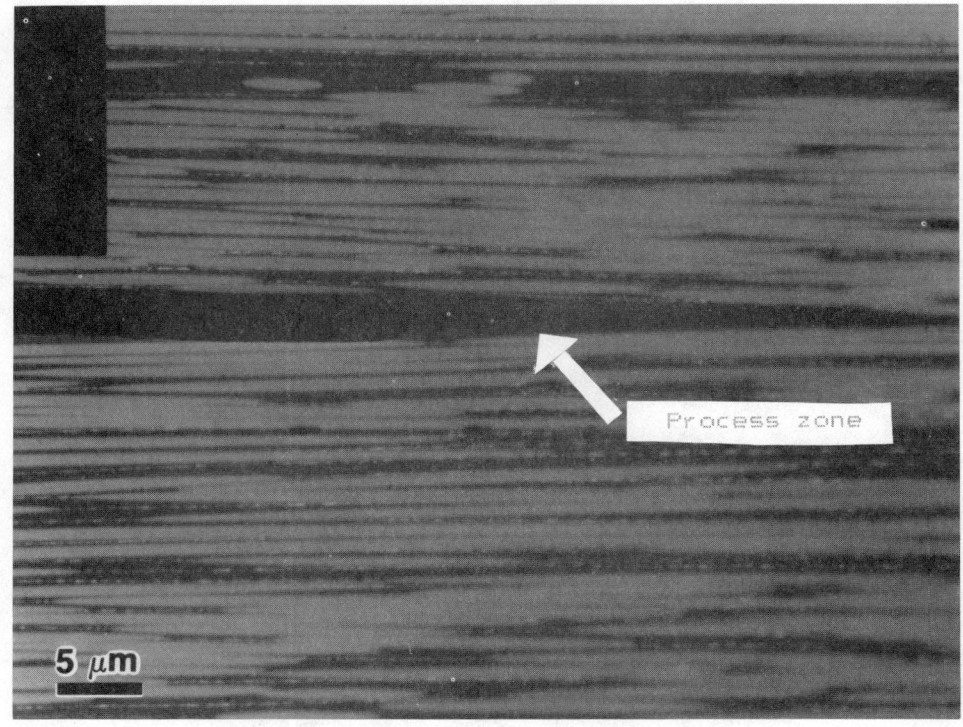

FIG. 5—*Process zone ahead of original delamination crack at 93°C (200°F).*

where

a = crack length,
A, n = material parameters, and
ΔG = effective energy release rate.

An important aspect of this relationship between cyclic crack growth rate and G or K is that the energy release rate or the stress intensity factor is a driving force for the growth of the fatigue crack.

The room temperature data are represented in Fig. 6 using the relation in Eq 1. Analysis of SCG behavior is considered more complex. The elevated temperature behavior observed (Fig. 4) exhibiting slow crack growth is similar to the relaxation-controlled process zone growth behavior in polymers. This phenomenon was extensively studied by Williams and Marshall [7] and Passaglia [8] for time-dependent conditions. In order to model this behavior for the elevated temperature fatigue problem as in our case, we will adopt the following approach.

In many polymers, the failure process ahead of the crack (craze-plastic zone) has been modeled by a line zone as proposed by Dugdale [9]. For an infinite plate with an initial crack length, a, and an applied stress, σ, the process zone length (Fig. 7) is given by

$$\alpha = \frac{\pi}{8} \frac{K^2}{\sigma_c^2} \qquad (2)$$

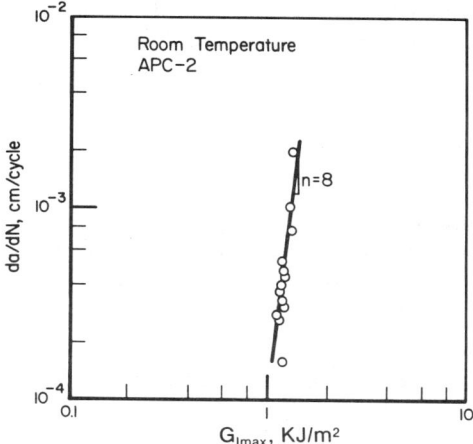

FIG. 6—*Fatigue crack growth rate versus strain energy release rate for specimens tested at 23°C (73°F).*

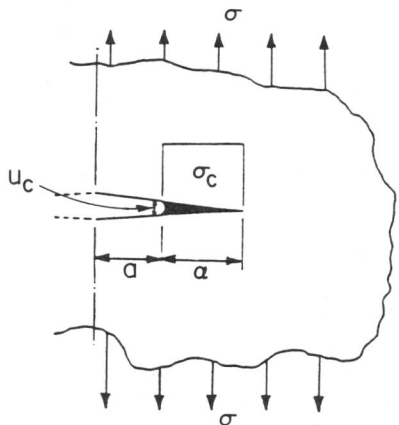

FIG. 7—*Dugdale zone ahead of crack.*

where

K = stress intensity factor, and
σ_c = stress in the process zone (craze-plastic zone).

For low values of critical stress intensity factor, K_c, below initiation, fracture does not occur. Instead, the crack-tip process zone may propagate, with the crack itself remaining stationary. Within the process zone, it is expected that there is variation of σ_c stress. The Dugdale model requires that the zone stress σ_c be constant. Also, since it is reasonable to assume for many polymers that the time dependence may be written as power laws [7], we will postulate similar relation for combined cycle and temperature dependence. For the elevated temperature case discussed in this paper, the time-dependent aspect is present since viscoelastic effects will be enhanced at elevated temperatures. Considering the relaxation of stresses within the process zone due to cyclic loading at elevated temperature similar to the development by Williams and Marshall [7], we can write the following

$$\sigma_c(N) = \sigma_0 N^{-m} \tag{3}$$

where

σ_0 = initial yield strength,
m = material parameter (a combined measure of cycle and temperature-dependent response of yield stress), and
N = number of fatigue cycles.

It is expected that there would be a variation of stress along the process zone, reducing from the unrelaxed value at the process-zone tip to the relaxed value at the crack tip after N cycles. The average process zone stress may be approximated by

$$\bar{\sigma}_c = \frac{1}{N} \int_0^N \sigma_c dN = \frac{\sigma_0 N^{-m}}{1 - m} \tag{4}$$

This approximation is made by ignoring all cycle dependent and viscoelastic history effects.

A relaxation controlled process may now be described from Eqs 2 and 4, so that

$$\alpha = \frac{\pi}{8} \frac{K^2}{\sigma_c^2} = \frac{\pi}{8} \frac{K^2}{\sigma_0^2} (1 - m)^2 N^{2m} \tag{5}$$

and the process zone growth may be described as

$$\frac{d\alpha}{dN} = 2m(1 - m)^2 \frac{\pi}{8} \frac{K^2}{\sigma_0^2} N^{2m-1} \tag{6}$$

In either case (Eqs 5 and 6), K refers to the original stress intensity factor, since the crack is stationary until the process zone reaches its critical length.

From log $d\alpha/dN$ versus log N curves, the value of m is obtained as illustrated in Fig. 8. Once m is obtained, Eq 6 may be used to predict life, N, if the critical process zone length is known. Although there was no effort in life prediction in this investigation, a future effort is planned.

To clarify and to determine the failure mechanisms and their relationship to the delamination growth behavior, scanning electron microscopy was used.

Figures 9 and 10 display fracture surfaces of the material in the stable (early in life) and unstable (late in life) crack growth regions of the curve at room temperature. In the stable crack growth region (early part of life), some ductility (that is, deformation and fracture morphology on the fracture surface) is readily evident with virtually no ductility in the region of unstable crack growth. At 93°C (200°F) in the stable process zone growth region, ductility is dramatic (Fig. 11), while in the unstable crack growth region (later part of life), the ductility is again lowered (Fig. 12). In comparing fracture surfaces generated at different temperatures, it appears that the level of deformation on the 93°C (200°F) fracture surface in the unstable region of the curve approximates that of the 23°C (73°F) fracture surface in the stable region of the curve. Alternating regions of crack stability and instability in PEEK laminates have been recently documented in delamination under Mode I loading conditions [9].

The source of ductility in the PEEK system is best understood in light of the mechanism of deformation in semicrystalline polymers. Semicrystalline materials have an amorphous phase

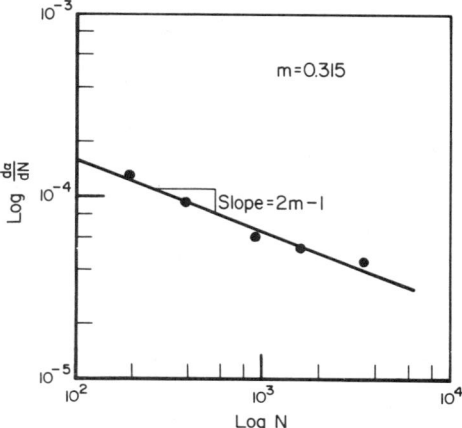

FIG. 8—*Process zone growth rate versus fatigue cycles for elevated temperature (93°C) tests.*

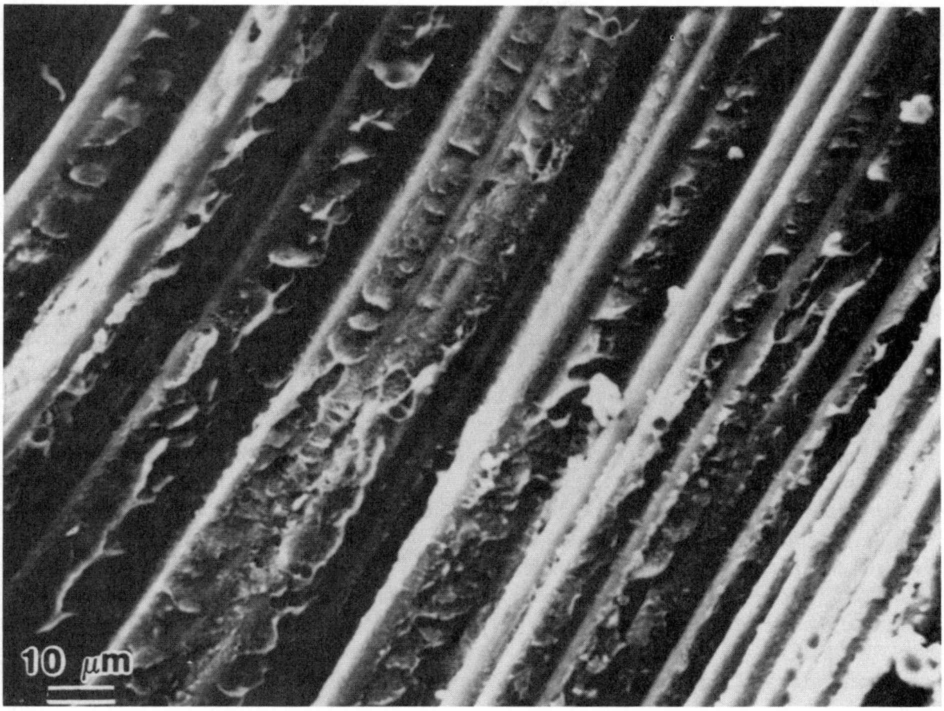

FIG. 9—*Fracture surface in stable crack growth region at room temperature showing moderate ductility.*

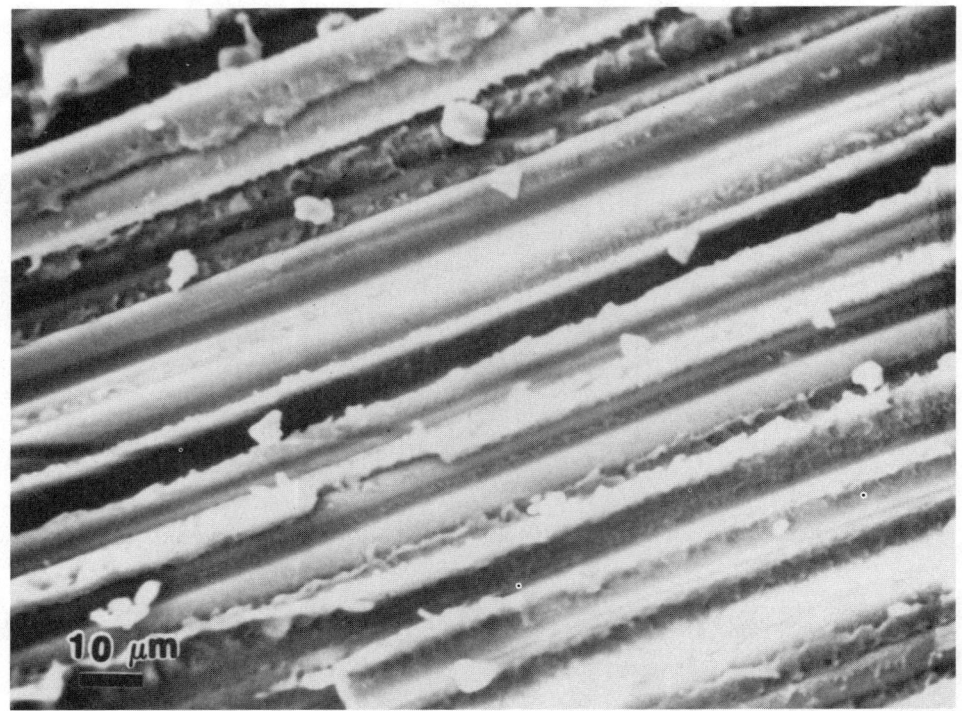

FIG. 10—*Fracture surface in unstable crack growth region at room temperature showing little ductility.*

and an ordered crystalline phase. A simplified schematic of this arrangement is shown in Fig. 13. Although the exact conformation of polymer chains in the crystalline phase is scientifically as yet unresolved, in this model it is assumed that these chains fold on each other.

When solidified from the melt, the crystals form platelets or lamellae which usually nucleate radially from a central point to form larger spherical structures known as spherulites. If a tensile load is applied normal to the face of the lamellae, the tie molecules connecting lamellae begin to stretch as shown in Fig. 14(*a*). At a certain point, however, they can be pulled out no further (Fig. 14(*b*)). At this time, the lamellae break up into smaller units (Fig. 14(*c*)). According to this model, as advanced by Peterlin [*10*], these so-called "mosaic blocks" are directly incorporated into a new fiber morphology (Fig. 14(*d*)). These fibers are thus the source of the observed ductility on the fracture surface.

In order for this mechanism of fiber deformation to be operative in the PEEK system, molecular mobility, especially in the tie molecules, is necessary. Although high molecular mobility is not normally associated with polymeric material tested below their glass transition temperatures (T_g of PEEK = 137°C), in reality ductility is highly rate and temperature dependent. At lower strain rates and higher temperatures, the molecular mechanisms favoring ductility in viscoelastic materials can take place (as they do in other polymeric materials tested below T_g such as nylon and polycarbonate) [*11*]. At higher temperatures and longer times, tie molecules can have time to move in semicrystalline systems, allowing reorganization of lamellae into fibers.

Elevated temperature and lower rates of deformation will tend to enhance ductility on the fracture surface of PEEK-based composites as is clear from Figs. 9 through 11. The very long

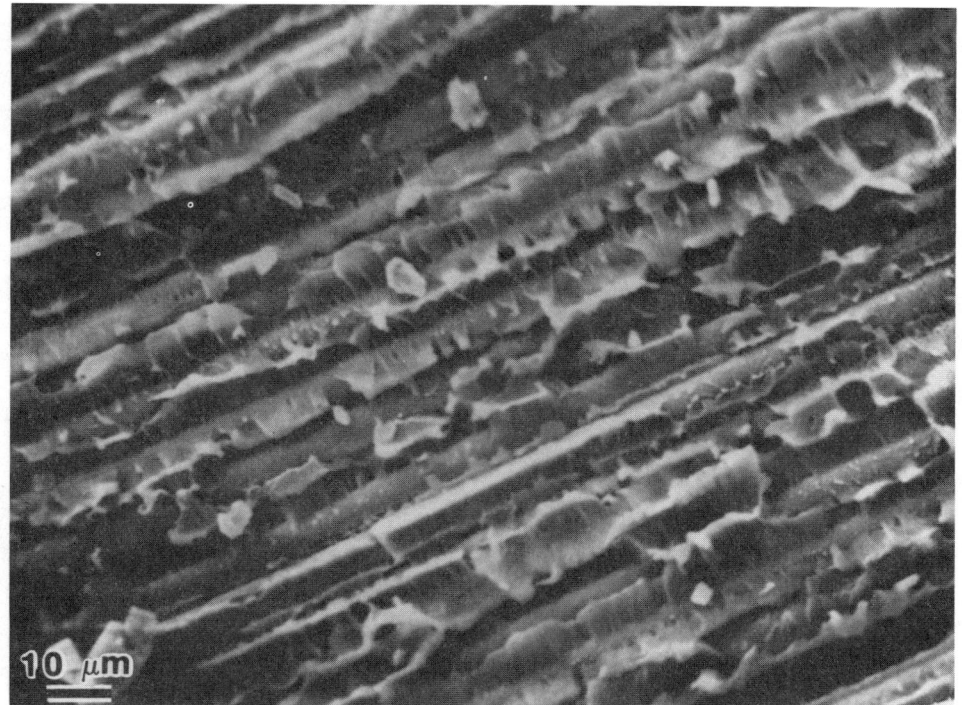

FIG. 11—*Fracture surface in stable crack growth region at 93°C (200°F) showing high ductility.*

uniform fibril formation on the specimens tested at 93°C (200°F) tends to stabilize the crack front, most likely through craze formation ahead of the crack tip. As a result, the region of crack stability extends well beyond the same region of the curve for the material tested at room temperature. Correspondingly, in the region of crack instability, less ductility is evident since molecular mobility is lower at higher deformation rates.

In the case of crack instability at room temperature (Fig. 10), the relative brittleness of the material under these test conditions is manifested in a most interesting way. When this specimen is viewed at higher magnification (Fig. 15), radial patterns are evident. These patterns clearly suggest delamination fracture through the center of the spherulites. As explained above, spherulites consist of lamellae which are also radially oriented. On the basis of a similar fracture surface feature, Bandopadhay and Brown [12] inferred that environmental stress cracking of polyethylene preferentially occurs between lamellae. Lovinger and Davis [13] indicate that lamellar orientation in PEEK is uniplanar, with lamellae preferentially oriented on edge. This lamellar organization would tend to favor the interlamellar failure suggested here.

The brittle fracture feature in PEEK thus suggests that rather than tie molecules stretching and causing lamellae to break apart as shown in Fig. 14, the tie molecules themselves break, leaving the lamellae themselves largely unaffected as shown in Fig. 16. By way of contrast, the very ductile deformation of the matrix for the specimen tested at 93°C (200°F) in the region of process zone growth is also shown at about the same magnification in Fig. 17. Thus, the relative ductility or brittleness on the fracture surfaces of these specimens can be described on a molecular level as a competition between the fibril deformation shown in Fig. 13 and the interlamellar separation shown in Fig. 15.

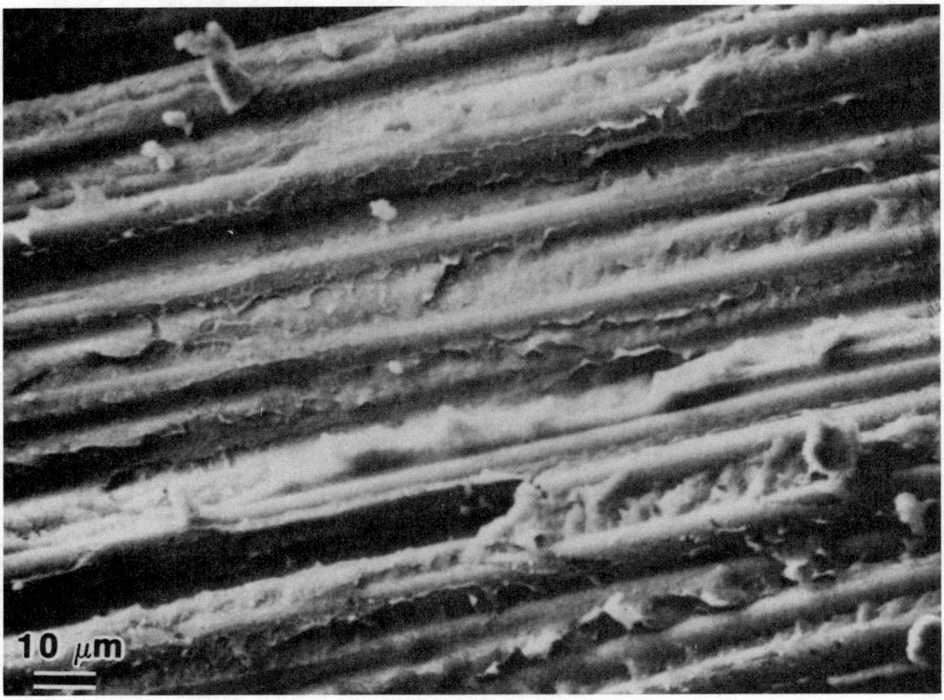

FIG. 12—*Fracture surface in unstable crack growth region at 93°C (200°F) showing moderate ductility.*

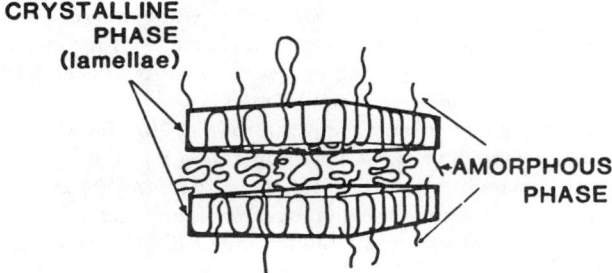

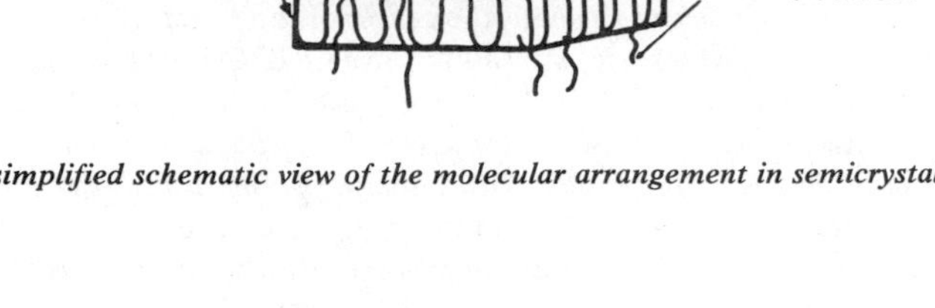

FIG. 13—*A simplified schematic view of the molecular arrangement in semicrystalline polymers.*

Conclusions

Based on the results of this investigation, the following conclusions can be drawn:

1. The Mode I cyclic delamination growth characteristics at elevated temperature in advanced thermoplastic composites are complex. The decreasing delamination growth rate early in life followed by slow growth rate until rapid delamination growth is significantly different

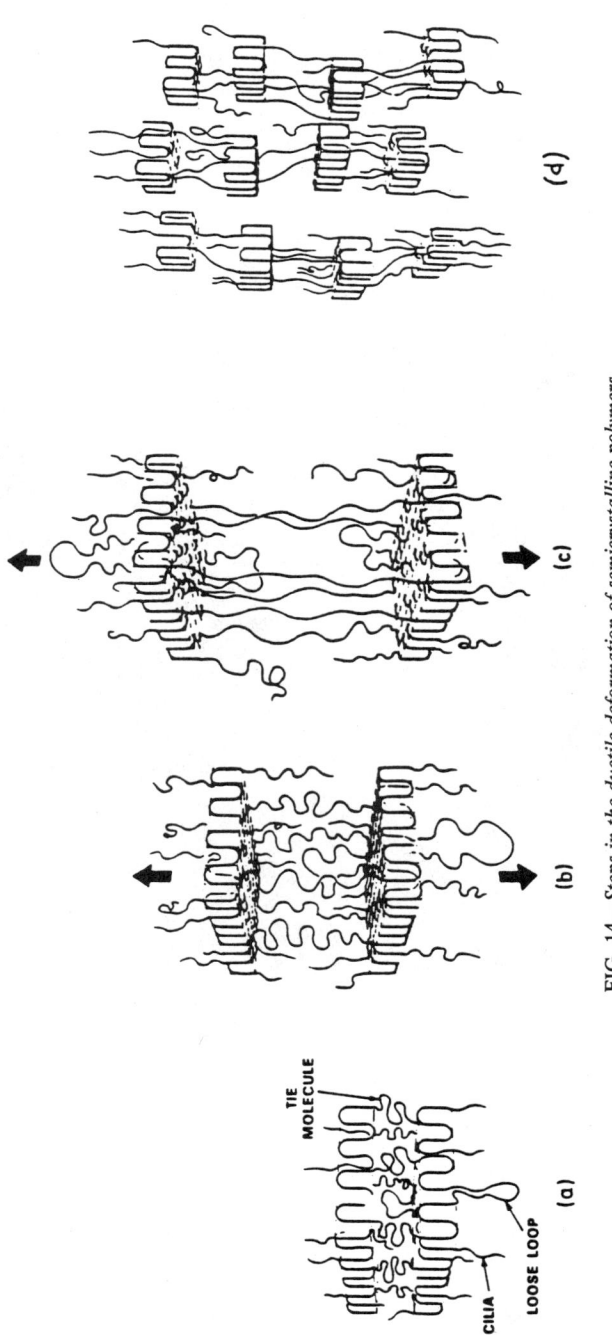

FIG. 14—*Steps in the ductile deformation of semicrystalline polymers.*

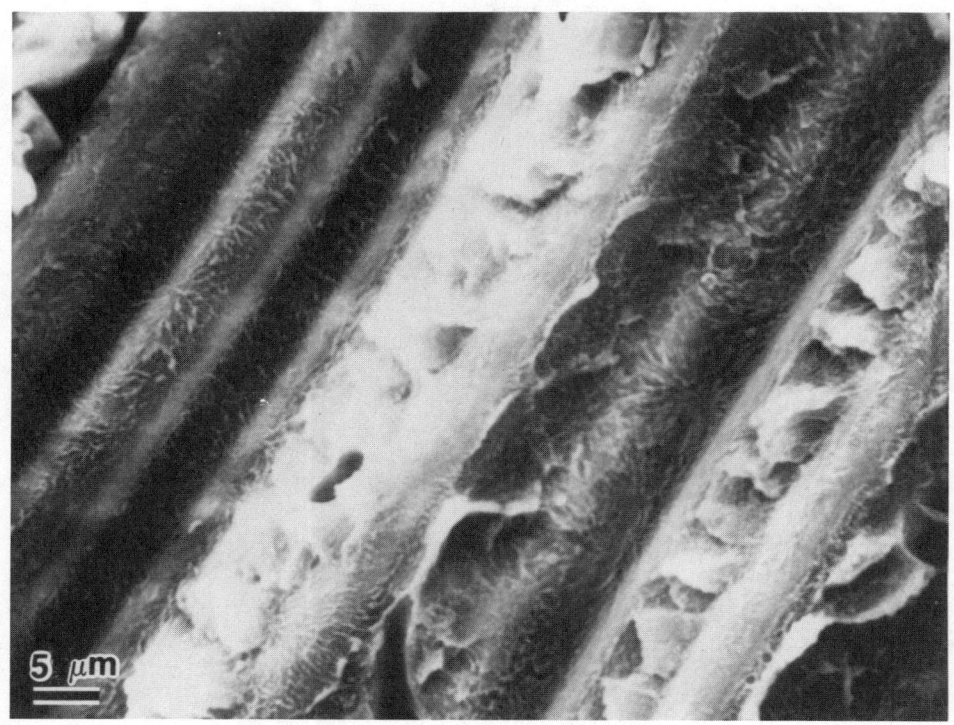

FIG. 15—*Higher magnification view of fracture surface in Fig. 9, showing radial patterns suggesting interlamellar failure.*

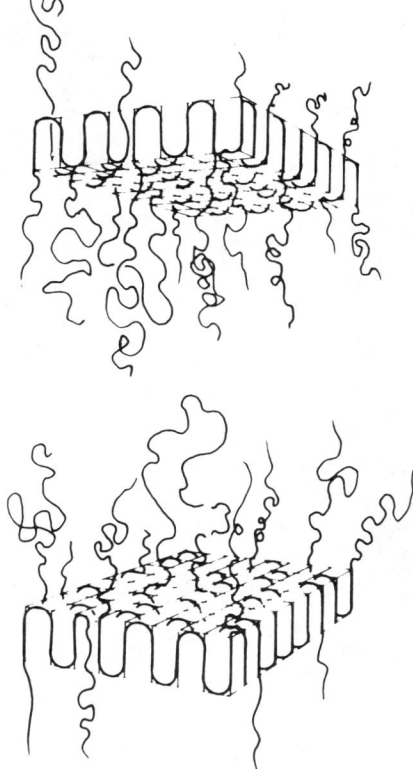

FIG. 16—*Final step in the brittle fracture of semicrystalline polymers.*

than the behavior at room temperature. The delamination growth behavior at room temperature is as expected under load-controlled condition.

2. The complex delamination process zone (craze-plastic zone) growth behavior at elevated
temperatures is modeled using the concept of relaxation controlled growth. The process zone
growth rate is related to original stress intensity factor, K, a material parameter, m, initial yield
stress, σ_0, and fatigue cycles, N. The relationship can be used (Eq 6) to predict the life, N, up to
the breakdown of process zone under cyclic loading at elevated temperatures.

3. Ductility is enhanced in PEEK-based composites at elevated temperature. Also, the original delamination crack is found to stabilize through process zone formation ahead of it. This
increases the life of the composite. Rapid delamination growth at the later stages of fracture is
associated with less ductility due to the lower molecular mobility at higher deformation rates.

4. The relationship between ductility and brittleness on the fracture surface can be described
as a competition between fibril deformation and interlamellar separation. This understanding
provides an explanation on a molecular level to the observations of macroscopic delamination
growth response.

References

[1] *Delamination and Debonding of Materials, ASTM STP 876,* W. S. Johnson, Ed., American Society
for Testing and Materials, Philadelphia, 1985.

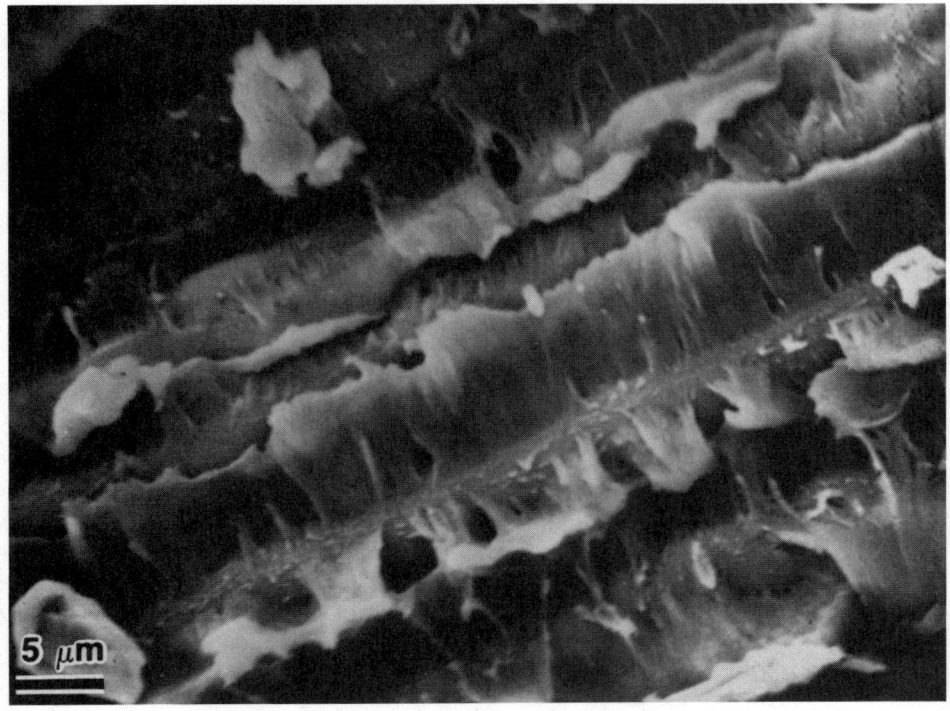

FIG. 17—*Higher magnification view of fracture surface in Fig. 10, showing significant fiber formation suggesting intralamellar failure.*

[2] Gillespie, J. W., Carlson, L. A., Pipes, R. B., Rothchilds, R., Trethewey, B., and Smiley, A., "Delamination Growth in Composite Materials," Center for Composite Materials, University of Delaware, Newark, DE, 1985.

[3] Martin, C. C., O'Connor, J. E., and Lou, A. Y., "Polyphenylene Sulfide High Performance Composites," *SAMPE Quarterly*, July 1984, pp. 12–16.

[4] Brady, D. G., "Aerospace Discovers Thermoplastic Composites," *Materials Engineering*, September 1986, pp. 41–44.

[5] Newaz, G., unpublished data on Mode I Delamination Toughness of Thermoplastic Composites, Battelle Columbus Division, Columbus, OH, 1987.

[6] Mall, S., Yun, K. T., and Kochhar, N. K., "Characterization of Matrix Toughness Effect on Cyclic Delamination Growth in Graphite Fiber Composites," ASTM Conference on *Fatigue and Fracture*, Cincinnati, April 1987.

[7] Williams, J. G. and Marshall, G. P., "Environmental Crack and Craze Growth Phenomena in Polymers," *Proceedings of the Royal Society of London*, A.342, 1975, pp. 55–77.

[8] Passaglia, E., "Relaxation of Stresses in Crazes at Crack Tips and Rate of Craze Extension," *Polymer*, Vol. 23, 1982, pp. 754–760.

[9] Crick, R. A., Leach, D. C., Meakin, P. J., and Moore, D. R., "Interlaminar Fracture Morphology of Carbon Fibre/PEEK Composites," *Journal of Materials Science*, Vol. 22, 1987, p. 2094.

[10] Peterlin, A., "Morphology and Fracture of Drawn Crystalline Polymers," *Journal of Macromolecular Science—Physics*, Vol. B8, 1973, p. 83.

[11] Matsuoka, S., "Non-Linear Viscoelastic Stress-Strain Relationships in Polymeric Solids," *Failure of Plastics*, W. Brostow and R. D. Corneliussen, Eds., Hanser Verlag, 1986.

[12] Bandopadhay, S. and Brown, H. R., "Environmental Stress Cracking and Morphology of Polyethylene," *Polymer*, Vol. 19, 1978, p. 589.

[13] Lovinger, A. J. and Davis, D. D., "Electron Microscopic Investigation of the Morphology of a Melt-Crystallized Polyaryletherketone," *Journal of Applied Physics*, Vol. 58, 1985, p. 2843.

K. Benjamin Su[1]

Delamination Resistance of Stitched Thermoplastic Matrix Composite Laminates

REFERENCE: Su, K. B., **"Delamination Resistance of Stitched Thermoplastic Matrix Composite Laminates,"** *Advances in Thermoplastic Matrix Composite Materials, ASTM STP 1044,* G. M. Newaz, Ed., American Society for Testing and Materials, Philadelphia, 1989, pp. 279–300.

ABSTRACT: This paper shows how to improve delamination resistance in composite laminates by the use of thermoplastic matrix resins and by stitching. Discussed are fatigue crack growth characterics in thermoplastic matrix composites as well as the differences between the crack growth processes in thermoplastic and thermosetting matrix composites under cyclic fatigue and monotonically increasing loads. Further improvement of delamination resistance achieved by stitching shows a 20 to 30% increase in the critical crack-opening load, a 20 times higher fatigue threshold, and a two-orders-of-magnitude increase in fatigue life. Fractography evidence is presented to illustrate the role of the stitching yarns in resisting delamination. A fracture model was constructed to simulate delamination crack propagation in the presence of stitching yarns. The effects of geometric stitching parameters, as well as intrinsic and extrinsic yarn properties, were simulated with this fracture model. Finally, practical issues on the effective and economical application of stitching to composite structures are discussed.

KEY WORDS: thermoplastic matrix composites, 3-D composites, stitching, toughness, delamination, interlaminar fracture, fracture model, crack closure, crack propagation, fatigue, stress intensity factor, critical strain-energy release rate, plastic deformation, deformation process

Delamination in composite laminates can be initiated at relatively low applied loads. Growth of delamination cracks leads to the rapid deterioration of mechanical properties and may cause catastrophic failure of the composite structure.

There are two basic approaches to improve delamination resistance in composite laminates. The first approach is to replace the matrix with tough resin systems. This approach is represented by the current interest in toughened epoxy resins and the new generation of thermoplastic matrix resins. Thermoplastics such as polyetheretherketone, polyimides, polyphenylene sulfide, and polyamides all show significant improvement in interlaminar fracture toughness of the composites by about one order of magnitude over traditional epoxy matrix systems [1–6].

The second approach to improve delamination resistance is through mechanical means such as providing fiber reinforcements in the weak out-of-plane direction. Stitching and the use of three-dimensional (3-D) fiber structures are the two common methods to accomplish this goal [7–9].

In the case of 3-D fiber structures, the laminated construction of the composites is completely eliminated, and so is the associated problem of delamination. However, modern high-speed manufacturing processes and suitable feed materials as well as associated engineering and design expertise are still largely in the development stages. The use of 3-D composites at this time

---

[1]Senior research engineer, E. I. du Pont de Nemours and Co., Inc., Wilmington, DE 19898.

is limited to certain specialty items. On the other hand, stitching, a special case of 3-D reinforcement, does not greatly alter the original laminate structure and thus does not require extensive modification of the traditional material systems and fabrication processes. If conducted properly, the stitched composite should retain most or all of the properties and functionalities of the original composite laminate but with greatly enhanced delamination resistance.

Discussions in the first part of this paper focus on recent developments in our continuing investigation into the toughening mechanisms of thermoplastic matrix composite systems. New evidence is presented to support the fracture model proposed in an earlier publication [5]; data are presented on the fatigue crack growth characteristics of thermoplastic and thermosetting matrix composites. The remaining part of the paper discusses further enhancement of the delamination resistance of thermoplastic matrix composites through the use of stitching. Experimental data and an analytical model are presented to develop a fundamental understanding of the strengthening mechanisms provided by the stitching yarns. Principles that can be applied to the effective design and use of stitched laminates are also presented.

Thermoplastic Matrix Composite Laminates

Material Systems

The thermoplastic matrix resin discussed in this paper is a model semicrystalline polyamide based on bis(para-amino-cyclohexyl methane). This matrix resin is designated as J1-polymer, to distinguish it from the amorphous polyamide J2-polymer used in many commercial thermoplastic matrix composite products offered by Du Pont. Unidirectional composite laminates were made from fully impregnated unidirectional AS4-graphite/J1-polymer tapes. The fabrication of laminates and specimens are detailed elsewhere [5]. A different type of sample was used in the stitched laminate study. These laminates were made from J1-polymer film and plain-weave graphite fabric via a fabric/film stacking technique, the details of which are discussed in a later paragraph.

Mode I Interlaminar Fracture and Toughening Mechanisms

The Mode I interlaminar fracture of unidirectional AS4-graphite/J1-polymer composites was investigated extensively and reported [5]. Its Mode I interlaminar fracture toughness represented by the critical strain-energy release rate, G_{Ic}, is 2.01 kN/m (11.5 lbf/in.) compared with the 0.26 kN/m (1.5 lbf/in.) of AS4-graphite/3501-6 epoxy laminate. This dramatic increase in interlaminar fracture toughness is attributed to energy dissipation through large amounts of plastic deformation in the thermoplastic J1-polymer during fracture (Figs. 1 and 2). As illustrated in Fig. 3, the plastic deformation process extends from the plastic zone in front of the crack tips to the trailing deformation process zone on the fractured surfaces. Figure 4 shows further evidence of the extensive plastic deformation in the matrix in progress. Upon application of external load, classical ductile failure processes such as the nucleation of microvoids proceed in the plastic zone in front of the crack tip; further matrix deformation in the form of void growth and plastic drawing of the resins are apparent also on the fracture surfaces behind the crack tip.

Fatigue Crack Propagation

The investigation of Mode I fatigue crack propagation was conducted on 36-layer unidirectional AS4-graphite/J1-polymer and AS4-graphite/3501-6 epoxy laminates. Width-tapered double cantilever beam (DCB) fracture specimens instead of straight DCB specimens were used in the test (Fig. 5). A four-to-one tapering ratio was used in our specimens. The critical applied

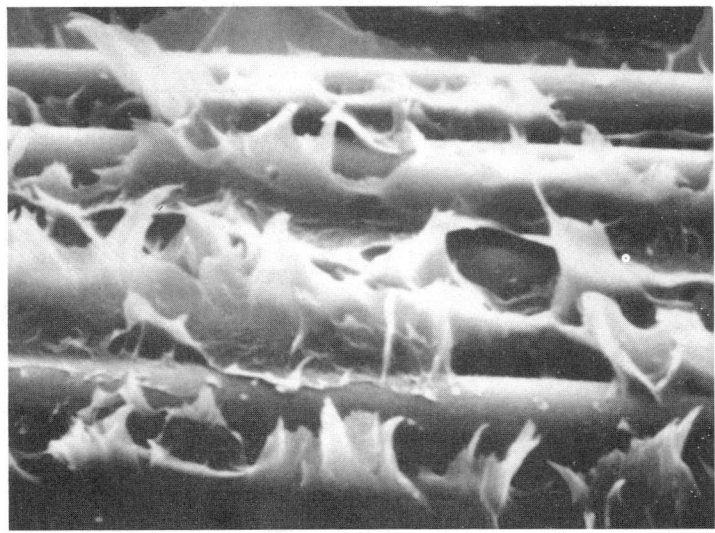

FIG. 1—*Mode I fracture surface of a unidirectional AS4-graphite/J1-polymer laminate. Note the extensive plastic deformation in the matrix resin.*

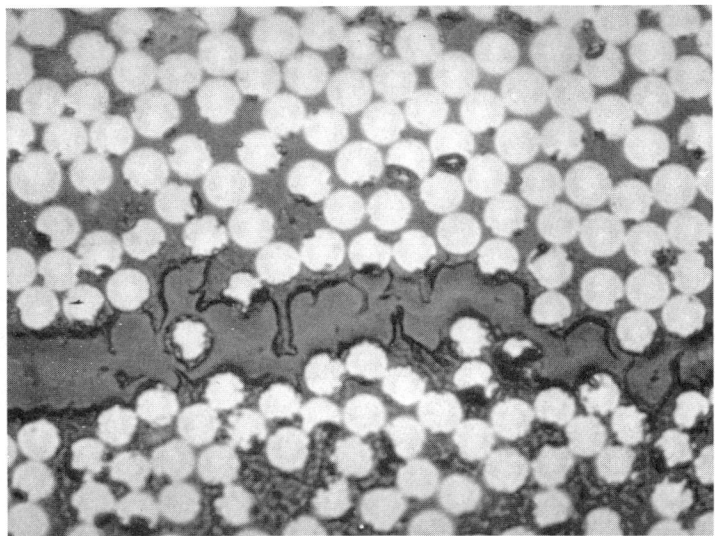

FIG. 2—*Mode I interlaminar crack of a unidirectional AS4-graphite/J1-polymer laminate as viewed along the direction of the fiber axis. Note the extensive drawing of matrix resins from the fracture surfaces.*

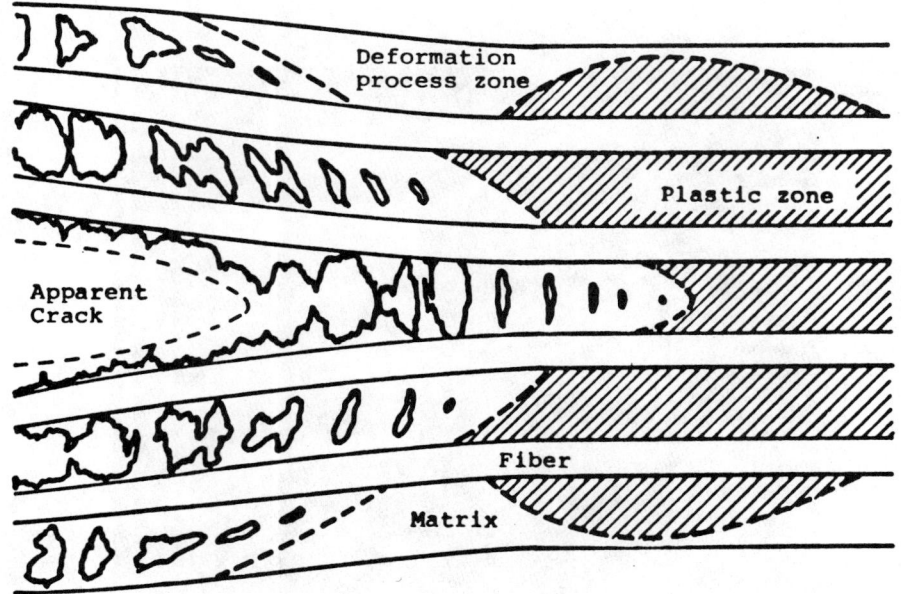

FIG. 3—*A physical model of the Mode I interlaminar fracture in thermoplastic matrix composite laminates* [5].

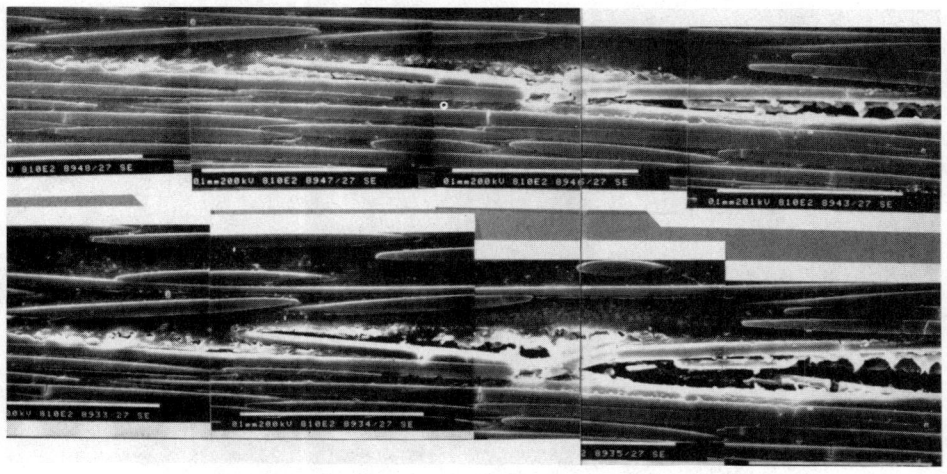

FIG. 4—*Mode I fracture initiation in progress in a AS4-graphite/J1-polymer DCB specimen. Micrograph at the top shows the crack tip region before load was applied. Micrograph at the bottom was taken as the crack was wedge-opened. Note the micro-voids nucleated in the plastic zone in front of the crack tip and the resin deformation on the crack surfaces behind the crack tip.*

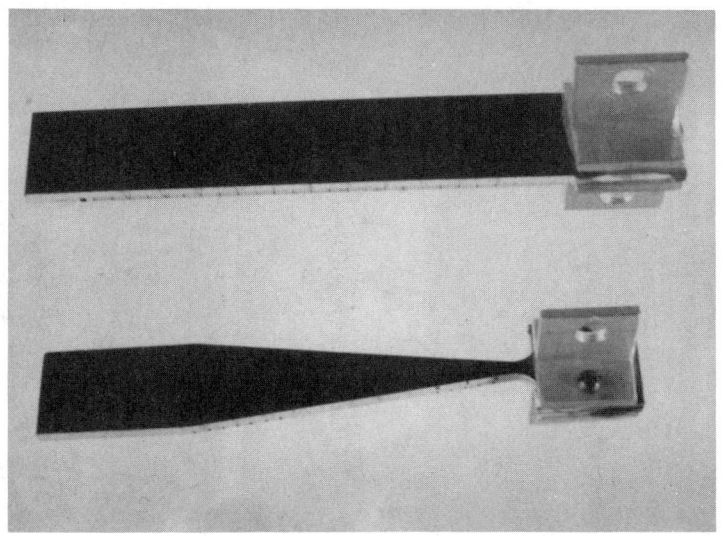

FIG. 5—*Straight and width-tapered DCB specimens.*

load, P_c, required to propagate the interlaminar crack is constant at all crack lengths for a tapered DCB specimen, which makes it especially convenient for investigating fatigue crack propagation under constant crack driving forces. The tests were run at maximum load levels set at selected fractions of the critical load. The R ratio (P_{min} to P_{max} load ratio) used ranged from 0.25 to 0.5; the higher R ratios are used when cracks propagated to some appreciable length and significantly increased the compliance of the specimen. Adjustments to raise the minimum load level were needed to keep up with the rather high test frequency (10 Hz) used.

The critical load, P_c, for crack propagation is 142 N (32 lbf) for the AS4-graphite/J1-polymer width-tapered DCB specimens compared with the 53 N (12 lbf) for specimens made from AS4-graphite/3501-6 epoxy. This value represents a 2.5 times higher load-carrying capacity for the thermoplastic J1-polymer composite laminates than for the conventional epoxy matrix composite system. The results of the fatigue crack propagation tests are shown in Fig. 6.

Crack growth is insignificant in the thermosetting AS4/3501-6 system at P_{max} smaller than one third of its critical load. Crack growth becomes noticeable and the crack growth rate increases rapidly when P_{max} increases beyond one half of P_c. As P_{max}s approach P_c, the fatigue crack will grow at a stable rate for a small number of cycles after initiation, begin to propagate unstably, and quickly cause the complete fracture of the specimen.

For the AS4/J1-polymer system, crack growth is noticeable when the P_{max}s are about one third of the critical load. The crack growth is stable throughout the entire test, even for tests conducted at P_{max}s approaching P_c. At such high P_{max} levels, it was observed that, if the specimen survives the first few hundred cycles, continuing crack growth will proceed at a stable rate and will not result in sudden and catastrophic failure, as in the case of the AS4/3501-6 specimens. It should also be noted that the load level at which fatigue crack growth becomes noticeable in the thermoplastic AS4/J1-polymer laminate is already above the critical load level in the thermosetting AS4/3501-6 laminate.

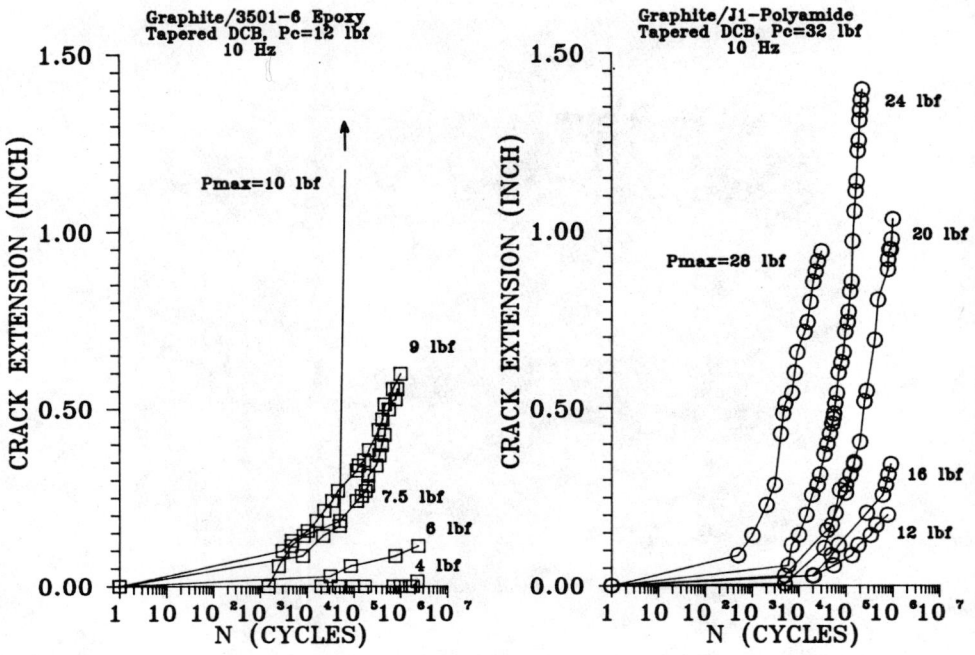

FIG. 6—*Fatigue crack propagation in AS4-graphite/3501-6 epoxy and AS4-graphite/J1-polymer unidirectional width-tapered DCB specimens.*

It is likely that a significant amount of heat is generated at the crack tip at the rather high test frequency of 10 Hz, especially at high loads. The AS4/J1-polymer laminate will dissipate more heat than the AS4/3501-6 laminate because there is more plastic deformation in the thermoplastic J1-polymer than in the thermosetting 3501-6 epoxy resin. Temperature rise in the specimen will cause softening in the matrix resin and results in the reduction of the critical load.

Stitched Thermoplastic Matrix Composite Laminates

Sample Preparation

The stitched composite laminates used in this investigation were prepared from plain-weave graphite fabric (Fiberite W-322) and 3.81 by 10^{-5} m (1.5 mils) thick J1-polymer film. Fabric instead of continuous tow was used because of its easier handling in the many processing steps involved in preparing the stitched laminates.

First, a single layer of graphite fabric was sandwiched between two J1-polymer films in a press at 90°C and 1.586 MPa (230 psi) for 15 min to produce a lightly tacked but not fully impregnated soft "preform." Full consolidation of this preform was deliberately avoided to facilitate easy stitching and to minimize fiber/fabric damages in the later stitching step.

As illustrated in Fig. 7, sixteen layers of this soft preform were cut into 0.152 by 0.152 m (6 by 6 in.) squares and stacked up in a $(0, +45, -45, 90, 90, -45, +45, 0)_s$ quasi-isotropic arrangement. A 0.0254-m (1-in.) wide Teflon-coated glass fabric was inserted between the two center

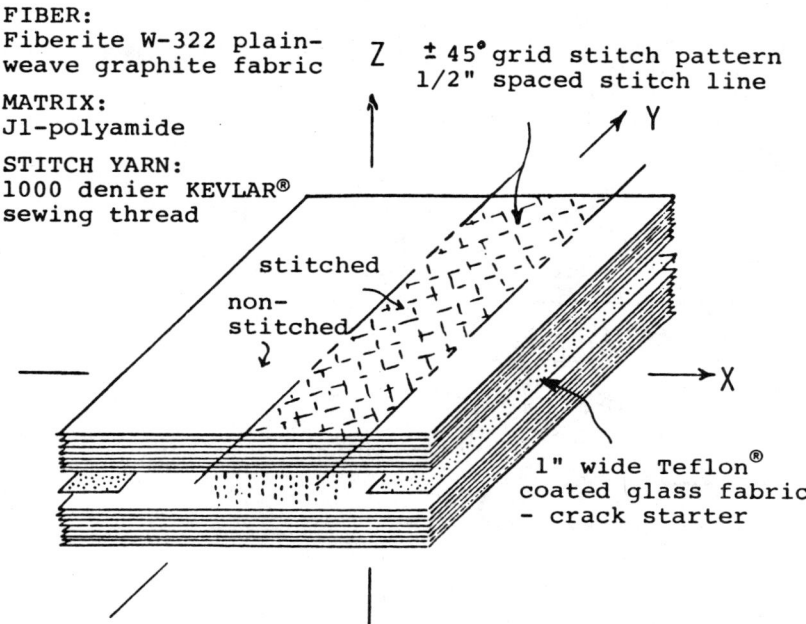

FIBER:
Fiberite W-322 plain-
weave graphite fabric

MATRIX:
J1-polyamide

STITCH YARN:
1000 denier KEVLAR®
sewing thread

Z ± 45° grid stitch pattern
   1/2" spaced stitch line

Y

stitched

non-
stitched

X

1" wide Teflon®
coated glass fabric
- crack starter

FIG. 7—*Stitch construction of the laminates used in investigating delamination resistance of stitched composites.*

layers at both ends of the stack. Stitching was performed on one half of the area by an industrial sewing machine. A 1000-denier Kevlar aramid sewing thread was used for the stitching. The stitch pattern is a ±45° orthogonal grid with 0.01247-m (0.5-in.) spaced stitch lines.

A fair amount of stiffness was required for the DCB fracture specimen, especially when used for characterizing fatigue crack propagation. Therefore, the 16-layer quasi-isotropic stitched preform was sandwiched between two 16-layer 0°/90° soft preform stacks to obtain additional stiffness. This 48-layer structure was then compression molded at 300°C and 15 MPa (2200 psi). The molded laminate was cut into 0.0762 by 0.0254 m (3 by 1 in.) strips and was attached with aluminum end tabs to make "short" DCB specimens (Fig. 8).

Mode I Fracture Characterization

The experimental investigation of the stitched laminates was conducted in 1982 before we completed the systematic investigation of the interlaminar fracture in thermoplastic matrix composites using the energy balance approach to measure G_{Ic}. At the time of the experiment, the interlaminar fracture toughness of the stitched laminates was characterized by the compliance approach, as well as the stress intensity factor K_I approach with several different K_I calibration formulae. Therefore, only data on compliance, critical load, and critical crack-opening displacement (COD) were recorded for the tests, which are insufficient to calculate G_{Ic} accurately by the energy balance method.

The stress intensity approach is used to report the results of the interlaminar fracture characterization on the stitched laminates. Exact solutions of the stress intensity factor, K_I, are avail-

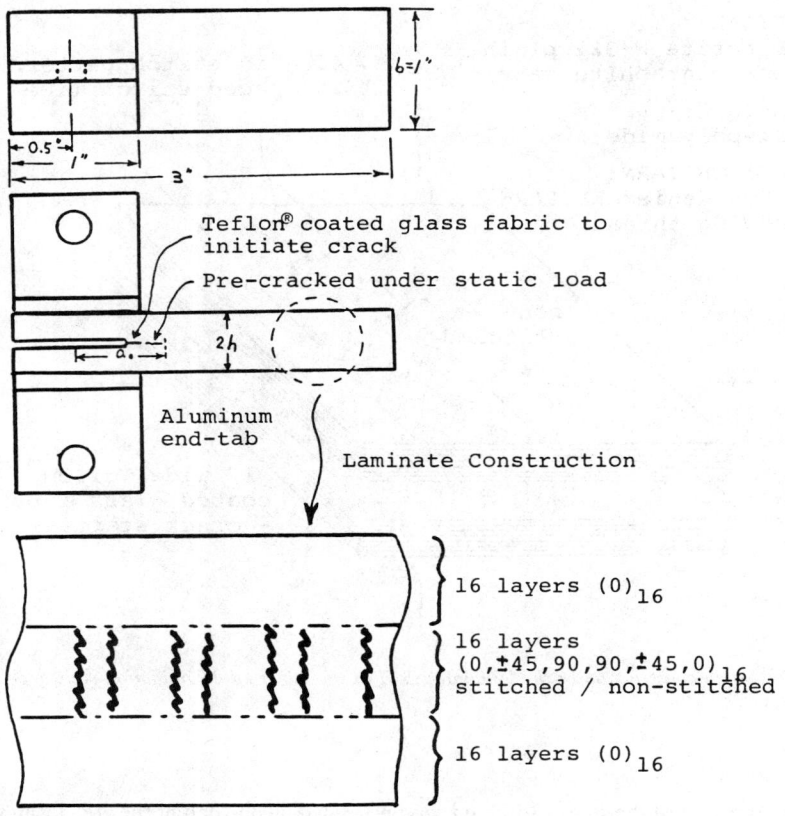

FIG. 8—"Short" DCB specimen used in stitched laminate study. Note that only the middle 16-layer was stitched; the other layers were used to increase the stiffness of the specimen.

able for the isotropic and homogeneous DCB specimens [10–13]; they give very similar results in their respective applicable ranges. An approximate form of the Fichter's solution is used [12] as follows

$$K_I = \left(\frac{12}{h}\right)^{1/2} \cdot P \cdot \left[\left(\frac{a}{h}\right) + 0.6728 + 0.0377 \cdot \left(\frac{h}{a}\right)^2\right] \tag{1}$$

where

$2h$ = thickness of the DCB specimen,
a = crack length, and
P = applied load.

The fracture criterion is $K_I = K_{Ic}$ as P reaches the critical load, P_c. The critical strain-energy release rate, G_{Ic}, then is calculated by employing the linear elastic fracture mechanics relationship of the following

$$G_I = (1 - \nu)^2 \cdot (K_I^2/E) \tag{2}$$

where

v = Poisson's ratio, and
E = Young's modulus.

The tensile modulus in the longitudinal direction of the DCB specimen, E_{11}, is used to substitute for Young's modulus, E, in Eq 2. The validity of applying these formulae, which are derived for a linear elastic isotropic DCB specimen, to the anisotropic composite DCB specimen was verified experimentally on a typical unidirectional AS4-graphite/J1-polymer DCB specimen. The results in Table 1 show that the G_{Ic} values calculated from Eqs 1 and 2 agree very well with G_{Ic} values obtained from the energy balance calculation.

Results on the Mode I interlaminar fracture characteristics of AS4-graphite/J1-polymer stitched laminates and their nonstitched counterparts are compared in Fig. 9. The critical load and the associated K_{Ic} and G_{Ic} in the stitched laminates are 25 to 30% higher than those of the nonstitched laminates. G_{Ic} was calculated from K_{Ic} using the measured value of $E_{11} = 40$ GPa ($5.8 \cdot 10^6$ psi) and a Poisson's ratio of 0.1. For the nonstitched specimens, the K_{Ic} and the derived G_{Ic} are relatively independent of crack length and have the respective values of 3.85 MPa$\sqrt{m}$ (3.5 ksi$\sqrt{in}$.) and 0.35 kN/m (2.0 lbf/in.).

The G_{Ic} value was significantly lower than that of the unidirectional laminate. There is evidence that this inequality was due to differences in the material form (fiber tow versus woven fabric), differences in laminate construction (unidirectional layup versus quasi-isotropic layup with 0°/90° backup layers), and variations in resin distribution inside the laminates. The amount of resin deformation was found also to be small on the fracture surfaces of the stitched laminates. K_{Ic} and G_{Ic} of the stitched specimens increased with the crack length because the number of stitch yarns that contribute to resist crack growth increases as the crack propagates. This strengthening effect is discussed in a later section of this paper.

Resistance to Fatigue Crack Propagation

Figure 10 shows the results of fatigue crack propagation in the stitched and nonstitched laminates. Six to eight specimens from each batch were tested at the same P_{max} and P_{min} levels.

TABLE 1—*Comparison of G$_{Ic}$s obtained from direct energy balance method and from K$_{Ic}$ using a linear elastic isotropic DCB beam formula (Eqs 1, 2). The sample is a unidirectional AS4-graphite/J1-polymer DCB specimen.*[a]

Crack Length, Inch	K_{Ic} ksi (in.)$^{0.5}$	Calculated from K_{Ic}; G_{Ic}, lb/in.	Direct Use of Energy Method, G_{Ic}, lb/in.
1.050	12.551	7.96	9.17
1.350	12.927	8.44	8.61
1.733	13.696	9.47	8.45
2.117	14.522	10.65	9.59
2.417	14.888	11.19	11.22
2.650	14.456	10.55	9.44
3.750	14.460	10.56	11.28
	Average:	9.83	9.68
		(1.721 kN/m)	(1.695 kN/m)

[a]Parameters:
$E_{11} = 123$ GPa ($17.9 \cdot 10^6$ psi).
$v_{12} = 0.31$.
$h = t/2 = 2.019 \cdot 10^{-3}$ m (0.0795 in.).

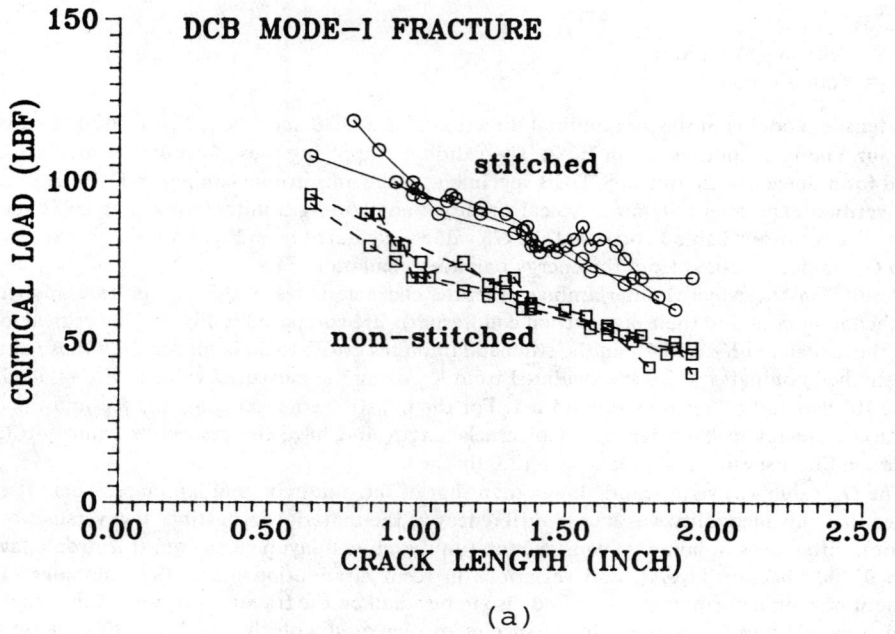

(a)

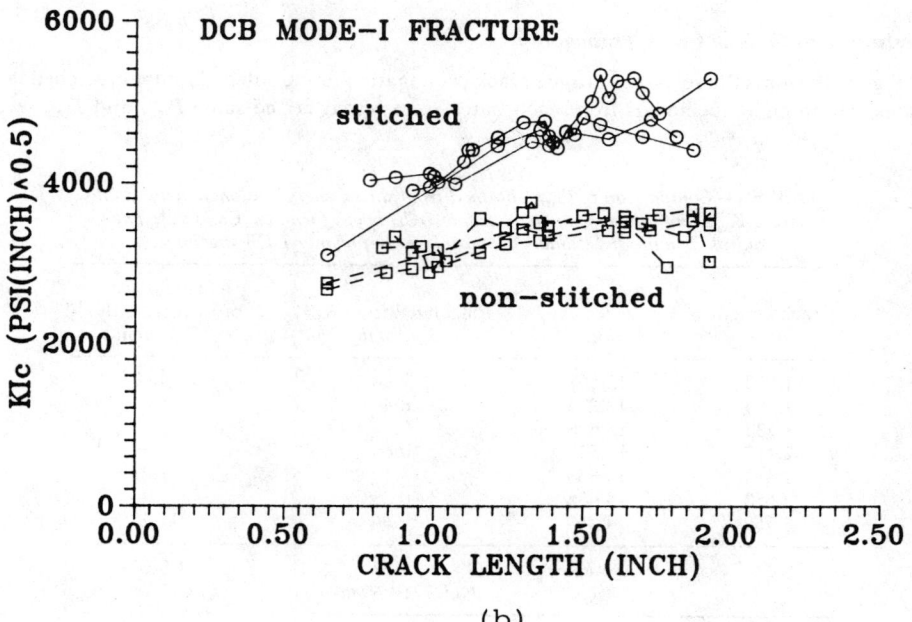

(b)

FIG. 9—*Mode I interlaminar fracture characteristics of stitched and nonstitched AS4-graphite/J1-poly-mer DCB specimens.*

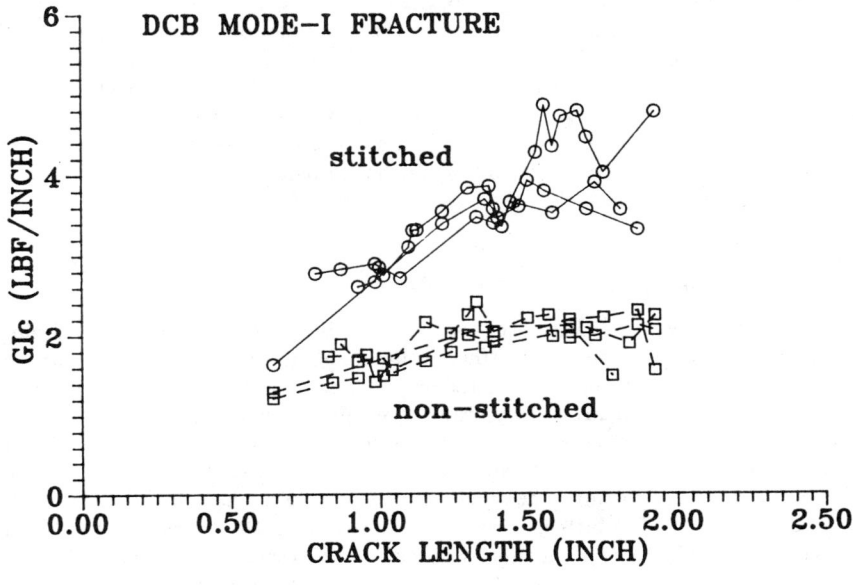

(c)

FIG. 9—*(Continued).*

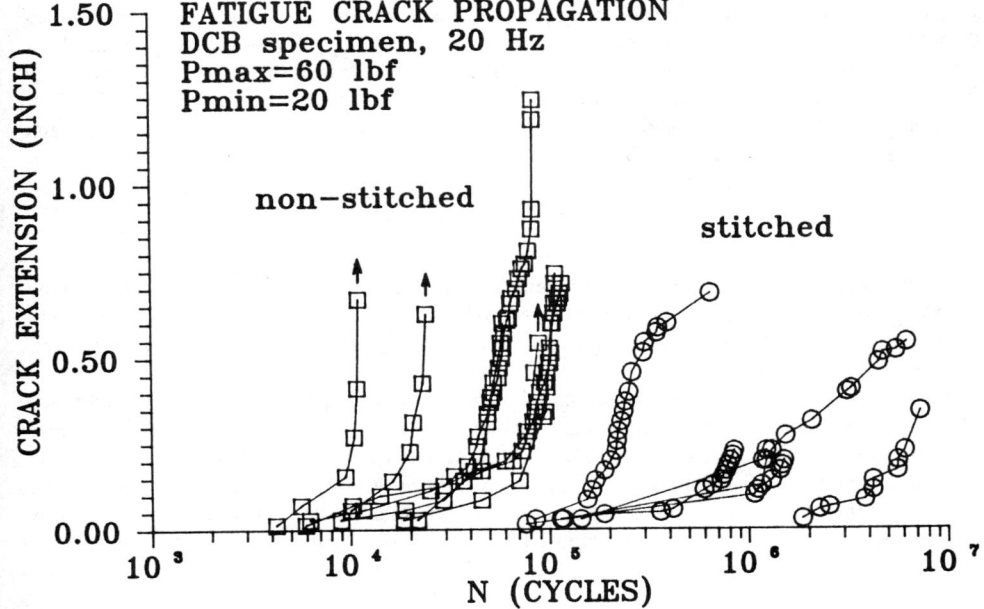

FIG. 10—*Fatigue crack propagation in stitched and nonstitched AS4-graphite/J1-polymer DCB specimens.*

Because straight DCB specimens were used in this case, the $K_{I\max}$ and $K_{I\min}$ levels increased as the crack advanced. In general, we observed that the number of fatigue cycles to reach the threshold for crack growth in the stitched laminate was about one order of magnitude more than that required for its nonstitched counterpart. Beyond the fatigue threshold, crack growth in the stitched specimens progressed about two orders of magnitude more slowly than that of the nonstitched specimens. A significant amount of data scattering was expected, especially for the stitched specimens, since the locations of the stitches in relation to crack length varied from specimen to specimen.

Roles of Stitching Yarn

When the local stress field in front of the crack tip begins to intensify upon loading, the stitched yarns, because of their high stiffness compared with that of the surrounding matrix resins, begin to carry most of the load and reduce the stress intensity in the surrounding resins. Therefore, higher applied loads will be required to initiate interlaminar crack propagation through the matrix resins and resin/fiber interfaces. In addition, the high-tenacity aramid stitching yarn does not fracture readily when encountering the crack front. Most of these yarns will survive, largely intact, behind the crack tip and provide a crack-closure force to the two opposite crack surfaces. The farther away these unbroken stitching yarns are left behind the crack front, the greater contributions they will make to the crack closure process.

Contributions from the surviving stitching yarns in resisting delamination are especially evident in the case of fatigue crack growth. Figure 11 shows micrographs of a stitched yarn bundle in a DCB specimen fractured under monotonically increasing loads; the multiple splits in the fractured fiber ends resemble those of fibers fractured in a single filament tension test.

On the other hand, micrographs (Fig. 12) of stitched yarn bundles in fatigue-tested DCB specimens show many shear bands on the fiber surface, evidence that these fibers have been through repetitive tensile-compression load cycles. Microcracks nucleate and grow in the shear bands and eventually break the filaments. In many cases, the fracture surfaces of individual filaments have a cone-and-cup type of appearance, typically that of ductile fracture. This evidence shows that most of the tough aramid stitching yarns survived the encounter with the crack front and continued to contribute to the increase in crack-growth resistance.

Analytical Model—Residual Strength Prediction

Many parameters influence the effectiveness of the delamination resistance provided by stitching. An analytical model was constructed to evaluate these parameters and to conduct case simulations to assist in the design of stitching experiments and of real 3-D reinforced structures.

Illustrated in Fig. 13 is an analytical model of interlaminar crack growth in a stitched laminate. The model employs the superposition principle applicable to linear elastic fracture mechanics. The selection of the system boundary allows us to treat the unbroken stitching yarns as pairs of externally applied forces acting on the two opposite crack surfaces inside the system. The equivalent stress intensity factor, $K_{I(eq)}$, for the system is the sum of $K_{I(ext)}$, contributed by the applied load P, and of $K_{I(int)}$, contributed by the unbroken stitching yarns as follows

$$K_{I(eq)} = K_{I(ext)} + K_{I(int)}$$

$$= P \cdot Y(a, h) - \sum_{i=1}^{n} P_i \cdot Y(a_i, h) \tag{3}$$

(a)

(b)

FIG. 11—*Fractured stitching yarn in a stitched DCB specimen tested under monotonically increasing load. Note the multiple splits along the fiber axis on fiber ends.*

(a)

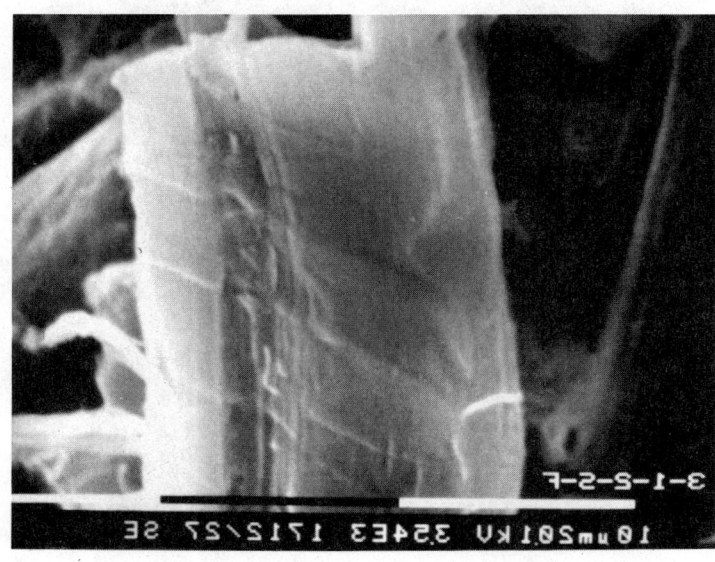

(d)

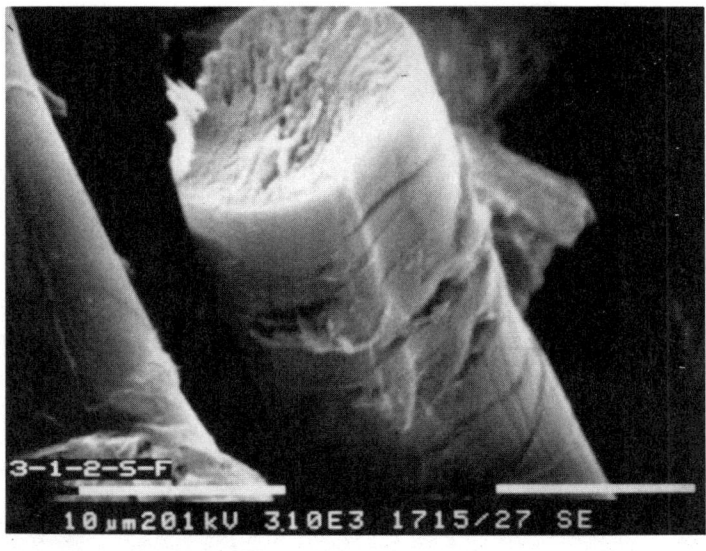

(c)

FIG. 12—*Fractured stitching yarns in a stitched DCB specimen tested under cyclic fatigue load. Note the many shear bands and the micro cracks that nucleated in the shear bands. Note also the cone-and-cup type of ductile fracture fiber ends.*

The crack will propagate when $K_{I(eq)}$ reaches the K_{Ic} of the material system. For our sparsely stitched laminate, the interlaminar crack still propagates predominantly through matrix resins, just as in the nonstitched specimens. Therefore, it is reasonable to assume that the criterion for crack propagation in the stitched laminate is the same as the nonstitched laminate, that is, $K_{Ic} = 3.85$ MPa$\sqrt{m}$ (3.5 ksi$\sqrt{in.}$). The residual strength at any crack length, which is the critical load required to continue propagating the crack, can be calculated by rearranging the equation as follows

$$P_c = \frac{1}{Y(a, h)} \cdot \left(K_{Ic} + \sum_{i=1}^{n} P_i \cdot Y(a_i, h) \right) \tag{4}$$

The magnitude of the reaction forces in each of the unbroken yarns is also needed to solve for Eq 4. These forces are estimated by first obtaining, from experimental data, the crack-opening displacement (COD) at the load point for all crack positions. COD values at individual yarn locations are calculated using beam formula and the above-measured load point COD. The forces in the stitching yarns then are calculated from the load-displacement characteristics of the yarn and the estimated COD at the yarn locations.

Case Simulations

Geometric stitching parameters that are inputs to the case simulations are shown in Fig. 14. In all of the simulations conducted, the total amount of stitching yarns was kept constant at the

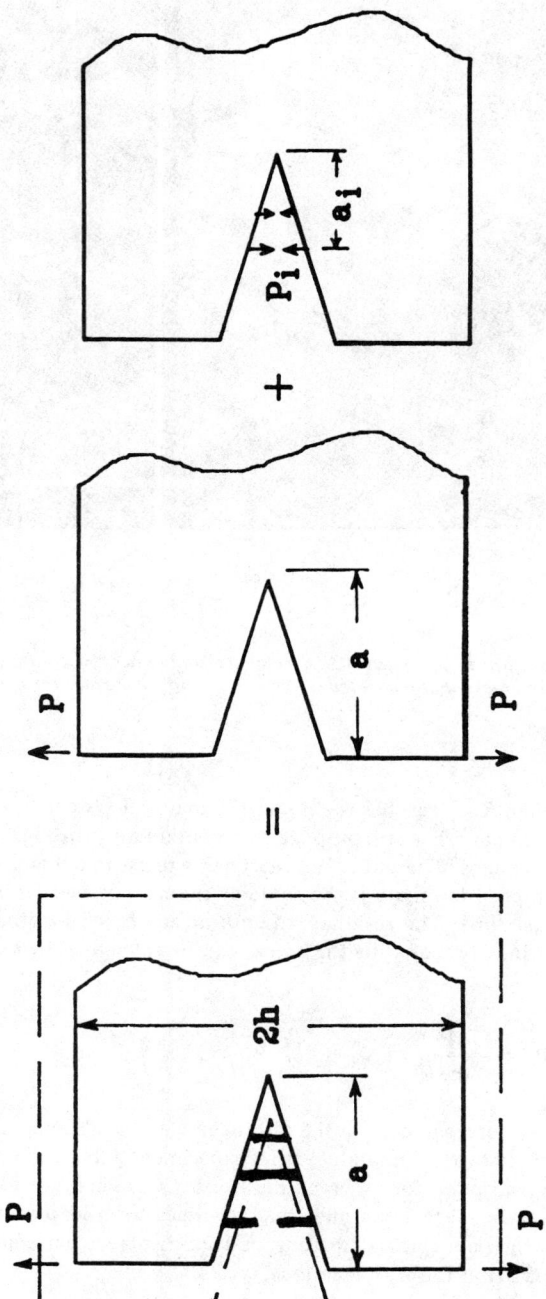

FIG. 13—Fracture model for a stitched DCB specimen with unbroken stitching yarns that provide crack-closure forces on the fracture surfaces.

STITCHING PARAMETERS

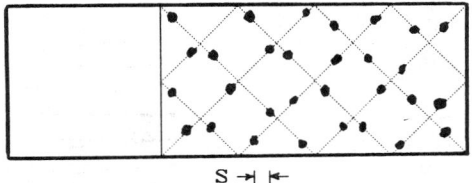

	CONTROL CASE
Yarn Size:	1000 denier
Row Spacing - S:	0.125 inch
Yarns/Row - N:	2/row

FIG. 14—*Geometric stitching parameters simulated in the residual strength analysis of stitched DCB specimen.*

YARN PROPERTY

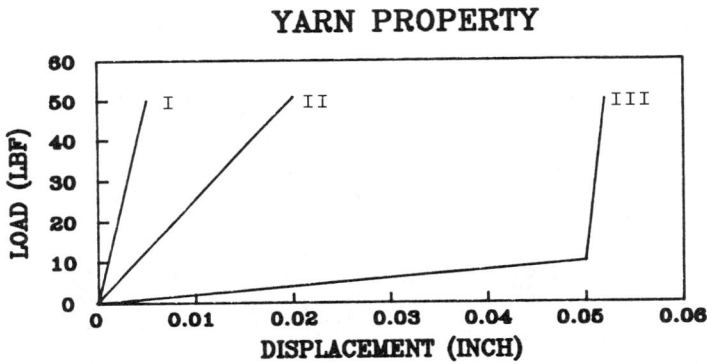

I. Straight stitching yarns
II. Twist in yarns. Off-angle alignment
III. Slacks in stitching yarn

FIG. 15—*Stitching yarn behavior simulated in the residual strength analysis of stitched DCB specimen.*

level used in the stitched laminate experiment. The other important parameter investigated in the simulation is the load-displacement characteristics of the stitching yarn. The break strength of the 1000-denier aramid yarn is about 222 N (50 lbf) in a yarn tension test. Based on COD measurements and microscopy evidence, the break extension of the stitching yarn in the DCB specimen is estimated to be about $1.27 \cdot 10^{-4}$ m (0.005 in.). In reality, the stitching yarns in the laminates are not straight and will not exhibit the deformation characteristics represented by the single-yarn test.

Several factors contribute to the nonideal yarn behavior shown in Fig. 15. First, there are twists in the sewing threads, and the stitched yarns usually do not line up perfectly in the direction normal to the laminate. Under these conditions, the yarns will exhibit lower stiffness than

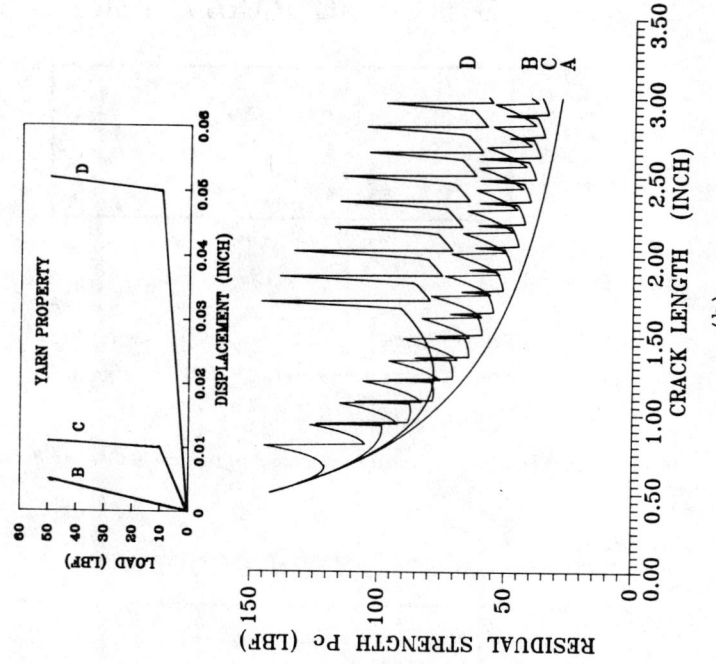

(a)

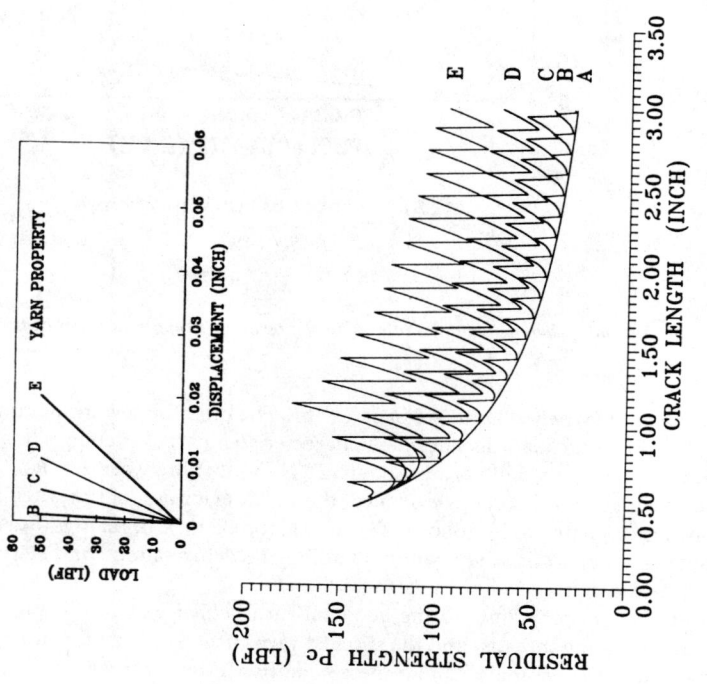

(b)

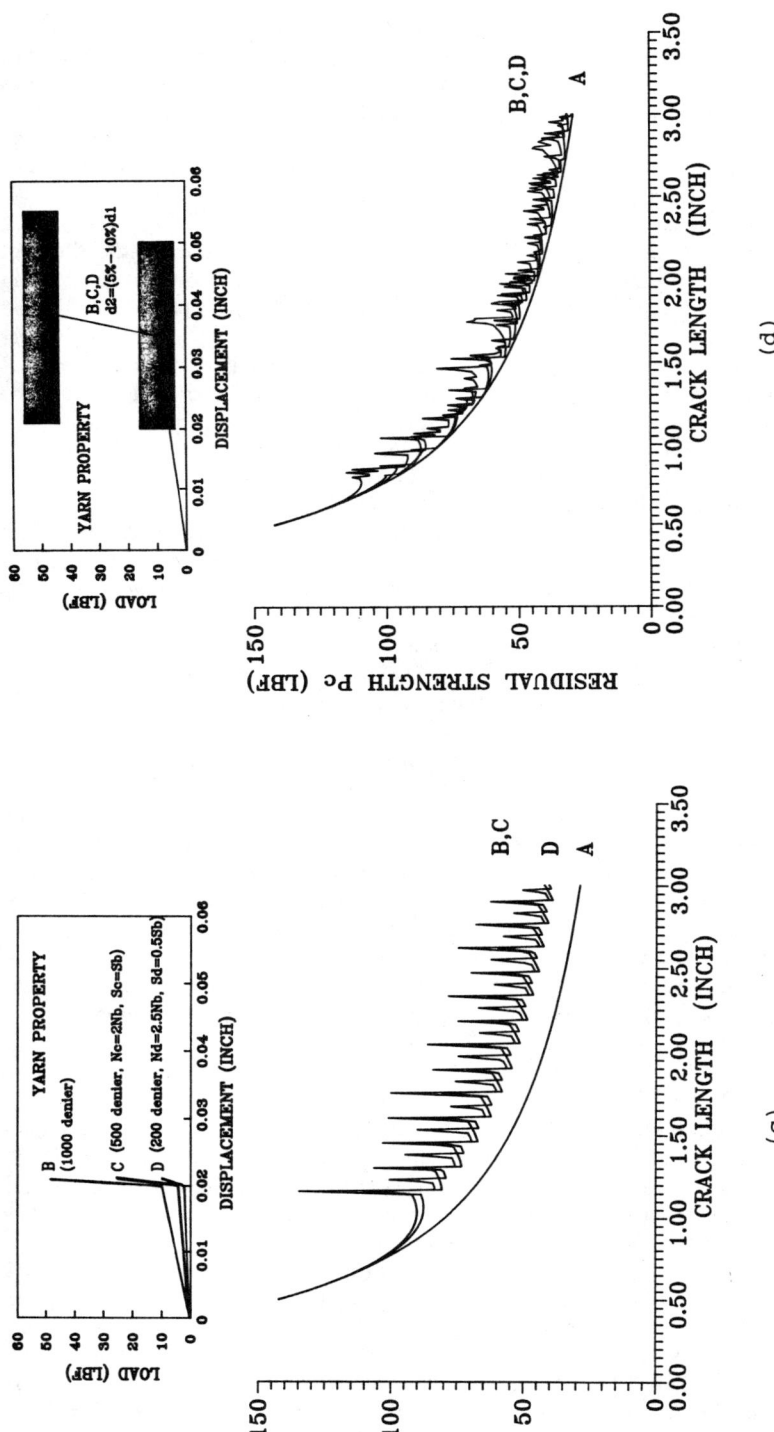

FIG. 16—Results on case simulations of residual strength in stitched DCB specimens. Note the effects of (a) yarn stiffness, (b) slacks in the stitching yarn, (c) yarn size and geometric parameters, and (d) randomly distributed stitch locations and yarn properties.

they do in the ideal behavior. Second, there are slacks in the stitches. These slacks are introduced mainly by the debulking effect during compression molding, and, to a lesser extent, as the result of the stitching action. These artificially produced slacks are locked into the structure when the resin solidifies. The slack yarns are likely to exhibit a two-part stiffness characteristic after they are released by the passing crack: a low-stiffness part associated with the untangling of the slacks; and a high-stiffness part associated with stretching of the fibers. The effects of these nonideal but nonetheless realistic yarn behaviors on delamination resistance were investigated in the case simulations.

Results of four case simulations are shown in Fig. 16. Two of the cases simulate the effects of stitching yarn properties and the other two cases simulate the effects of geometric stitching parameters. Curve A in all of the figures represents the base-line results of the nonstitched control specimen.

Simulation (a) examines the effects of yarn stiffness on the residual strength of the DCB specimen as the crack propagates. The results show that the stiffer the stitching yarns, the earlier they begin to contribute to the increase of delamination resistance in the specimen. For tougher yarns that have longer break extensions, the increase in residual strength will not take effect immediately as the crack starts to propagate. However, a greater strengthening effect is achieved once the yarns begin to stretch to their full extent.

Simulation (b) investigates the effect of slacks in the stitch yarns on delamination resistance. A small amount of strengthening is observed when the slacks are first released by the propagation of the crack. Residual strength increases sharply when the slacks are completely released, and the yarns react to further extension with their intrinsic material response. The longer the response is delayed, that is, the more slacks in the stitching yarn, the more pronounced is the strengthening effect obtained. However, excessive delay may allow extensive crack growth and fast unstable crack propagation before the stitching yarns begin to provide strengthening to the structure.

Simulation (c) examines the effect of geometric factors that include yarn size, row spacing, S, and the number of stitches per row, N. Smaller stitching yarns with closer row spacings produce a smoother transition to adjacent yarns, allowing them to become effective in the strengthening process. They also raise the mean level of the strengthening effect provided by the earlier part of the yarn stretching process. On the other hand, large yarn bundles, with all fibers being stretched at about the same time, provide a higher instantaneous strengthening effect.

Simulation (d) comprises three separate runs with randomly generated value for stitch locations and yarn properties that fall within a selected range. Compared with the previous cases, the increase in residual strength by the randomly stitched yarns does not seem to be as effective as in the cases where groups of yarn contribute together in a regular repetitive sequence.

Applications to Stitched Laminate Design

These experiments and analyses on the stitched composite laminates provide some insight into the effective and economical use of stitching and other forms of 3-D reinforcements in composites structures. First, complete through-thickness stitching is not necessary for improving delamination resistance. It is difficult to completely eliminate fiber damage in through-thickness stitching. Fiber damage will cause reduction in the in-plane properties of the composites. Also, complete through-thickness stitching is technically difficult to perform on very thick laminates. In most potential stitching applications such as joints, overlaps, and structures with bifurcated geometries, areas that are prone to delamination are usually quite easy to identify. Therefore, only the few plies adjacent to the suspected crack-prone interfaces need be stitched to enhance its delamination resistance; the rest of the plies can be left free of any stitching-induced damage. Second, it is generally not necessary to provide stitching or other forms of 3-D reinforcement to the entire laminated structure; in the case of resisting delamination, stitching need only be applied at the near-edge areas, which can be achieved easily with relatively simple and inexpensive equipment.

Intrinsic as well as extrinsic properties of the stitching yarn are both important factors in resisting delamination crack growth. High-modulus fibers are effective in sharing the load with the matrix resin. They greatly reduce the stress intensity within the matrix ahead of the crack tip, thus improving delamination resistance. Even greater delamination resistance can be achieved by using tough fibers that can survive the encounter with the crack front and continue to provide a reacting crack-closure force on the crack surfaces. The magnitude of this crack-closure force depends not only on the intrinsic strength and modulus of the fiber, but also on extrinsic properties provided to the yarn through yarn texturing and stitching processes. These parameters can be manipulated to create the most effective 3-D reinforcement for the specific application.

Conclusions

Evidence presented in this paper further confirms the important role plastic deformation processes in thermoplastic matrix resins play in improving the delamination resistance of composite laminates. The fatigue investigation shows that at relatively low load levels compared with their respective critical loads, an interlaminar fatigue crack is easier to grow in thermoplastic matrix composites but not in the traditional thermosetting epoxy matrix composites. However, fatigue crack growth in the thermoplastic matrix composites is always very stable even when load levels approach their critical load, whereas, fast unstable crack propagation is always the case for epoxy matrix composites when subjected to high fatigue loads. More importantly, the improved interlaminar fracture toughness of the thermoplastic matrix composites has significantly increased the critical load of the laminate structure, such that the load level at which fatigue crack growth becomes noticeable in the thermoplastic matrix composites has already exceeded the critical load of the same structure made from traditional thermosetting matrix composites.

This study also quantifies the advantages of using stitching to further improve the delamination resistance of composites. It should be emphasized that the extrinsic properties of the stitching yarn are just as important as the yarn's intrinsic properties in enhancing delamination resistance. Materials and processing parameters, as well as geometric parameters of the stitches, can be adjusted to produce the most effective stitching results.

Acknowledgments

The author wishes to thank his colleague, Dr. P. Popper, for valuable discussions on 3-D fiber structures. Thanks are also due to I. N. Carroll, B. M. Moran, J. F. Pedrick, F. J. Best, and R. A. Pazdalski for their assistance in conducting the experiments.

References

[1] Bascom, W. D. Bitner, J. L., Moulton, R. J., and Siebert, A. R., "The Interlaminar Fracture of Organic-Matrix, Woven Reinforcement Composites," *Composites,* Vol. 11, 1980, pp. 9–18.

[2] Bradley, W. L. and Cohen, R. N., "Delamination and Transverse Fracture in Graphite/Epoxy Materials," *Proceedings of the Fourth International Conference on Mechanical Behavior of Materials,* 1983.

[3] Carlile, D. R. and Leach, D. C., "Damage and Notch Sensitivity of Graphite/PEEK Composite," *Proceedings of the Fifteenth National SAMPE Technical Conference,* 1983, pp. 82–93.

[4] O'Connor, J. E., Ma, C. M., and Lou, A. Y., "Polyphenylene Sulfide: Resin, Prepreg and High Performance Composites," *Proceedings of the 39th Annual Conference of Reinforced Plastic/Composites Institute,* 1984, Section 11-E, pp. 1–5.

[5] Su, K. B., "Mechanisms of Interlaminar Fracture in a Thermoplastic Matrix Composite Laminate," *Proceedings of the Fifth International Conference on Composite Materials*, 1985, pp. 995–1006.

[6] Wedgewood, A. R., Su, K. B., and Narin, J. A., "Toughness Properties and Service Performance of High Temperature Thermoplastics and Their Composites," *Proceedings of the Nineteenth International SAMPE Technical Conference*, October 13-15, 1987, pp. 454–467.

[7] Ko, F. K., "Development of High Damage Tolerant, Net Shape Composites Through Textile Structural Design," *Proceedings of the Fifth International Conference on Composite Materials*, 1985, pp. 1201–1210.

[8] Klein, A. J., Ed., "Braids and Knits: Reinforcement in Multidirections," *Advanced Composites*, Sept./Oct. 1987, pp. 36–48.

[9] Popper, P. and McConnell, R., "A New 3-D Braid for Integrated Parts Manufacture and Improved Delamination Resistance—The 2-step Process," *Proceedings of the 32nd International SAMPE Symposium and Exhibition*, 1987, pp. 92–103.

[10] Kanninen, M. F., "An Augmented Double Cantilever Beam Model for Studying Crack Propagation and Arrest," *International Journal of Fracture*, Vol. 9, 1973, pp. 83–92.

[11] Kanninen, M. F., "A Dynamic Analysis of Unstable Crack Propagation and Arrest in the DCB Test Specimen," *International Journal of Fracture*, Vol. 10, 1974, pp. 415–430.

[12] Fichter, W. B., "The Stress Intensity Factor for the Double Cantilever Beam," *International Journal of Fracture*, Vol. 22, 1983, pp. 133–143.

[13] Foote, R. M. L. and Buchwald, V. T., "An Exact Solution for the Stress Intensity Factor for a Double Cantilever Beam," *International Journal of Fracture*, Vol. 29, 1985, pp. 125–134.

Author Index

B

Bankert, R. J., 73
Burrett, P., 183

C

Carlile, D. R., 199
Chen, J. R., 213
Chervenak, J. G., 146
Coulter, J. P., 14

D

Dan-Jumbo, E., 113
Drews, M. J., 50

E

Edie, D. D., 50
Ellison, M. S., 50

G

Gantt, B. W., 50
Guceri, S. I., 14

H

Hackett, R. M., 62
Hahn, H. T., 183
Hinkley, J. A., 33, 251
Hoa, S. V., 213
Hoffman, R. R., 146
Howes, J. C., 33

J

Jerina, K. L., 183
Johnston, N. J., 251

L

Lambropoulos, N. D., 73
Leach, D. C., 199

Leung, C.-L., 5
Liao, T. T., 5
Lickfield, G. C., 50
Lin, S., 213
Loos, A. C., 33
Lucas, J. P., 231
Lustiger, A., 264

M

Moore, D. R., 199
Myers, F. A., 154

N-O

Newaz, G. M., 264
O'Brien, T. K., 251
Odegard, B. C., 231

P

Palazotto, A., 91
Prasad, S. N., 62

R-S

Ramey, J., 91
Shephard, M. S., 73
Simonds, R. A., 133
Sternstein, S. S., 73
Stinchcomb, W. W., 133
Su, K. B., 279
Sun, C. T., 113

T-Z

Tye, R. P., 146
Yung, J.-Y., 264
Zahlan, N., 199
Zhou, S. G., 113

Subject Index

A

ABAQUS F.E., 74, 86
Adhesion, 257
Aluminum flake, 149
    conductivity, 153
Anisotropy, 73, 78, 88
Aramid sewing yarn, 285
    nonideal behavior in stitching, 295
ASTM standards
    proposed heat flow technique, 147
    D 412-51T: 35
    D 695: 200
    D 790: 200
    D 2344: 200
    D 3039: 200
    D 3418: 200
Autohesion, 34
    bond strength development, 34
    temperature dependence, 41

B

Backward finite difference method, 6
Boundary-fitted coordinate systems (BFCS), 20
Brittleness, 258

C

Carbon fiber, 50, 148, 183
Chevron notch short bar/rod method, 236
Coatings, on carbon fiber, 50
    distribution, 59
    evaluation, 51
    matrix polymer redistribution, 59
    powder deposition, 52
    process, 53
    thickness, 61
    uniformity, 53, 54
Compact tension tests
    fracture toughness test, 35
    healing test, 38
Composites—APC-1
    tension and compression strength, 96
Composites—APC-2/AS-4, carbon/poly-
        etheretherketone (PEEK) laminates
    brittle fracture, 273

compression test, 205
crack instability, 273
creep, 201, 206
critical stress intensity factor, 269
cyclic loading, 208
deformation, 270
delamination crack growth, 266
delamination growth, 264
ductility, 270
elastic properties, 202
fatigue crack growth, 267
fatigue tests, 201, 208
fiber deformation, lamellae, 272
fiber fracture, 205
fracture morphology, 270
instrumented falling weight impact test, 205
interlaminar fracture toughness, 203
Mode I delamination toughness, 265
Paris law, 267
short-beam test failure, 202
slow crack growth, 266
stiffness and strength properties, 200, 202
stress, process zone, 270
temperature profiles, 268
thermal expansion coefficients, 203
toughness tests, 200, 203
Composites—APC-2, carbon/polyetherether-
        ketone (PEEK) laminates
    dynamic viscosity, 8
    one-dimensional heat flow, 5
    physical properties, 7
    process model, 5
    temperature profiles, 9
    thermal diffusivity, 7
    thermal properties, 7
Composites—APC-2, graphite/polyether-
        etherketone (PEEK) laminates, 91
    bonding characteristics, 101
    compression tests, 92, 98
    interlaminar matrix deformation, 102
    intralaminar fracture, 106
    intralaminar deformation, 106
    intralaminar separation, 99
    material properties, 93
    notch strength, 94

strength prediction, three parameter method, 95
temperature effects on strength, 92
Composites—AS-4-graphite/J1 polymer, 279
  crack growth, 283
    critical strain energy release rate, 280, 286
    delamination resistance, 290
      effects of yarns slacks on, 298
      fiber damage, 298
    geometric stitching parameters, 293
      strengthening effects, 298
    interlaminar crack growth, 290
    load displacement characteristics, 295
    Mode I fatigue crack propagation, 280, 287
    Mode I interlaminar failure, 280
    Mode I interlaminar fracture, 287
    plastic deformation processes, 279
    residual strength, 293
      effects of yarn stiffness on, 298
    stiffness, 285
    stitching, 280, 285
    stress intensity factor, 285
    yarn behavior, 295
Composites—AS4/polycarbonate laminates, 252
  adhesion, 257
  double cantilever beam test, 253
  edge delamination, 253
  interlaminar toughness, 260
  matrix deformation, 257
  modulus, 255
  ply properties, 255
  stiffness, 255
  thermal expansion, linear coefficient of, 257
Composites—AS4/polyetheretherketone (PEEK) laminates, 133
  compressive strength, 137
  damage region, 141
  damage tolerance, 133, 145
  delamination, 141
  failure mode, 144
  fatigue lives, 134, 142
  fatigue loading, 136
  fatigue tests, 136
  fiber fracture, 142
  monotonic mechanical tests, 136
  radiographs during cyclic loading, 141
  residual strength, 139
  residual tensile strength, 143
  specimen stiffness, 137
  stiffness ratio, 138
  tensile strength, 137
Composites—AS4/polyimidesulfone blend laminates, 252
  adhesion, 257

double cantilever beam test, 253
edge delamination, 253, 261
fiber bridging, 261
interlaminar toughness, 262
modulus, 255
ply properties, 255
stiffness, 255
thermal expansion, linear coefficient of, 257
Composites—AS4/TPI blend laminates, 252
  adhesion, 257
  double cantilever beam test, 253
  edge delamination, 253, 261
  fiber bridging, 261
  interlaminar toughness, 262
  modulus, 255
  ply properties, 255
  stiffness, 255
  thermal expansion, linear coefficient of, 257
Composites—BMI toughened laminate, 113
  fatigue behavior, 113
  fatigue life, 115
  in-plane shear modulus, 114
  stress-strain, 114
Composites—Graphite-epoxy
  fatigue life, load frequency effect on, 113
  fatigue tests, 114
  finite element pultrusion model application, 68
  in-plane shear modulus, 114
  laminates, 68, 113, 245
  temperature effect on strength, 93
  tension and compression strength, 97
  toughness, 133
Composites—Graphite/polysulfone, 3
  critical strain energy release rate, 39
    factors affecting, 47
  double cantilever beam fracture, 37
  healing function, 45
  healing test, 38
  refracture toughness, 38
Composites—IM6/PEEK, graphite/polyetheretherketone laminates, 231
  crack tip stress, 234
  critical strain energy release rate, 239
  fiber bridging, 240
  fracture mechanisms, 240
  fracture surface morphology, 241
  fracture toughness, 237, 239
    chevron notch short bar/rod method, 236
  interlaminar fracture, 239
  laminate construction, 233
  Mode I delamination fracture toughness, 232

moisture absorption, 236, 238
short bar hybrid test specimen, 231
stress distribution, 236
Composites—IMP6/APC-2 laminates, 113
crack propagation, 124
failure modes, 127
fatigue life, 119
fatigue life prediction, 126
fatigue tests, 115, 121, 128
frequency-dependent fatigue model, 126
load-frequency effect on fatigue, 113
material properties, 115
shear modulus, 114
stiffness variation during cyclic loading, 116
temperature increase summary, 130
tension-compression fatigue tests, 117
test temperature increase history, 115
Composites—Polyamide-imide (PAI), carbon fiber reinforced, 184
creep tests, 185
fatigue tests, 185
fiber orientation distribution, 186
orientation angles, 187
fracture morphology, 183, 192, 196
modulus, 185
spatial fiber distribution, 183
ultimate tensile strength, 185
Composites—Polyetheretherketone (PEEK), carbon fiber reinforced
creep tests, 185
fatigue tests, 185
fiber orientation distribution, 186
orientation angles, 187
fracture morphology, 183, 192
modulus, 185
spatial fiber distribution, 183
ultimate tensile strength, 185
Composites—Polyphenylene sulfide (PPS), carbon fiber reinforced, 154
constant deformation, 155
creep tests, 166, 176, 185
crystallinity, 158, 166
dynamic mechanical tests, 155, 158
fatigue tests, 185
fiber orientation distribution, 186
orientation angles, 187
flexure tests, 155, 166
fracture morphology, 183, 192
frequency sweeps, 166
modulus, 185
relaxation tests, 166, 176
spatial fiber distribution, 183
stiffness, 176
temperature scans, 158
thermal response, 155

torsional tests, 155
ultimate tensile strength, 185
Composites—Polyphenylene sulfide (PPS), glass fiber reinforced, 213
crystallinity, 214
debonding, 227
glass transition temperature, 214
microstructures, 226
modulus, elastic, 222
moisture absorption, 217
moisture effects on samples, 226
moisture resistance, 213
molding, prepregs, 217
molding, stampable sheets, 217
stress-strain curves, 219
tensile strength, 219, 225
Composites—Ryton, carbon fabric reinforced, 155
constant deformation, 155, 175
creep tests, 166, 175
crystallinity, 158, 176
dynamic mechanical tests, 158
relaxation tests, 166, 175
stiffness, 176
temperature scans, 158
Composites—XAS/polycarbonate laminates, 252
adhesion, 257
double cantilever beam test, 253
edge delamination, 253
interlaminar toughness, 260
matrix brittleness, 258
matrix deformation, 257
modulus, 255
ply properties, 255
stiffness, 255
thermal expansion, linear coefficient of, 257
Compression molding, 15, 26
Compression tests, 92, 98, 205
Compressive strength, 137
prediction, 95
Computational domain equations, 22
Computational fluid dynamics, 14
Conduction-convection matrix, 66
Conductivity, composite packaging for electronics, 146
Constant loading boundary condition, 18
Contact point relocation method, 25
Crack growth rate, 266, 283
Crack propagation, 75, 124, 279, 287
Crack tip stress, 75, 86, 234
Creep tests, 166, 176, 185, 201, 206
Critical strain energy release rate, 45, 239, 251, 277, 280
factors affecting, 47

Crystallinity, 158, 166, 176, 214
Cyclic loading, 140, 208

D

Damage region, 141
Damage tolerance, 33, 133, 145
Darcy's laws, 17
Debonding due to moisture, 227
Deformation
   constant, 155
   interlaminar matrix, 102
   intralaminar, 106
   laminate, 117
   matrix, 257
   viscoelastic, 34
Delamination, 141, 264
   crack growth, 266
   edge, 253, 258, 261
   fracture toughness, 73
   resistance, 279, 290
      effects of yarn slacks on, 298
Density, 7
Diffusion, 35
   modeling, 41
   moisture, 236
   self-diffusion coefficient, 42
Dirichlet boundary conditions, 21
Double cantilever beam (DCB) composite
    test, 37, 45, 251
   fracture toughness parameters, 89
   specimens, 76
Dynamic viscosity, 8
   temperature profiles, 9

E

Eastobond FA252, 51, 54
Electrical conductivity, 151
Electrical resistance heating and powder de-
    position apparatus, 52
Electronics packaging, 146
Electrostatic powder deposition, 51

F

Failure mode, 127, 144
Fatigue
   behavior, 113
   frequency dependent model, 126
   life, 115, 134
   load frequency effect on fatigue life, 113
   life prediction, 126
   tests, 114, 121, 136, 185, 201, 208
Fiber
   bridging, 45, 83, 240, 258, 261
   damage, 298
   deformation, lamellae, 272
   deposition process, 52
   distribution, 183, 186

   filler, 146
   matrix debonding, 258
Fibrous preform, 26
Filler, fiber, 146
Finite difference techniques, 6
Finite element method, 63
   conduction-convection matrix, 66
   model, 74
   model results, 76
   weighted residuals, 65
Flexure tests, 155, 166
Fracture, fiber, 142, 205
Fracture mechanisms, 240
Fracture properties
   sensitivity to flaw microstructure, 86
Fracture morphology, 183, 192, 196, 270
   surface, 192, 241
Fracture toughness, 45, 91, 133, 237
   chevron notch short bar/rod method, 236
   compact tension test, 35, 37
   double cantilever beam test, 36
   fracture toughness parameters, 89
   interlaminar, 46, 203, 239, 251, 279
   moisture effects on, 231
   specimens, 76
Frequency-dependent fatigue model, 126
Frequency sweeps, 166

G

Galerkin method, 65
Geometrical transformation coefficients, 23
Geometric stitching parameters, 293
   strengthening effects, 298
Glass transition
   temperature, 214
   zone, 154
Graphite-polysulfone composites, 3

H

Healing
   function, 45
   isothermal model, 43
   lower toughness, 47
   nonisothermal, 41
   polymer flow during, 47
   rehealing, 46
   tests, 35, 40
Heat capacity, 7
Heat flow technique, 147
Heat transfer
   coefficient, 41
   model, one-dimensional, 5, 62, 64
High temperature testing, 91

I

Imposed orthogonality method, 25
Impregnation configurations, 27

Impregnation front, resin, 19
  movement, 24
Impregnation processes
  resin, 14
  thermoplastic on carbon, 50
Injection molding, 183
Instrumented falling weight test, 205
Interfacial tests, 35
Interlaminar crack growth, 290
Interlaminar fracture, 46, 203, 239, 251
Interlaminar matrix deformation, 102
Intralaminar deformation, 106
Intralaminar fracture, 106, 203
Intralaminar separation, 99
Interlaminar toughness, 260
Isothermal healing model, 46
Isotropic model, 75
  stress distribution, 77

L

Laminates (*see also* Composites—types)
  heat transfer, 6
  system comparison, 97
  temperature effects on strength of, 92
LARC-TPI (Langley Research Center—Ther-
    moplastic Polyimide), 51
  powder, 54
Lewis-Nielson semitheoretical prediction
    methods, 146
Line zone model for failure process, 268
Load displacement characteristics, 295
Load-frequency effect on fatigue, 113

M

Matrix brittleness, 258
Mechanical tests, dynamic, 155, 158
Mesh generator, finite quadtree, 75
Mode I delamination fracture toughness, 232,
    265
  temperature effects on, 265
Mode I delamination growth, 264
Mode I failure, 73, 280
  anisotropy, 78, 88
  characterization, 82
  displacement analysis, 75
  fiber bridging, 83, 89
  isotropic model, 75
  orthotropic analysis, 78
  stress distribution, 74
Mode I fatigue crack propagation, 280, 287
Mode I fracture, 285
Mode I interlaminar fracture, 231, 280, 287
Mode I strain energy release, 239
Mode II failure, 73
  anisotropy, 85, 88

characterization, 79
crack length effect on shear stress distribu-
    tion, 81
crack propagation, 85
isotropic analysis, 79
orthotropic analysis, 79
stress distribution, 74, 79
Modeling
  autohesion, 34
  delamination resistance, 290
  diffusion, 41
  finite element, 74
  finite element results, 67
  frequency-dependent fatigue, 126
  heat transfer, 5, 62
  interlaminar crack growth, 290
  isotropic, 75
  Lewis and Nielson, 146
  line zone for failure process, 268
  nonisothermal healing, 44
  numerical grid generation, 14
  one-dimensional heat flow, 5
  one-dimensional heat transfer, 62
  orthotropic, 74
  porous medium flow, 16
  process, 5
  pultrusion process, 62
  resin flow, 16
  thermal conductivity, 147
  viscoelastic constitutive, 8
Modulus, 185, 255
  elastic, 222
Moisture absorption, 217, 232, 236, 238
Moisture resistance, 213
Mold die movement, 26
Monotonic mechanical tests, 136

N

Neat resin/fibrous preform interface, 18
Negligible internal resistance (NIR) heat
    transfer solution, 40
Nonisothermal healing model, 44
Notch strength, 94
Numerical grid generation, 14
Nylon 6/6, 146

O

Orthotropic modeling, 74
  stress distribution, 77

P

Paris law, 267
PET, glass-fiber reinforced, 184
Plastic deformation processes, 279

Ply properties, 255
Polyester coated carbon fiber, 53
Polymer distribution in composites, 55
Polypropylene, glass fiber reinforced, 184
Porous medium flow, 16
Powder deposition on carbon fibers, 51
   coating uniformity, 53
   continuous process, 52
   distribution, 59
   evaluation, 53
Process simulation (*see* Modeling)
Pultrusion, 62
   heat transfer model, 64
   rate, 64

R

Relaxation tests, 166, 176
Residual strength, 139, 293
Residual tensile strength, 143
Resin film stacking, 15, 26
Resin flow, 14
Resin impregnation, 14
   boundary fitted coordinate systems, 20
   fibrous preform interface, 18
   fibrous preform interface front, 19, 24
Resin interface, neat and impregnated, 18
Resin movement, 24
Resin, neat
   autohesive bond strength development, 35
   temporal distribution of depth, 27
Resin, PEEK, dynamic viscosity, 13
Resin, polysulfone, 33
   autohesive bond strength development, 37
   compact tension test, 37
   critical strain energy release rate, 39
   healing test, 38
   refracture toughness, 38
Resin systems, 50
Resin, thermosetting, 50, 62
   pultrusion process, 63
Resin toughness, 251
Resin transfer molding, 15

S

Shear modulus, in-plane, 114
Short bar hybrid test specimen, 231, 233
Short-beam test failure, 202
Slow crack growth, 267
Stiffness and strength properties, 133, 176, 200, 255, 285
   moisture effects on, 213, 232
Stiffness ratio, 138
Stiffness variation during cyclic loading, 116
Stitching, 280, 285
Strain energy, 84
Strain energy release rate, 45, 74, 239, 251, 280

factors affecting, 47
Strength development, interply, 36
Strength prediction method, 95
Stress distribution, 73, 236
Stress intensity factor, 269, 285
Stress, process zone, 270
Stress-strain curves, 114, 219

T

Tack measurements, 35
Temperature profiles, 268
Tensile strength, 137, 219, 225
   residual, 143
   ultimate, 185
Tension-compression fatigue tests, 117
Tension/compression loading, 133
Tension strength prediction, 95
Thermal conductivity, 7, 146
   model, 147
Thermal diffusivity, 7
Thermal expansion, 189, 203, 257
Thermal profiles, 5
Thermal properties, 7
Thermal response, 155
Thermodynamic pressure, 19
Thermoplastic coating
   carbon fibers, 51
   distribution, 59
   evaluation, 51
   melting, 51
   powder deposition, 52
   process, 53
   uniformity, 53, 54
Thermosetting resins
   pultrusion, 62
   systems, 50
Three dimensional fiber structures, 279
Three parameter strength prediction method, 95
Torsional tests, 155
Toughness rankings, 262
Toughness tests, 200, 203
Tungsten powder, 149
Two-dimensional Darcy flow, 14

V

Viscoelastic constitutive model, 86
Viscosity, dynamic, 8
   temperature profiles, 9

W

Weighted residuals method, 65
Wetting, 35
WLF temperature dependence, 42